TESTING RESEARCH HYPOTHESES
WITH THE GENERAL LINEAR MODEL

TESTING RESEARCH HYPOTHESES WITH THE GENERAL LINEAR MODEL

Keith McNeil,
Isadore Newman,
and Francis J. Kelly

Southern Illinois University Press
Carbondale and Edwardsville

99 98 97 96   4 3 2 1

Excerpts in this book are from Keith A. McNeil, Francis J. Kelly, and Judy T. McNeil, *Testing Research Hypotheses Using Multiple Linear Regression* (Carbondale: Southern Illinois University Press, 1975). Excerpts in chapters 10 and 13 are from F. J. Kelly, K. A. McNeil, and I. Newman, "Suggested Inferential Statistical Models for Research in Behavior Modification," *Journal of Experimental Education* 41 (1973): 54–63, and from K. A. McNeil, "Meeting the Goals of Research with Multiple Regression Analysis," *Multivariate Behavioral Research* 5 (1970): 375–386, and are used by permission of the journals.

Library of Congress Cataloging-in-Publication Data

McNeil, Keith A.
   Testing research hypotheses with the general linear model / Keith
McNeil, Isadore Newman, and Francis J. Kelly.
     p. cm.
   Includes bibliographical references (p. -) and index.
   1. Statistical hypothesis testing. 2. Linear models (Statistics)
I. Newman, Isadore. II. Kelly, Francis J., 1926– . III. Title.
QA277.M34   1996
519.5'6—dc20   95-32920
ISBN 0-8093-2019-3 (cloth)   CIP
ISBN 0-8093-2020-7 (pbk.)

CONTENTS

Preface   ix

1. Introduction to the General Linear Model   1
Conceptual Orientation of the Text   1
Research Orientation of the Text   2
Statistical Orientation of the Text   3
Purposes of Research   5
Complex Determinants of Behavior   9
Pictorial Representation of Accounting for Variance   15

2. Hypothesis Testing   17
The Population of Interest   18
Analysis of Variance   19
R^2: The Proportional Estimate of Sums of Squares   28

3. Vectors and Vector Operations   32
Introduction to Vectors   32
Elementary Vector Algebra Operations   36
Useful Properties of Vectors   38
Linear Combinations of Vectors   41
The Unit Vector   44
Pieces of Information   45

**4. Research Hypotheses That Employ Dichotomous
Predictor Variables**   48
Dichotomous Predictors, Continuous Criterion   48
Research Hypotheses Involving k Groups   58
Orthogonal Comparisons—Post Hoc   61
Orthogonal Comparisons—A Priori   69
Dichotomous Predictor, Dichotomous Criterion   73
Advantage of the GLM Approach over Correlation and ANOVA   77

5.   Research Hypotheses That Employ Continuous Predictor Variables   **79**
Research Hypotheses Requiring a Single Straight Line of Best Fit   **79**
Summary of F and R^2 **88**
A Second Illustrative Example of a Single Straight Line of Best Fit   **89**
Mathematical Calculation of the Single Straight Line of Best Fit   **94**
Venn Diagram   **95**
Geometric Interpretation   **95**
Relationship to Other Statistical Techniques   **95**
The Use of Applied Research Hypotheses: Single Straight-Line
    Computer Problem   **95**
A Note about the Unit Vector in Computer Solutions   **96**

6.   Multiple Continuous Predictors   **98**
Two Continuous Predictors   **98**
Testing the Overall Predictability of the Model   **105**
Testing One Variable's Contribution   **110**
Multiple Continuous Predictors: Comparing Two Models of Behavior   **113**

7.   Interaction   **116**
Interaction Between Dichotomous Predictors   **116**
Interaction Between a Dichotomous Predictor and a Continuous
    Predictor   **127**
Interaction Terms as Covariates   **140**
Interaction Between Two Continuous Predictors   **140**
Comparison of Categorical and Continuous Interaction   **143**

8.   Statistical Control of Possible Confounding Variables   **149**
Control of Confounding Variables Through Theory   **149**
Control of Confounding Variables Through Research Design   **149**
Confounding Variables under Statistical Control   **151**
Introduction to Power   **168**

9.   Nonlinear Relationships   **174**
Homoscedasticity and Heteroscedasticity   **174**
Fitting Expected Nonlinear Functional Relationships   **178**
Rescaling Observed Nonlinear Relationships   **197**

10.   Detection of Change   **216**
Ascertaining Causality   **216**
Directional Research Hypotheses   **217**
Random Assignment   **217**
One-Group Models   **218**
Two-Group Models   **223**
Matched Groups   **231**
Matched Pairs of Individuals   **231**
Point Change Measuring Multiple Subjects   **232**
Curvilinear Relationships That Account for the Data Better than
    Stimulus Change   **235**

Functional Change in Individual-Organism Research 237
Suggested Inferential Statistical Models for Research in
      Behavior Modification 239

11. **Miscellaneous Questions about Research That Regression Helps Answer** 252
Samples and Populations 252
Criterion and Predictor Variables 254
The Dollar Cost of Prediction 256
Attenuation, Replication, and Cross-Validation 257
Traditional Sequencing of Hypotheses 258
Sequencing of Hypotheses 259
Interpretation of Weights 264
Missing Data 265
Outliers 266
How to Beat the Horses 267

12. **Application to Evaluation** 268
Evaluation of Compensatory Education Programs 268
Applications to Other Evaluation Situations 274
Policy Capturing 283

13. **The Strategy of Research as Viewed from the GLM Approach** 287
Meeting the Goals of Research with the GLM 287
Newton States the Problem 287
The Most General Regression Model 294
The Goal of Control in the Behavioral Sciences 297
A Proposed Research Strategy 302
Reporting 307
Further Reading 310

Appendixes
A. Data Set of Sixty Subjects 313
B. SAS Discussion 315
C. Activities with Circles and Squares 318
D. SAS Statements for the Various Applied Research Hypotheses 324
E. ANOVA Source Tables 338
F. Equivalency of F-test Formulae 340
G. Power Tables 342
H. Policy-Capturing Activity 348
I. Microcomputer Setups for Selected Applied Research Hypotheses 354
J. Index for SAS 361

References 363

Index 370

PREFACE

The technique of multiple linear regression has been accepted by the research community since 1975. Therefore, we do not devote much space to a defense of the equivalence of correlational and analysis of variance procedures with multiple linear regression. We have also chosen to use the term *general linear model (GLM),* although in some places we have used the terms *multiple linear regression technique, multiple linear regression approach*, and *multiple regression* synonymously.

The first three sections of chapter 1 provide a conceptual, research, and statistical orientation to the entire text. The remainder of chapter 1 provides the rationale for the utility of a conceptual model of behavior, along with one such model that can be used to identify predictor variables. We strongly suggest that readers familiar with the GLM read these three sections before delving into the more advanced material. We invite readers who are relatively unfamiliar with the GLM to read the first eight chapters before branching off into topics that are of immediate interest or seem too irresistible to delay.

We have found that the computer examples can provide additional insight and reinforcement for the concepts. The initial computer examples contain the bare essentials needed to obtain the answers to the Research Hypotheses. Subsequent computer examples provide more detail (such as labels, plots, and alternative ways of obtaining results). In all cases, the minimum computer instructions are indicated. Many assertions are made in the text, and often these assertions are backed up with the necessary computer statements. The reader is referred to Appendix B for the computer lines that will substantiate those assertions. All computer examples are performed on the data set in Appendix A, which is available from either Keith McNeil, whose e-mail is KMCNEIL @ NMSU.EDU, or Isadore Newman, whose e-mail is INEWMAN @ UAKRON.EDU.

TESTING RESEARCH HYPOTHESES
WITH THE GENERAL LINEAR MODEL

1

INTRODUCTION TO THE GENERAL LINEAR MODEL

CONCEPTUAL ORIENTATION OF THE TEXT

Limited Mathematical Approach

The approach we use in this text relies on vectors instead of matrices. Few formulae are presented; we focus instead on conceptual understanding. With the ready availability of computers, researchers do not have to calculate their statistics, although they do need to be able to use computers. Therefore, the emphasis in this text is on developing one's own statistical models based on the Research Hypothesis and then having the confidence to interact with a computer program to obtain the results of that test. The text also emphasizes the interpretation of the results.

Research Guiding the Statistical Effort

It is our firm belief that statistical procedures should be used to answer researchers' questions. Most statistical texts present a vast array of statistical techniques but do not facilitate the understanding of the advantages and problems of the various techniques. We take the position that the Research Hypothesis should be stated first and that the hypothesis dictates all the remaining statistical, computer, and interpretational effort.

Directional and Nondirectional Hypotheses

Although in most statistical texts directional and nondirectional hypotheses are discussed, few emphasize the appropriate interpretations and necessary adjustments when using the F test. In this text, we emphasize the role of each of the two hypotheses and encourage the directional test whenever there is reason to use it.

Nonzero Hypotheses

Although most authors of statistical texts discuss the possibility of asking a nonzero Research Hypothesis, most do so only casually and do not encourage the researcher to consider seriously the benefit of such a hypothesis. When a researcher refers to effect size, the researcher is using another technique (usually post hoc): answering what should have been answered through a nonzero Research Hypothesis. Reporting of R^2 is argued in this text to be a vehicle for understanding research. The larger the effect size, the larger the R^2, all other things being equal.

RESEARCH ORIENTATION OF THE TEXT

Research Hypothesis as the Basis for Inferential Statistics

The reason inferential statistics exist is to answer Research Hypotheses. Inferential statistics are of no other value, and that is why we always capitalize the term *Research Hypothesis*. The Research Hypothesis must come first, and it should not be limited by the statistics that the researcher knows. The most compelling reason for using the general linear model (GLM) approach is that the vast majority of Research Hypotheses can be tested with the GLM approach. Thus the researcher is freed from the shackles of asking only a limited number of Research Hypotheses and can spend more effort asking the Research Hypotheses that are of real interest. All parametric statistical procedures investigating a single-criterion variable are computational simplifications of the GLM and therefore can be tested with the GLM. In addition, many nonparametric procedures are subsets of the GLM. Where in the text some specialized statistical techniques are identified, the purpose is to assist those researchers familiar with these techniques, not to perpetrate the misguided impression that it is important to view each of these techniques as separate, necessary entities. Because they still appear in the literature is the only compelling reason to recognize them. The statistical test should be in the background, at the service of answering the Research Hypothesis. For instance, one should *not* say, "I did a *t* test." One should say, "In order to answer the Research Hypothesis of interest, I did a *t* test."

Focus on the Criterion Instead of on the Predictor Variables

Many researchers forget that they are trying to understand their criterion variable and therefore place more emphasis on their predictor variables. Some researchers develop new treatments or procedures before they can account for a substantial amount of the variance in the criterion variable. If a researcher does not know what accounts for the variance in the criterion, how can intervention strategies be developed? The emphasis in this text is on accounting for the criterion variance by reporting the obtained R^2. We encourage researchers to transform the existing statistical tests of significance in the literature to R^2 values. Indeed, many metaevaluations do just that (Rosenthal, 1984). Researchers are encouraged to increase the accounted-for variance by including other information in the regression model—either with other variables or with interacting or nonlinear functions.

Multiple Predictor Variables

A single predictor usually will not account for all of the criterion variance. Most social science researchers consider their constructs to be complexly determined yet employ very simple statistical models to test those constructs. On the other hand, carefully chosen predictor variables and well-thought-out relationships between those predictor variables and the criterion may require only a few variables to reach the ultimate goal of accounting for all of the criterion variance. (Chapter 13 provides an example of one predictor variable accounting for all of the criterion variance.)

Predictors Probably Needed from Various Areas

Many researchers feel that there is an inherent limit to the amount of variance that can be accounted for, and they often stop their search for predictability much too soon. Our position is that a researcher should attempt to obtain an R^2 close to 1.0, and in order to do that, one may need to obtain predictors from various groupings or areas. One such grouping is provided at the end of this chapter. Some statistics authors say two or three predictor variables will result in the maximum amount of variance accounted for. Our position is that an R^2 of 1.0 is theoretically attainable, but the researcher just may need to consider additional variables. However, in research on humans, the measuring instruments usually do not meet a very high level of precision, and that foils our attempt to obtain an R^2 of 1.0. An R^2 of .60 or lower is typically reported, and that is too low.

Predictors as "Pieces of Information"

It is valuable for the researcher to think of each variable as another "piece of information." There are two advantages to using this concept. First, the researcher may come to realize what other "information" is known about the subjects under study. Second, this concept leads to the calculation of degrees of freedom, which is usually difficult with statistical procedures other than the GLM.

Looking at predictors as pieces of information helps to answer the following questions: What information can I get and at what cost? Is this really new information? If I have already considered the two subtests, is the total test a new piece of information? If I have already decided who is in my sample, and I know that I am going to have two gender groups, do I need to consider both male and female as new pieces of information?

STATISTICAL ORIENTATION OF THE TEXT

The Generalizability of the General Linear Model

Almost all the statistical procedures encountered in an introductory statistics class, or a correlational course, or an analysis of variance (ANOVA) course are computational simplifications of the GLM. As such, the questions answered by these procedures also can be answered by the GLM. Indeed, all parametric and most non-parametric Research Hypotheses investigating one criterion variable can be answered by the GLM. In addition, many Research Hypotheses that have not had computational simplifications developed for them can be tested by the GLM. Therefore, the reader of this text will be encountering a flexible, generally applicable statistical procedure that can replace most of the other procedures.

Before the widespread availability of the computer and statistical computer packages, the GLM was computationally difficult to set up, and that is why statisticians developed computational simplifications. Because the computer can do routine and complex operations easily and quickly, we can conveniently use the computer to do the necessary calculations. The researcher is then free from wondering if the "correct" statistical analysis is known or if the desired Research Hypothesis can be stated in such terms that one of the few statistical tests known by the researcher can be used to answer that hypothesis. Thus researchers can now let the desired Research Hypothesis wag the tail of the statistical procedure instead of the tails of the known statistical tests "wagging" the way the Research Hypothesis is stated.

Continuous Versus Categorical Representation of the Variable

Traditionally, there have been two distinct statistical camps—one dealing with continuous variables (the correlation camp), and one dealing with categorical variables (the ANOVA camp). One benefit of the GLM is that neither the statistical technique nor the computer care whether we want the variable to be treated as continuous or categorical. How to represent the variable is a decision for the researcher to make. That decision depends on the Research Hypothesis, not on the statistical tools in the researcher's toolbox.

We, the authors, are biased toward continuous variables for two reasons. First, most variables do occur naturally in a continuous fashion. For convenience, we often artificially dichotomize a continuous variable, producing such results as a hire or no-hire decision, a tall or short label, or a like or dislike attitude. Where one specifies the cut point often becomes problematic and usually differs from situation to situation.

Second, we strongly believe that functional relationships occur in a continuous fashion. Behavior usually does not change abruptly as a result of being either below or above a certain cut point. If it does change, then it is highly unlikely that the a priori cut point is the "real" cut point. Initial consideration of continuous data still allows the investigation of the existence of alternative cut points.

Hypothesis Testing Versus Data Snooping

Many researchers have used the GLM as a "data-snooping" tool, specifically when they conduct a stepwise regression analysis. The emphasis of this text is on the hypothesis-testing use of the GLM. The reader needs to know when the two uses are appropriate: Data snooping is valuable in the service of hypothesis generation, while comparison of models is valuable in the service of hypothesis testing.

A Priori Comparisons Instead of Post Hoc Comparisons

There is an enormous amount of literature regarding which statistical technique to use after a significant ANOVA has been found. These techniques are referred to as *post hoc comparisons* because they test hypotheses that were not originally stated. Given the preceding discussions, the reader will realize that in this text we make the case that there is little value in post hoc comparisons and much utility in a priori comparisons, comparisons that encourage the investigation of directional Research Hypotheses. The limited value of the post hoc comparisons will be discussed within the framework of data snooping.

Interactions

Most researchers, when thinking about what pieces of information account for the criterion variance, usually indicate that it depends. Whenever researchers use that language, they are referring to statistical interaction. Our position in this text is that interaction may well account for meaningful variance in many research areas. In traditional ANOVA, interaction is usually viewed as a source of variance that is not wanted, that complicates the situation, and that is not a meaningful question in its own right. We treat interaction as a possibly valuable piece of information that can be tested, just as any other piece of information can be tested.

Covariance Adjustment

When groups are initially different, the technique of covariance adjustment can be used. The covariate must, though, have variance within groups, otherwise the criterion cannot be adjusted for that covariate, and the covariate remains as a competing explainer of the results. For instance, if a researcher is studying the criterion of the time it takes to run a mile, then pretreatment body fat would be a reasonable covariate, as those with less body fat can usually run faster. If a researcher conducts one treatment at one altitude and the other treatment at another altitude, then each subject at the one altitude would have the same covariate score on the altitude measure. Thus altitude cannot be used as a covariate and remains as a competing explainer for whatever differences the researcher finds in the criterion. Similarly, if the two sites have, say, different levels of pollution, then the researcher cannot treat pollution as a covariate unless data were obtained on different days having different levels of pollution.

Clarification of Possible Statistical Confusions

It has been our experience during the past 25 years that students who understand the GLM understand the big picture of statistics. Many concepts that were skipped, placed on the back burner, or incorrectly learned are clarified with the GLM. Be prepared for an "Aha" experience if you have unresolved questions, such as the following:

• How many subjects should I have?
• How many variables should I have?
• What does *interaction* really mean?
• I think that the relation between my predictor and my criterion is not linear; how can I reflect that relationship?
• The GLM computer output provides the term *constant*; what does that mean?
• What are *degrees of freedom*, anyway?
• What is the formula for the degrees of freedom for my Research Hypothesis?
• My data do not meet the assumption of interval data; what do I do?
• What does *significance* mean?
• I found statistical significance, but do I have practical significance?

PURPOSES OF RESEARCH

Investigations into human behavior take many forms and are conducted for various reasons. The purpose of some is to describe, others to predict, and still others to improve (control) behavior. Underlying these behaviors are some expectations held by the investigator regarding a causal network behind observed behavior. Research designs and statistical procedures have been developed to help the researcher make decisions regarding the adequacy of the *description*, *prediction*, or *improvement* in relation to expected outcomes. These three functions are illustrated in the following simplified example.

Description

Description refers to reporting data that have been obtained and applies only to the particular people (or other entities) from whom those data come. In description, no inference is made to other data or to other people.

A social scientist may intuitively suspect that children from "lower-class" homes seem to have less success in school than children from "middle-class" homes. This research investigator, based upon his observations as filtered through his theoretical frame of reference, would expect children from lower-class homes to have lower grade point averages (GPAs) when compared with the GPAs of their classmates from middle-class homes.

To *describe* the expected relationship, the investigator will usually select a representative number of children from middle-class and lower-class homes (based upon some operational definition) and then collect data on the criterion (dependent variable) selected to represent school success (GPA in this case). Suppose that the observed data reveal that 50 children from lower-class homes have a mean GPA of 1.52 (on a five-point scale), and 50 children from middle-class homes have an observed mean GPA of 3.68. The researcher confirms his expectation: The children from lower-class homes do have lower GPAs on the average than children from middle-class homes. It seems that lack of preparation in the home causes less school success. Further inspection of the individual GPAs may reveal that there is some overlap between the two groups. Some children from lower-class homes have higher GPAs than some children from middle-class homes. In fact, some children from lower-class homes have higher GPAs than the average child from a middle-class home, and conversely, some children from middle-class homes score lower than the average child from a lower-class home. Lack of preparation attributed to home activity does *not always* cause poor school success; there is some overlap of observed scores among groups. Perhaps other predictor (or independent) variables also influence school success; this matter will be discussed shortly.

Another way to *describe* these data is to report the strength of the relationship between home preparation and school success using a squared bivariate correlation coefficient (R^2). This value (R^2) represents the proportion of the total observed GPA variance that is explained by group membership. Suppose in this case that $R^2 = .45$. Then the percentage of observed variance (variation of a subject's GPA from the overall mean GPA) that is explained by home background is 45%. The large mean differences and the percentage of explained variance between groups may be sufficient for the investigator tentatively to accept *some* causal relationship between home preparation and GPA for this specific group of children.

Prediction

The process of using data from a smaller group (sample) of people (or measurable entities of any kind) to make estimates about a larger group (population) from which such data have not been gathered is called *prediction*. Our position in this text is that a researcher uses data to establish a prediction only *after there are descriptive data to support that prediction.* For example, a second researcher may read the descriptive data on the relative school success of lower- and middle-class children that were reported by the researcher described above. This second researcher may have noted that schoolroom activities approximate the activities in the middle-class homes, and that lower-class homes provide few of the classroomlike activities. The lack of preparation in the lower-class home could cause these children to be at a competitive disadvantage to the prepared middle-class child.

In a situation such as the one just given, the investigator usually desires to do more than describe a particular set of children; typically, he wants to generalize from the sample of children he measured to a theoretical population of similar children. The researcher then chooses children for the study who adequately represent the two theoretical subpopulations (children from lower-class homes and middle-class homes), and then the information from that sample of children can be used to generalize to all children from lower- and middle-class homes.

To know what generalization to make, one must figure out how likely it is that the observed difference between means (or the R^2 value) would occur due to sampling variation. Suppose that this researcher finds with his sample that the home preparation and school success has an R^2 value of .47, and the mean GPA values are 1.59 for the lower-class children and 3.73 for the middle-class children. The R^2 value of .47 can be subjected to a statistical test to figure out how likely it is that the observed mean differences can be attributed to sampling variation (discussed in chapter 4). Given this sample of 100 children, the observed R^2 and mean differences are highly unlikely to be due to sampling variation. On the basis of a statistical test on these two means, the investigator would conclude that home preparation apparently causes differential school performance. If a new set of children from the two types of homes were selected, based on this researcher's predictive study, he would *predict* (or estimate), with some confidence, lower GPAs for children from lower-class homes when compared with the GPAs of children from middle-class homes. The researcher may tentatively conclude that the type of home "causes" the level of GPA. A definite inference of causality would have to wait for manipulation.

The causal interpretation just given is based on the theoretical network developed by the investigator. The theory, bolstered by observation of differential home experiences in lower-class and middle-class homes, led to specific expectations. As with most research, other causal factors may be operating. The causal interpretation is always subject to modification based upon subsequent information.

Perhaps the difference between children from lower- and middle-class homes is due to nutrition or to expectations of teachers. These possible competing explainers have not been controlled in the present study. Some possible competing explainers can be eliminated by the design of the study; some can be statistically controlled; most must be logically eliminated by the researcher.

Please note that determining causality is a process of logic and research design and not a process of statistics. Statistical verification is a necessary but never sufficient condition for causal interpretation. We present more on causality in later chapters. Knowing the causers allows a meaningful change in the situation, our next subject.

Improvement

Improvement involves making a change in a given situation that has been expected and verified to be a "cause." Many applied behavioral scientists want to improve the lot of humans and will attempt to manipulate the environment to bring about a controlled outcome. For example, the researcher discussed above can *predict* that lower-class children will have lower GPAs than middle-class children. But he may wish to "upset" the prediction and do something to change the situation so that the children for whom he would have predicted low GPAs will have higher GPAs.

He cannot, however, be successful in improving the situation until he is certain of the cause of the low GPAs. The researcher found that home preparation was related to GPA, but changing a child's home preparation *may not* raise his GPA (if social class itself is not the cause).

Improvement cannot be determined until the variables that are found to be predictors and are suspected of being causes are *manipulated* in a sample and found to be related to improvement. Some variables, such as gender, cannot even be considered to be manipulatable. Other variables will be difficult to manipulate. Variables that are manipulatable may cause a change in other variables that are under consideration or were previously thought to be exogenous. Additionally, some subjects may resist being "controlled." Finally, even if the researcher can change subjects on the predictor variable, the desired change in the criterion may not occur. It is an empirical world, and we will never know until we manipulate.

Manipulation to Search for Causal Factors

Suppose the researcher investigating the performance of the children from lower- and middle-class homes wanted to upset the predicted "poor" school success for the children from lower-class homes (LCH). He therefore must manipulate (change) one of the factors suspected of causing poor school success. He may first select 100 children who represent the same theoretical population of children from LCH and on some random basis select from these 100 LCH children 50 children for an intervention program and 50 children to act as a natural control group. If home experience is the relevant causal variable, then a change in the home environment—introducing classroomlike experiences—theoretically should result in higher GPA performance than expected. There is nothing sacred about choosing 100 subjects; more subjects would have provided a more powerful test of the hypothesis, and fewer than 100 would have provided a less powerful test of the hypothesis.

If these classroomlike experiences in the home were provided for the 50 experimental LCH children and not for the 50 control LCH children, a GPA difference between these two groups favoring the experimental group should theoretically be found. When the data on the criterion are examined, it is found that the mean GPA for the experimental group is 2.98, and for the control group the mean GPA is 1.54 (see Figure 1.1). The bivariate R^2 of .30 and its associated F value are very unlikely to be due to sampling variation. Apparently, the improved home experiences cause better school performance; the improved GPA is 2.98, compared to what might have been expected for these children on the basis of the above descriptive study (mean GPA was 1.52 for LCH children). However, note that the improved GPA does not equal 3.68, the GPA observed for children from middle-class homes (MCH) in the earlier study. The investigator has apparently isolated some of the causal factors underlying lack of success, but some additional influence or set of influences are also operating. It could be that the experimental treatment was not as strong as the natural MCH environment or that another variable or set of variables were operating to influence the GPA. Are children from MCH a bit brighter? Are they more motivated for success? The data to answer these questions are lacking in the present study.

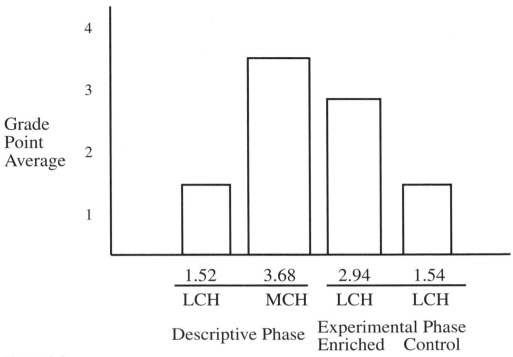

Figure 1.1.
GPA means for children from lower-class homes (LCH) and middle-class homes (MCH) obtained in the descriptive phase and in the experimental (improvement) phase.

The investigator also may note that the control group's GPA of 1.54 was close to the GPA of 1.52 observed for children from LCH in his initial study. The difference is so small that he would likely conclude that the two samples represent the same population. With these sample values (1.52 and 1.54), he still does not know the population GPA mean, but he might assume that it is close to the average (1.53) of the two sample means.

COMPLEX DETERMINANTS OF BEHAVIOR

The simple example just presented was provided to illustrate the *descriptive*, *predictive*, and *improvement* aspects of research. At the present level of theoretical sophistication in behavioral research, such bivariate (single predictor and single criterion) research surely would be labeled naive. From past research it is "known" that there are many other variables, such as test anxiety, measured intelligence, peer expectations, content of the material to be learned, and past special learning (e.g., as measured by standardized achievement tests), that enter into broad performance constructs, such as GPA. Furthermore, these variables may not be additively related to performance. For example, Castaneda, Palermo, and McCandless (1956) showed that the way anxiety level affected a student's performance of a task depended upon task difficulty. Given a difficult task, low-anxious students tended to do better than other students; but given an easy task, high-anxious students tended to do best. This effect is called a *difficulty-by-anxiety interaction*. Statisticians use the term *interaction* when the effect of one variable *depends* upon another variable.

Besides the interaction discussed above, one might expect task difficulty to depend upon the relative brightness of the performing student. A difficult task for the average student may not be difficult for the exceptionally bright student. If one were to extend the Castaneda et al. study to include a task difficulty-by-anxiety-by-"intelligence" interaction, one would expect to explain more of the criterion variance (and have less unexplained criterion variance).

Given this complex state of affairs, it seems that researchers who want to *describe*, *predict*, and *improve* need complex theoretical models and flexible statistical procedures that accurately reflect the complex models. The GLM, as presented in this book, is one flexible statistical procedure that meets the needs of a broad spectrum of behavioral researchers.

A Conceptual Model

The details and various applications of this flexible statistical procedure, the GLM, are presented in the rest of this text. But we first present a conceptual model that provides a way of considering variables that relate to the behavior the researcher is investigating. One of the major reasons for the grouping of variables discussed below is to stimulate the search for multiple variables, since consideration of multiple variables is usually necessary to account for variance in any criterion variable. The variable groupings presented here are applicable to behavioral research; other groupings could be developed for this or other areas of research.

A critical review of the behavioral literature reveals that the variation on the criterion is caused by a network of interrelated predictor variables. These predictor variables can be arbitrarily grouped into three categories: (a) *person variables*, (b) *focal stimulus variables*, and (c) *context variables*.

Person Variables

Most areas of research into human behavior have substantiated that there are pervasive differences between individuals. In attempting to account for differing performances, one would want to specify the *person variables*, or the characteristics of the behaving individual that may influence her degree of success.

Focal Stimulus Variables

There are many ways to measure any construct that a researcher wishes to consider a criterion variable. The ways in which those measures differ from one another are the *focal stimulus variables*; they are the characteristics of the instrument or task that is used to measure the criterion behavior.

In attempting to account for differing performances, one would want to specify dimensions of the task to be completed that may influence an individual's degree of success.

Context Variables

Many stimuli unrelated to the task can influence an individual's behavior. Such expectancies are neither part of the task nor within the person; they are part of the *context*, the situation, in which the criterion behavior occurs. In a sense, the context provides information regarding "payoff" or reinforcement associated with performance.

In an educational setting, context variables that might interact with student characteristics to influence response acquisition might be the physical condition of the classroom, the reward system for good grades, and peer expectations. (If a student's friends value school success, he might work even if he does not value school achievement.)

Symbolic Representation of the Conceptual Model

In view of the preceding discussion, one can say that *complex behavior* of a person is a function of (a) *person characteristics*, (b) *focal stimulus characteristics*, and (c) *context characteristics*. The preceding statement can be reduced to a quasi-mathematical model as a shorthand notation:

$$Y = f(P, S, C)$$

where:
   Y = the criterion;
   P = person characteristics;
   S = focal stimulus characteristics;
   C = context characteristics; and
   f = the functional relationship of the variables in the three classes (P, S, C) as
       they relate to Y.

The equation can be read: The criterion behavior of a number of individuals is a function of the characteristics of these individuals, the focal stimulus characteristics, and the context characteristics.

Please do not panic at the sight of symbols. Symbol notation is used to simplify the expressions so they can be presented in the form of quasi-mathematical models, a practice that becomes extremely useful as this text goes on. There is no need to remember the particular symbols used here, or for that matter to understand fully the variables and examples used in this chapter. They are presented to show how one might go about grouping and exploring variables that may contribute to prediction. Indeed, some variables are difficult to classify. Accurate classification is not the goal; increasing the amount of accounted-for variance is the goal. The whole conceptual model for the study of complex behavior with examples of person, focal stimulus, and context variables is presented below.

An illustration of variables in the three predictor categories of the comprehensive model, for the criterion of math achievement follows.

$$(Model\ 1.1)\ Y = f(P, S, C)$$

where:
   Y = scores on a math achievement test for a set of individuals;
   P = person variables;

Examples of person variables for this criterion:

1. convergent thinking ability (as measured by most standardized intelligence tests)
2. divergent thinking ability (as measured by tests of creativity)
3. symbol aptitude
4. motivation
5. sex role identification (masculinity-femininity)
6. past learning relevant to success on the task

S = focal stimulus variables;

Examples of focal stimulus variables (characteristics of the criterion measure) for this criterion:

1. length of the test
2. ordering of items by difficulty
3. numerical items versus "word problems"
4. types of mathematics problems contained in the test
5. number of operations required by each item

C = context variables;

Examples of context variables for this criterion:

1. peer expectancies
2. adult expectancies
3. physical plant (light, noise, heat)
4. reward conditions

The functional relationships among the three classes of variables and the criterion (Y) can take many forms depending upon the research expectations of the investigator. The above quasi-mathematical model contains examples for each of the three classes that might explain observed behavior. The examples would, of course, vary for any given criterion behavior. One task of research is to quantify the variables and then attempt to account for the observed criterion variation in terms of some weighted combination of the predictor variables. It is premature to delineate what functional relationships might exist between (P, S, C) and Y. We provide the following example to give some ideas about how one might operationalize the three categories.

Accounting for Complex Behavior

Consider the task of predicting complex behavior, such as success on a job-training program. If the training program is costly, it would be worthwhile to predict the successful and unsuccessful trainees in terms of cost effectiveness (to avoid training many individuals who are unlikely to be successfully trained).

In view of the proposed conceptual model just presented, several aspects of the situation should be investigated. One might first examine the set of behaviors that are related to the criterion (Y), the observed terminal behavior. What are the task characteristics (represented by S) that might be relevant to the criterion? Such an examination might give the investigator a few notions regarding relevant human characteristics (represented by P) needed as prerequisite skills (e.g., are specific abilities or special previous learning necessary for training success?). What are the conditions (represented by C) surrounding the learning and testing setting? Is there peer pressure for success? Are the trainers placing pressure on the trainees? If so, will trainee anxiety be relevant? What are the intermediate payoff schedules? Must the trainee work through the program before some reinforcement is given? Will the need to achieve be relevant to the reinforcement schedule? If he has failed often in the past, does he expect to fail in this new task? If so, can the training context be manipulated to minimize these effects? Familiarity with the research literature should provide additional suggestions regarding the relevant variables that account for complex behavior. Indeed, the whole research endeavor focuses on selection of variables based upon theoretical expectations and past empirical findings regarding the relationships among sets of variables.

One might hypothesize that the desired terminal behavior (criterion) of the training program is related to spatial abilities (SA), anxiety level (Anx), ability to manipulate symbols (Sy), the amount of written work the job to be trained for will require (Wr), the length of the training program (Ln), and the number of trainees in the group (Nt). Note that the first three variables are person variables; the fourth variable is a focal stimulus variable; and the fifth and sixth variables are context variables. This set of variables can be cast as:

$$Y = f(SA, Anx, Sy, Wr, Ln, Nt)$$

where:
>   Y = observed terminal behavior of the trained individuals; and SA, Anx, etc., are defined above.

The function sign (f) implies that there are some mathematical functions that express the relationship of the predictors to the observed criterion. Types of commonly used functions are additive, multiplicative (allowing for interaction, discussed in detail in chapter 7), squared (second-degree polynomial, discussed in detail in chapter 9), square root, cubic, and trigonometric functions (e.g., sine and cosine).

One may assume that the scores on the training task (Y) are a sum of the six weighted predictor variables (a weighted additive function). Equation 1.1 expresses this function:

$$Y = a_1 SA + a_2 Anx + a_3 Sy + a_4 Wr + a_5 Ln + a_6 Nt \qquad (1.1)$$

The weights $a_1, a_2, \ldots, a_6$ might be chosen rationally or be empirically derived using some mathematical solution. In the domain of any science, researchers have seldom found a set of weights that satisfies the equality (=) expressed above. The reason for lack of equality is that theory, measurement, and expressed functions typically are not perfect. Errors of prediction are made due to incomplete or erroneous theory, inadequate measurement tools, and lack of ability to adjust relationships perfectly. What this means, then, is that the task of behavioral research is to develop more comprehensive theories, better measurement, and more appropriate quantitative procedures in the effort to minimize errors of prediction.

The equality expressed in Equation 1.1 is an ideal that is not observed in most research endeavors. There will usually be errors of prediction. That is, no matter how "good" the weights are, the sum of the weighted predictors will hardly ever be equal to the criterion score. Equation 1.1 can be adjusted to acknowledge errors of prediction and therefore satisfy the expressed equality. Equation 1.2 reflects the inclusion of the error component that satisfies the expressed equality.

$$Y = a_1SA + a_2Anx + a_3Sy + a_4Wr + a_5Ln + a_6Nt + E_1 \qquad (1.2)$$

Or, equivalently:

$$Y = \hat{Y} + E_1$$

where:

E_1 = the difference between the predicted score, which is the sum of the weighted scores ($\hat{Y}$), and the observed terminal performance (Y); $E_1 = (Y - \hat{Y})$.

The GLM procedure derives the set of weights ($a_1, a_2, \ldots, a_6$) to minimize the sum of the squared differences between the observed criterion scores (Y) and the predicted criterion scores ($\hat{Y}$). The squared discrepancy between an observed score and a predicted score can be expressed symbolically as $(Y - \hat{Y})^2$, or as E_1^2. The sum of those squared errors across all subjects, or $\Sigma(E_1)^2$, has interesting properties that satisfy statistical distributions (e.g., the *F* statistic). Thus, with some manipulation, the least squares solution can be used inferentially for decision-making purposes. Chapters 2 and 4 treat statistical inference in greater detail.

Accounting for Complex Behavior with Alternate Functions

Suppose the weighted additive function expressed in Equation 1.2 yielded a large $\Sigma(E_1)^2$ for the group of *N* individuals. (The sum of the squared errors of prediction for the individuals is large.) Some of this error might be due to selection of inappropriate functions. Knowledge of past research might suggest to the investigator that the additive function is not adequate. She might suspect that Wr is geometrically related to Y. That is, the criterion scores are approximately the same for very low to middle Wr levels; but the criterion scores increase rapidly as Wr scores increase from medium to high. Furthermore, the investigator might expect anxiety (Anx) and length of training (Ln) to interact, such that anxiety might adversely influence terminal behavior, especially when the training period is long. These two circumstances can be reflected by adding two additional variables to Equation 1.2: a squared function of Wr (Wr^2, reflecting the geometric expectation) and a multiplicative function for anxiety and length of training (Anx * Ln, reflecting the interaction expectation). The expanded equation, Equation 1.3, would be:

$$Y = a_1SA + a_2Anx + a_3Sy + a_4Wr + a_5Ln + a_6Nt + a_7Wr^2 + a_8(Anx * Ln) + E_2 \quad (1.3)$$

where:

E_2 = a new error value ($Y - \hat{Y}$).

The error (E) is subscripted differently in Equation 1.3 because the error values for each subject will likely be different due to the two additional variables used for prediction in Equation 1.3. Note that the asterisk (*) will be used to indicate multiplication throughout this text.

If the eight variables in Equation 1.3 represent the functional relationship more adequately than the six variables in Equation 1.2, then the predicted behavior ($\hat{Y}$) should be closer to the observed behavior (Y) in Equation 1.3 than that found using Equation 1.2; thus Equation 1.3 should yield a smaller $\sum(Y - \hat{Y})^2$.

PICTORIAL REPRESENTATION OF ACCOUNTING FOR VARIANCE

Let us now take another look at the prediction of the criterion variable presented in Equation 1.2. Multiple predictor variables were included because it was suspected that the criterion was complexly caused. If only spatial abilities (SA) had been used as a predictor, a certain degree of predictability might have been observed, say $R^2 = .20$ (which is an overlap of 20% between predictor and criterion variance) as in Figure 1.2. Each of the Venn circles in Figure 1.2 represents one unit of variance.

The unshaded area of the criterion represents 80% of the criterion variance, which is unaccounted for. Often this variance is labeled as *error variance* and considered to be due to measurement error. The unshaded area, though, may be due to omitting the relevant (and necessary) predictor variables. Consequently, this error is also often referred to as *model specification error*. Adding another predictor variable to the prediction scheme (say, Sy, ability to manipulate symbols) might produce the situation depicted in Figure 1.3.

Note in Figure 1.3 that the second predictor (Sy) also accounts for approximately 20% of the criterion variance; but because the two predictor variables are correlated with each other, the new predictor adds only 10% new overlap with the criterion.

Predictor variables will most likely be correlated in the real world. The concern is not how highly they are correlated; the concern is how much of the criterion variance they account for together. Has the set of predictor variables come close to the goal of accounting for 100% of the criterion variance? If not, then what other

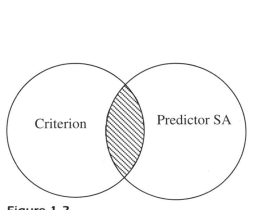

Figure 1.2.
A pictorial representation of a situation in which the predictor accounts for 20% (the shaded area) of the criterion variance.

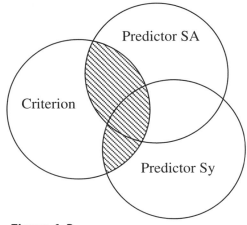

Figure 1.3.
A pictorial representation of a situation in which the predictors are correlated and account for 30% of the criterion variance.

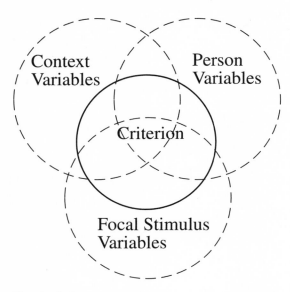

Figure 1.4.
Pictorial representation of the comprehensive model presented in Model 1.1.

predictor variables could be considered? There are many other within-person variables, but including other within-person variables may yield diminishing returns; that is, there is criterion variance that is accounted for by variables other than within-person variables. A criterion variable will most likely be a function of relevant focal stimulus variables, context variables, and person variables. Figure 1.4 represents the ideal situation (accounting for 100% of the criterion variance), with the three sets of predictor variables being represented by dashes because each set would not always be expected to account for the same amount of criterion variance. Deciding which variables to use from each set is based on theory or the knowledge base, and the proportion of criterion variance accounted for by each set is always an empirical question to be answered by the data.

2
HYPOTHESIS TESTING

In chapter 1 we suggested that research design and statistics are tools to help an investigator make decisions regarding the adequacy of *description*, *prediction*, and *improvement* of behavior. When testing expectations, one seeks to generalize from a small group of individuals to other individuals who are similar. We present a brief discussion of sampling theory for review purposes. Sampling theory is the basis for the steps in hypothesis testing presented below.

Steps in Hypothesis Testing

1. Statement of the *Research Hypothesis*: a statement about the population that the researcher is hoping to support. (The Research Hypothesis allows for a multitude of population values—e.g., $\mu > 5$ and $\mu_1 - \mu_2 \neq 0$.)
2. Statement of the *Statistical Hypothesis*: a statement about the population that is antithetical to the Research Hypothesis. (The Statistical Hypothesis specifies one population value—e.g., $\mu = 5$ and $\mu_1 - \mu_2 = 0$.)
3. Statement of *alpha*: the risk (probability) the researcher is willing to make in rejecting a true Statistical Hypothesis. (One minus alpha is called the *confidence level*.)
4. Collection of the data from a representative sample of the population.
5. Calculation of the *test statistic* (t, F, chi-square, etc.).
6. Determination of significance.
   With computer printout:
      a. If the reported probability is less than or equal to the stated alpha, reject the Statistical Hypothesis and accept the Research Hypothesis.
      b. If the reported probability is larger than the stated alpha, fail to reject the Statistical Hypothesis and fail to accept the Research Hypothesis.
   With tabled values:
      a. If the test statistic is greater than or equal to the tabled value, reject the Statistical Hypothesis and accept the Research Hypothesis.
      b. If the test statistic is less than the tabled value, fail to reject the Statistical Hypothesis and fail to accept the Research Hypothesis.

THE POPULATION OF INTEREST

In a typical research situation, a *population* is a collection of individuals about whom one is seeking information. The population information to be estimated is assumed to have a *true* value. For example, theoretically there is a true mean value ($\Sigma X/N$) for the height of all men in the United States over the age of 21. This value is called a *parameter* (a population fact). There is a true value (another parameter) that represents the dispersion of the height of men from the true mean (population variance, $\Sigma(X - \bar{X})^2/N$).

With most populations of interest, it is usually impossible to find and measure every individual, so it is impossible to arrive at the parameter value. A researcher therefore selects a subset of the population of interest and measures the individuals in the subset to make a statement about the population.

A *sample* is a collection of individuals who are a subset of individuals representing the research population. The sample information to be collected is assumed to be an *estimate* of the population parameter.

Whenever a random sample is drawn, the mean of that sample is the best estimate of the population mean and is called an *unbiased estimate of the population mean*. On the other hand, the sample variance is called a *biased estimate of the population variance*. (An explanation of that statement can be found in other statistical texts.) While the sample variance, $\Sigma(X - \bar{X})^2/N$, is a biased estimate, the unbiased estimate of the population variance is $\Sigma(X - \bar{X})^2/(N - 1)$.

Both the sample mean and the sample variance are subject to sampling error. For example, assume that we are interested in a population of 100 men in a particular location. We could randomly select 20 men from this population, measure their heights, and calculate a mean value for these 20 men. This mean value is the best estimate of the average height of the population of 100 men that can be obtained from 20 men. Another sample of 20 men from this population can be randomly selected and measured, and a mean value of the measures can be calculated. The likelihood of these two means being exactly the same is very small. Yet one would expect them to be somewhat close in value. Furthermore, if one measured the total population and calculated the population mean, one would expect this parameter value and the two sample mean values to be very similar but not exactly the same. The difference between a population mean and a sample mean is assumed to be due to sampling error—of course, this may not be exactly true because there may be systematic errors of measurement in the sample. If one assumes there are no systematic measurement biases, then the random errors of measurement should cancel out, and the sample mean would approximate the population mean. Errors of measurement are treated extensively elsewhere and are beyond the scope of this chapter.

In real-life situations, the researcher is typically unable to obtain population parameters for the population of interest. For example, in a particular school system a researcher may be interested in the relative merits of Method "e" when compared with Method "f" for 7th-grade boys. He could give Method e to half the 7th-grade boys and give Method f to the other half. He could then use the mean scores on each of the two groups (on some test designed to measure success in the subject matter being taught) to *describe* the relative success of the two treatments *for these 7th-grade boys*. Subsequently, he would wish to use the data from these two groups to generalize to (or *predict* the success of these two methods for) a much broader population of students. In this example, he may wish to use the results of the methods for the two sample groups to generalize to 7th-grade boys in this school for the next, say, two years. He would be assuming that this year's 7th-grade class represents a random sample of the population of 7th-grade boys in the school district for the three years of interest (this year plus the next two years). Of course, something may happen in the community or school to change the composition of the next two years' 7th-grade boys, so that this year's boys do not represent that changed population. His confidence in his ability to predict, or generalize, would then be reduced by an unknown amount. His generalization might be totally invalid; without further data and analysis of that data, his original sample means are still the best information from which to predict the effects of the treatments.

The researcher wants to make a decision regarding which method to use for the next two years' 7th-grade boys. Since those two groups of boys are not yet 7th graders, he cannot expose them to the two methods and then decide. He therefore wants to use this year's 7th-grade boys to *predict* for the next two years of 7th-grade boys. Suppose there are 100 boys in the present year's 7th-grade class. Each boy can be assigned a number from 1 to 100; and using a table of random numbers, the researcher can assign 50 boys to Method e and 50 boys to Method f. This sampling procedure results in two treatment samples that are likely to be representative subsets of the total population being investigated. Random sampling does not guarantee that the samples are representative of the population, but it does minimize the chances that the samples are biased subsets of the population.

These boys can be subjected to the treatment to which they were assigned, and then the criterion performance can be measured on some task the methods were designed to teach. Suppose the following information is obtained: a mean value of 55 and variance of 25 for the sample receiving Method e, and a mean value of 35 and variance of 25 for the sample receiving Method f. Originally, the two groups represented the same population; as a consequence of the treatment, do they still represent the same population? Are the means of 55 for Method e and 35 for Method f different enough to consider Method e and Method f to be differentially effective? Or are the differences likely to be due to sampling variation? We now need to test whether the difference between the means is large enough to be considered "significant."

ANALYSIS OF VARIANCE

The ANOVA technique analyzes estimates of the population variance and yields a probability statement showing how likely it is that observed differences between means are due to sampling error. If the differences are highly unlikely to have been caused by sampling error (say 1 time in 1,000), one may be willing to conclude that, as a result of the two treatments, the two groups of boys now represent different populations. Risk and cost must enter into the decision of what level of probability to select. These matters will be discussed later.

The F statistic is based upon the ratio between two estimates of the population variance. In research designs there are various ways to estimate population variance, some sensitive to group mean differences and others that are insensitive to group mean differences. In our present example, two estimates are of interest: the within-group estimate of the population variance, and the among-group estimate of the population variance.

Within-Group Estimate of the Population Variance

The estimate of population variance using within-group information in analysis of variance is usually called *mean square within* (symbolized in many texts as MS_w). Our preference in this text is to refer to this variance estimate as the *within-variance estimate* and to use the symbol $\hat{v}_w$, where (a) the small v is used to represent the population variance; (b) the hat (ˆ) indicates an estimate; and (c) the subscript, w, indicates how the estimate is calculated (within group).

The $\hat{v}_w$ is an estimate of population variance based on how much the criterion scores of persons in each group differ from the mean score of their group and is calculated in several steps. First, the mean of one group is calculated, and that group mean is subtracted from each of the observed criterion scores (X) of the members of that group. Second, each of these discrepancy scores is then squared, and the squared scores are summed. This procedure is followed for each of the k groups. (In the above example, k is 2, the two groups being Method e and Method f.) The sums of squared discrepancy scores for all groups are then added (sum of squares within, or SS_w) and divided by the degrees of freedom within all the groups. The result is $\hat{v}_w$.

As typically presented, "degrees of freedom" is difficult to understand, though it is a simple concept. A more elaborate explanation of this matter is given later. A brief discussion is presented here because each estimate of the population variance has its own degrees of freedom.

A *variable* by definition can take on any of a range of numerical values. A variable is thus free to vary. Within Group e (Method e) there are 50 boys, and there are 50 observations on the criterion variable. Each of the 50 observations or scores is free to vary. Each score is considered free to vary because each represents one of the 50 boys and is therefore dependent on the performance of that boy.

After obtaining the 50 scores for Group e on the criterion variable, one can calculate the mean of those 50 scores ($\Sigma X/N$). When the differences between the observed scores and the mean are calculated, are all 50 difference scores free to vary? (Note that it is a property of the mean that the sum of the difference scores between the mean and the observed scores is equal to zero.) Start with any one score, and it is free to be any value; that is, the first score is free to vary—it can be any value. Also, each subsequent score through observation 49 is free to vary. But what about the 50th score? That score is *not* free to vary because each of the preceding 49 scores is specified, and the group mean is known. Given the 49 scores and the mean, there is only one value the 50th difference score can be. It is not free to vary.

The three scores of 4, 3, and 2 will help to explain the concept of degrees of freedom. The mean of these three scores is $\Sigma X/N = 9/3 = 3$. Now calculate the differences between each observed score and the mean. Given that the mean is 3, if one did not already know the actual scores, one might think that those three scores could take on an infinite number of values to get a mean of 3. If the first observed score is 4, then the difference score for that observed score is 4 - 3 = 1. (That observed score was free to vary; it happened to be 4.) If the second observed score is 3, the difference score for it is 3 - 3 = 0. That score was also free to vary. Now that it is known that two of the scores are 4 and 3, there is only one value the third score can be and still yield a mean of 3 for the three scores (and still have the difference scores sum to zero). That third score would, of course, be 2 because 2 - 3 = -1. Of the three scores, only two are free to vary once the mean is known. To generalize: When the mean is known, one less than the total number of scores is free to vary (thus, when researching variability within a group, degrees of freedom is the total number of subjects less one, or N - 1).

When the problem of calculating $\hat{v}_w$ was presented, it was given that for Method e, the mean score was 55, and the sample variance was 25; for Method f, the mean score was 35, and the sample variance was 25. One could go back to the 100 observed scores (50 for each group) and calculate $\hat{v}_w$ from the procedure described above. It is easier, however, to use the knowledge of the group means and variances. The variance for Group e (25) was calculated by finding the difference between each score and the group mean, squaring and summing the difference scores, and dividing the sum by the number of scores (50), or $\Sigma(X - \bar{X}_e)^2/N$. To obtain the unbiased estimate of the population variance ($\hat{v}_w$) one does not divide by N but by degrees of freedom. The degrees of freedom here is N - 2, where N is the number of subjects, and 2 is the number of groups. The separate sums of squares for each group are summed together before dividing by degrees of freedom, so the sum of squares for each group is needed. If the Group e variance was 25 and N = 50, then from the above formula one can see that the sum of squares for Group e must equal 1,250 (25 * 50). The variance and N are the same for Group f, so the sum of squares for Group f also must equal 1,250.

The $\hat{v}_w$ is calculated by adding the sum of squares of the k groups and then dividing by degrees of freedom. This problem contains two groups, and the sum of squares for each is 1,250. The total degrees of freedom is equal to the sum of the degrees of freedom for each group. In each group, a mean was calculated, so the degrees of freedom for each group is 50 - 1 = 49, and the total degrees of freedom would be 98. One also could arrive at the total degrees of freedom by looking at the total number of observations in both groups (N) and subtracting the number of means calculated (k), or in this case 100 - 2 = 98.

The $\hat{v}_w$ is therefore calculated in Equation 2.1 for the example problem as:

$$\hat{v}_w = \frac{\Sigma (X - \bar{X}_e)^2 + \Sigma (X - \bar{X}_f)^2}{N - k}$$

$$= \frac{1{,}250 + 1{,}250}{100 - 2} = \frac{2{,}500}{98} = 25.5$$

$$\hat{v}_w = 25.5 \tag{2.1}$$

This value (25.5) is the best estimate of the population variance because it is not influenced by group mean differences. It is best in the sense that repeated calculations on random samples from the same population would result in variance estimates that deviate the least from the actual population variance. If the group mean differences are small, then the estimate of the population variance using among-group information ($\hat{v}_a$ discussed next) should be close to the within-group estimate ($\hat{v}_w$).

Among-Group Estimate of the Population Variance

The among-group estimate of the population variance is symbolized here as: $\hat{v}_a$. The subscript denotes that it is an estimate using among- or between-group information.

The $\hat{v}_a$ is derived by subtracting the mean of all the boys, regardless of group membership (this is often called the *grand mean*, $\bar{X}_g$) from each of the k group means. For each group, this value is then squared and multiplied by the number of scores in the group. The product for each group is summed over groups (SS_a) and divided by degrees of freedom, which for this variance estimate is k - 1.

Degrees of freedom for $\hat{v}_a$ is equal to the number of groups (k) minus one (1). Following the logic given for $\hat{v}_w$, each group mean is an observation free to take on any value, but once the grand mean is calculated using these means, only k - 1 means are free to vary. In this example, $\bar{X}_e = 55$, $\bar{X}_f = 35$, $\bar{X}_g = (55 + 35)/2 = 45$. For this case, $\bar{X}_g$ can be derived by getting the average of the two means because the groups are of equal number. When groups differ in number, the above procedure is not applicable, but $\bar{X}_g$ can be obtained by summing all scores and dividing by N.

The $\hat{v}_a$ is therefore calculated in Equation 2.2 for the example problem as:

$$\hat{v}_a = \frac{[\,(\bar{X}_e - \bar{X}_g)^2 * N_e\,] + [\,(\bar{X}_f - \bar{X}_g)^2 * N_f\,]}{k - 1}$$

$$= \frac{[\,(55 - 45)^2 * 50\,] + [\,(35 - 45)^2 * 50\,]}{2 - 1} \tag{2.2}$$

The *F* Ratio

Two estimates of the population variance, $\hat{v}_w$ and $\hat{v}_a$, have just been discussed. Within-group estimates of the population variance are unaffected by group mean differences. Among-group estimates of the population variance are sensitive to group mean differences because they are based upon group mean variations from the grand mean. If there is no real difference between the two populations, then $\hat{v}_a$ will be a good estimate of the population variance and would be expected to be about equal to $\hat{v}_w$ (the best estimate of the population variance).

Consider the discussion of random sampling. When two samples are randomly drawn from a population, the sample means are expected to vary somewhat from each other. The difference between sample means is referred to as *sampling error*. The larger the number of subjects in a sample, the closer the sample means tend to be. Occasionally, however, large mean differences will be observed.

The *F* ratio represents the distribution of the ratios of two independent estimates of the population variance. It can be mathematically shown that the expected *F* ratio is 1.00 if the two samples are representative of a single population. The randomly selected groups will occasionally have large mean differences due to sampling variation. Therefore, an *F* ratio much greater than 1.00 would be obtained. Likewise, the randomly selected groups will occasionally have small mean differences due to sampling variation. In these instances, an *F* ratio less than 1.00 would

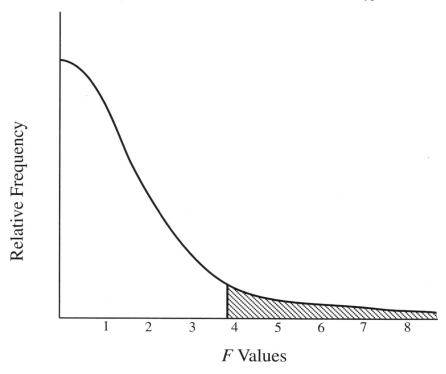

Figure 2.1.
An approximate distribution of F with 1 and 98 degrees of freedom.

be obtained. If the ratio of the two estimates is distributed as F, then for any set of degrees of freedom, the likelihood of an observed F ratio being due to sampling variation can be determined.

Figure 2.1 is an approximate distribution of F ratios with 1 (k - 1) and 98 (N - k) degrees of freedom. The area shaded to the right of 3.94 represents the 5% of the observed F ratios that will occur from the samples whose means are most discrepant. An F ratio of 3.94 or greater would be observed 5% of the time due to sampling variation. An F ratio of 6.90 or greater would be observed 1% of the time due to sampling variation. These values of 3.94 and 6.90 can be found in F tables in many statistics books. In the example with Method e and Method f, the estimated population variance using within-group data was $\hat{v}_w = 25.5$; and the among-group estimate was $\hat{v}_a = 10,000$. In Equation 2.3, the F ratio is:

$$F = \frac{\hat{v}_a}{\hat{v}_w} = \frac{10,000}{25.5} = 392.15; \, df = 1,98 \tag{2.3}$$

Statisticians tell us that an F of 392 with 1 and 98 degrees of freedom would be observed less than 1 time in 100 due to sampling variation. Since $\hat{v}_a$ is so very large in relation to $\hat{v}_w$, one can conclude that the difference between the two observed means (Method e mean was 55; and Method f mean was 35) is unlikely to be due to sampling variation. Most likely, the two treatments have resulted in two populations—

one of 7th-grade boys exposed to Method e and another population of 7th-grade boys exposed to Method f.

On the other hand, one must be aware that 1 time in 100 such a large mean difference is expected due to sampling variation. Suppose you are the decision maker regarding the selection of a method of instruction. What would you recommend? Remember, you have already stated a rival (statistical or null) hypothesis:

The mean of the criterion measures is the same for each treatment population.

When should one reject the Statistical Hypothesis? This depends upon the amount of risk one wishes to take. In the case just presented, the researcher asks, "If my observed sample mean differences are that 1-in-100 that result from random sampling variation, how much *will* that mistake cost? On the other hand, if the sample results are indicative of the population, how much *will it not* cost to act as if they are?" The probability that the present results came from a common population is very small. Whether this probability is tolerable must be determined by the researcher.

Figure 2.2a represents the two samples drawn from a theoretical population. When statistical significance is obtained as a result of differential treatment effects, one can say with some degree of confidence that the two samples no longer represent the same population with respect to the criterion measure of concern. Those two samples now represent two populations, Treatment e population and Treatment f population.

Figure 2.2b represents the other state of affairs. Statistical significance has not been obtained, and so the differential treatments have not produced a large enough difference to conclude that they have different effects on the criterion measure of concern.

Type I Error

The decision to reject or fail to reject the Statistical Hypothesis is essentially one of risk taking. Usually, the probability level of either 5 times in 100 or 1 time in 100 (alpha = .05 or alpha = .01) is the decision point used by researchers in the behavioral sciences; that is, they are willing to risk making 5 errors (or 1) out of 100 rejections of a true Statistical Hypothesis (a Type I error). The choice of alpha is up to the decision maker and the degree of risk she wishes to take. The level of alpha defines the probability of making a Type I error. Also, the decision maker must recognize that taking a conservative risk (e.g., 1 time in 1,000 instead of 1 time in 100) regarding Type I error increases the probability of making a Type II error.

Type II Error

The decision maker has two possible conclusions she can make from the Statistical Hypothesis testing procedure. She can reject the Statistical (rival) Hypothesis, or she can fail to reject the Statistical Hypothesis. One hopes that a correct decision is made, but because one never knows the population value(s), one can never be sure that the correct action has been taken. The Type I error (rejecting a Statistical Hypothesis when it should not be rejected) is discussed above. The researcher has complete control over the probability of error in the choice of the alpha level (when all assumptions are met). If alpha is .05, then 5% of the time the Statistical Hypothesis will be incorrectly rejected (if the Statistical Hypothesis is true).

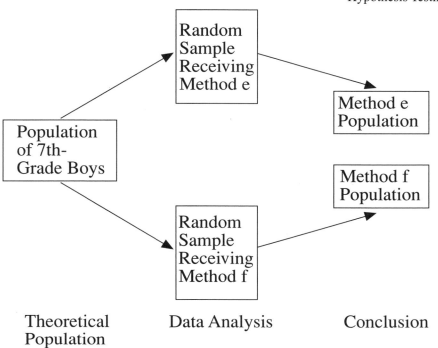

Theoretical
Population   Data Analysis   Conclusion

Figure 2.2a.
Conclusion, when statistical analysis *shows* significance.

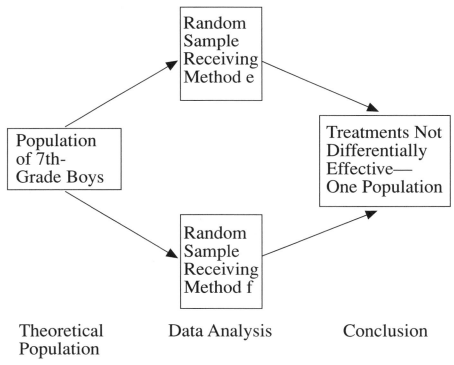

Theoretical
Population   Data Analysis   Conclusion

Figure 2.2b.
Conclusion, when statistical analysis *does not* show significance.

On the other hand, the other decision—failing to reject the Statistical Hypothesis—also may be in error (Type II error). If the Statistical Hypothesis is not true of the populations, but the sampled data do not provide enough evidence to reject it, then an error is made. The probability of the Type II error decreases as sample size increases, and it also decreases as the alpha (probability of a Type I error) increases. The probability of Type II error also decreases as the discrepancy between the statistically hypothesized population value and the true population value increases. The true population value is never known, therefore the probability of a Type II error being committed cannot be determined exactly. Because the consequences of making a Type II error are usually less devastating (e.g., failing to accept some new fact that is correct) than the consequences of making a Type I error (e.g., accepting a new fact that is incorrect), the alpha levels chosen by most researchers are low (e.g., .05, .01, or .001). The content of the Research Hypothesis should always determine the cost of making either a Type I or a Type II error. Though the probability of making a Type II error is difficult to calculate, the researcher needs to be aware of the concept to realize that when she fails to reject the Statistical Hypothesis she *may* be making an error. The calculation of Type II error through *power analysis* is discussed in chapter 8. (Keep in mind that power is equal to one minus the probability of making a Type II error.)

Assumptions Related to the *F* Test

A few assumptions must be met before the ratio between any two estimates of the population variance is distributed as F. There are four basic assumptions of concern.

1. All subjects in the treatment groups were originally drawn at random from the same parent *population*.
2. The variance (v) of the criterion measures is the same for each treatment *population*.
3. The criterion measure for each treatment *population* is normally distributed.
4. The means of the criterion measures are the same for each treatment *population*.

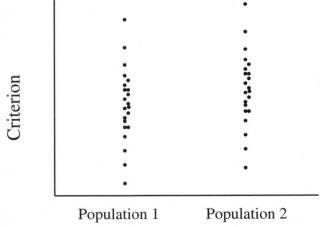

Figure 2.3.
Illustration of two populations that meet Assumptions 2 and 3 but not 4.

Figure 2.3 illustrates Assumptions 2 and 3, for the case where two populations are under consideration. Note that these assumptions refer to the populations. Since the populations are never available, the validity of the assumptions is never known for sure. Tests of significance are available to test Assumptions 2 and 3; but when using such tests, the researcher hopes to fail to reject the Statistical Hypothesis. This situation is opposite to the usual hypothesis-testing procedure and should require different values of alpha. The desire in Assumption 2 is to show that the *variances are not significantly different* and in Assumption 3 that the population distribution is not significantly different from a normal distribution. In both cases, the traditional hypothesis-testing process is reversed, in that an equivalent population state of affairs is desired. The Statistical Hypothesis, instead of the Research Hypothesis, is hoped to be true. If one wants to protect against falsely accepting this hypothesis, then alpha should be set at a much higher level—.25 is the level that some in the literature suggest, but we take a less liberal stance and encourage use of a much higher alpha, say .60. Because a small number of subjects leads to less likelihood of finding significance, this suggestion is particularly relevant when few subjects are available. If one really wants to establish "no difference," then one should obtain a large N size. Nonsignificance is particularly meaningful with large N.

Investigations into these assumptions have generally concluded that Assumptions 2 and 3 can usually be violated without seriously distorting the stated alpha level. The fourth assumption is really the Statistical Hypothesis (often referred to as the *null hypothesis*) being tested. The position we take in this text (and defend more fully in later chapters) is that attempting to obtain an R^2 close to 1.00, along with replicating findings, is a necessary and sufficient guard against any violation of assumptions.

The assumptions can be viewed in this fashion: When a significant F is obtained, it could be due to any of the four (or a combination) of the assumptions' not being true. If the researcher is reasonably sure that Assumptions 1, 2, and 3 are true, then Assumption 4 can be rejected. For example, with 1 and 100 *df*, an F of 6.90 or greater is observed 1 time in 100 due to sampling variation. If one specifies .01 as the amount of risk one will tolerate to reject a true Statistical Hypothesis, given an F larger than 6.90 and faith in the first three assumptions, one would reject the condition that the criterion means are equal for the two treatment populations. Woehlke, Elmore, and Spearing (1990) show how the assumptions underlying General Linear Model can be investigated in both SAS and SPSS. Some of those notions are incorporated in the SAS examples in this text.

Elaboration on Assumptions

The assumptions given above are applicable when one is dealing with treatment groups as in Figure 2.4a; however, they can be extended to cover a linear fit when the independent (predictor) variable is continuous. Assumptions 2 and 3, regarding equal variance and normal distribution, can be extended to the situation pictured in Figure 2.4b.

One may wish to establish a linear relationship hypothesis, such as "*There is a linear relationship between ability and performance*" (as depicted in Figure 2.4b). It is assumed that, for each data point on the continuum of the predictor variable of ability, the criterion values have equal variance and are normally distributed in the population. Given a sufficiently large sample (say, greater than 100 subjects per variable), it is not too serious if there is a violation of the assumptions of equal variance and normal distribution.

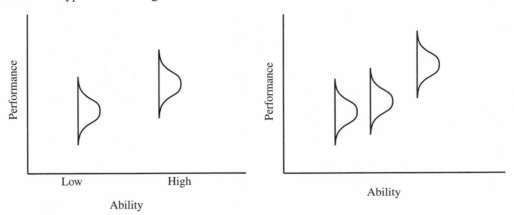

Figure 2.4a.
A relationship between two groups and the
criterion.

Figure 2.4b.
A linear relationship between two continuous
variables.

R²: THE PROPORTIONAL ESTIMATE OF SUMS OF SQUARES

The F ratio also can be derived using R^2, which is a proportional estimate of the total observed sums of squares for each source of variance. In the two-group case of Method e and Method f, an among-group sum of squares (which was 10,000) and a within-group sum of squares (2,500) was calculated. A total observed sum of squares also could have been calculated by subtracting the grand mean from each of the 100 student's scores, squaring the difference scores, and then summing these 100 squared scores. This total sum of squares (SS_t) would be 12,500, equaling the among sum of squares (10,000) plus the within sum of squares (2,500).

The range of R^2 is from .00 to 1.00 and is an expression of the proportion of the total sample criterion variance accounted for by the particular set of predictor information. The total sample variance can be calculated by dividing the total sum of squares by the number of subjects: Total variance = SS_t/N. The variance accounted for by knowing the group means can be calculated by dividing the sum of squares among by the number of subjects: Variance accounted for by knowing group means = SS_a/N. R_a^2 (which uses among-group data) is simply SS_a divided by SS_t. Therefore, R_a^2 is:

$$R_a^2 = \frac{SS_a}{SS_t}$$

and for this example, R_a^2 is equal to:

$$\frac{10,000}{12,500} = .80$$

In this example, SS_a is 80% of SS_t. The SS_a relies upon knowledge of which group (e or f) the subject is in; thus, knowledge of group membership accounts for 80% of the total observed sum of squares (SS_t). The reader should note that such a high R^2 is usually not associated with a treatment group.

R_w^2 (which uses within-group data) is SS_w divided by SS_t. Therefore, R_w^2 is:

$$R_w^2 = \frac{SS_w}{SS_t}$$

and for this example, R_w^2 is equal to:

$$\frac{2,500}{12,500} = .20$$

The sum of squares within reflects the variation within the groups that is not directly due to the two treatments. SS_w is 20% of the total sum of squares; thus 20% of the observed SS_t is due to unknown sources.

What does introducing the notion of R^2 provide that was not known when using estimates of the population variance ($\hat{v}_a$ and $\hat{v}_w$)? The R^2 notion provides a different perspective regarding variation and research designed to explain variation. Indeed, this different perspective is one of the fundamental reasons for this book.

The R^2 Perspective

Given any one sample from a population, the best source from which to estimate the population variance is the observed variability of the sample. In the sample of 100 subjects given Methods e and f there was a total sum of squares of 12,500. The estimated population variance ($\hat{v}_t$), based on the sample, was found by dividing SS_t by degrees of freedom (*df*). The grand mean had to be calculated to get SS_t; therefore, degrees of freedom must be N - 1. Once the grand mean is known, only 99 scores are free to vary; the last score is fixed; $\hat{v}_t$ then equals:

$$\hat{v}_t = \frac{SS_t}{N-1}$$

and for this example $\hat{v}_t$ equals:

$$\frac{12,500}{99} = 126.2$$

Using the R^2 perspective, given only a sample and a sample mean, 100% of the variance in the criterion scores is due to unknown sources. A researcher's goal is to find out what accounts for variance in the criterion—why some subjects score high, some low. Some of the variance might be due to different treatments, ability, motivation, and so on. The task of research is to seek out information that will account for all of the criterion variance, or for as much of it as is temporarily satisfactory.

In the case of the 100 7th-grade boys, an estimated population variance ($\hat{v}_t$) of 126.2 was observed. The researcher then brings into the picture the information that these boys were randomly assigned to two different methods of instruction. One might now wonder, "Does this treatment information explain some of the criterion variance?" As was shown above, R_a^2 was based upon knowledge of which boys received which treatment. It was found that this treatment information explained 80% of the total variance ($R_a^2 = .80$).

Now that it is known how much criterion variance can be accounted for by using knowledge of treatment groups, can it be determined how likely it is that this large an R2 value would be due to sampling error—as was done with the F ratio previously? As a matter of fact, it can; the F ratio can be calculated from R_a^2 and R_w^2 as well as from $\hat{v}_a$ and $\hat{v}_w$.

Using R_a^2 and R_w^2 (two proportional parts of the total sample variance), it can be determined whether the knowledge of group membership (Methods e and f) will explain a nonchance amount of the observed sample variance. The formula in Equation 2.4 for the F test can be used:

$$F_{(df_n, df_d)} = \frac{R_a^2 \,/\, df_n}{R_w^2 \,/\, df_d} \tag{2.4}$$

where:

R_a^2 = the proportion of unique observed variance due to knowledge of group membership;

R_w^2 = the proportion of unique observed variance due to unknown sources;

df_n = the number of group means free to vary once the grand mean has been calculated; and

df_d = the number of subjects whose criterion scores are free to vary once each group mean has been calculated.

As previously determined, R_a^2 equals .80, and R_w^2 equals .20. The degrees of freedom numerator, df_n, equals 2 - 1, or 1, because only one of the two group means was free to vary once the grand mean was calculated. The degrees of freedom denominator, df_d, is 100 - 2 because within each group the last (50th) score was not free to vary (the group mean plus the other 49 scores determined the last boy's score). Substituting these values into Equation 2.4, one obtains Equation 2.5:

$$F_{(df_n, df_d)} = \frac{R_a^2 \,/\, df_n}{R_w^2 \,/\, df_d} = \frac{.80 \,/\, 1}{.20 \,/\, 98} = \frac{.80}{.0020}$$

$$F_{(1,98)} = 392.15 \tag{2.5}$$

This F value is exactly the same as obtained by using $\hat{v}_a = 10,000$ and $\hat{v}_w = 2,500$.

What has been gained by using the R^2 approach? Not only can one determine how likely it is that the R_a^2 is due to sampling error, but one also gets an idea of what proportion of the sample variance can be explained by the piece of information, in this case group membership. Furthermore, the R_w^2 shows how much ignorance still exists concerning the source of the observed criterion variance.

For this example, $R_a^2 = 1 - R_w^2$, so the two points might seem redundant. This will not always be the case. The more general equation for F is developed in chapter 4.

Theoretical and Heuristic Benefits of R^2 in Analysis of Variance

The great amount of variance accounted for by the two methods in the above example is seldom observed in the behavioral sciences. Significant results often translate into extremely small R^2 values because large numbers of subjects can yield "statistically significant" results yet account for little variance. One study reported statistical significance with the R^2 value of .01. The authors likely thought that R^2 was similar to probability—the lower the better. One could paraphrase these authors' results as "knowing for sure about very, very little."

Heuristically, the use of R^2 keeps one's ignorance in full view because $1 - R^2$ is the proportion of variance yet to be accounted for. If theoretical expectations of causal relationships are empirically found to account for less than 50% of the criterion variance, then it seems that additional "relevant" variables must enter into the theoretical model. The determination of variables that are relevant variables is empirically a function of the criterion variable and the other predictor variables that are already in the predictive system. And, of course, such model building is in the spirit of chapter 1. Researchers should want as much as possible to reduce the unaccounted-for variance—the researcher's true enemy.

Many studies reported in journals give sufficient information to allow the reader to calculate the R^2 value. A preferable procedure would be one in which the researcher reported the R^2 in the published study, but this is not commonly done. R^2 is one way of representing effect size and as such should be reported in all research. Chapter 8 contains procedures that enable the reader to calculate R^2 values from information commonly found in published work. Appendix E contains an Analysis of Variance (ANOVA) Source Table that was produced from General Linear Model results.

3

VECTORS AND
VECTOR OPERATIONS

INTRODUCTION TO VECTORS

The general linear model can be most easily understood within the framework of vector notation. If a researcher is to develop a command over the GLM, she must become somewhat familiar with vector notation and vector operations. Once the researcher has grasped the basic components of vector algebra, she can construct for herself the models used to test a particular research question. Thus, knowledge of vector algebra, coupled with mastery of the GLM, allows the researcher to ask the research question in the particular way she wants and then to generate the statistical test. This approach may seem heretical to those researchers who were taught that they must frame their research question in a particular fashion or that only certain research questions are permissible.

Most people come into contact with vectors in their everyday lives; thus, the mastery of vector algebra is not really a difficult chore. Many properties of vector algebra are simply generalizations from the algebra with which most readers are already familiar.

Why Use Vectors?

A vector is an ordered set of numbers that allows data to be represented in a very concise fashion. While there are other kinds of vectors, for the purposes of this text, a vector is simply an ordered set of numbers. Once the conventions are learned and the symbols are identified, a large amount of data can be represented in a small amount of space. A familiar example of vectors is shown in Table 3.1.

Table 3.1 is one that may be familiar. The same kind of measurement has been made for each entity represented (i.e., the noon temperature has been recorded for each city listed). The collection of temperatures is called a *vector* because each member of the collection is, in fact, a number; and the members (or *elements*) are in some particular order. That is, the temperature of 20 degrees was observed in Chicago, and the temperature of 30 degrees was observed in St. Louis.

The data in Table 3.2 show another use of vectors, representing several measurements of a single entity. The information could have come from a biographical

Table 3.1

Example of a Single Vector: The Same
Measurement for Each of Several Entities

City	Noon Temperature (Degrees Fahrenheit)
Chicago	20
Miami	80
Pittsburgh	45
St. Louis	30

Table 3.2

Example of Several Measurements of a
Single Entity

Height (inches)	68
Weight (pounds)	150
Age (years)	25
Schooling (years)	19
Married (yes = 1; no = 0)	1
Children (total number)	1

data sheet. If one were given the following vector (Y) and told that this vector represents the same kind of information as presented in Table 3.2, then one would know that this individual is 72 inches tall, weighs 300 pounds, is 30 years old, has 12 years of education, is not married, and has no children.

$$Y = \begin{bmatrix} 72 \\ 300 \\ 30 \\ 12 \\ 0 \\ 0 \end{bmatrix} = \begin{matrix} \text{height of subject in inches} \\ \text{weight of subject in pounds} \\ \text{age of subject in years} \\ \text{years of education of subject} \\ 1 = \text{married; } 0 = \text{not married} \\ \text{total number of children} \end{matrix}$$

The data in Table 3.3 represent a combination of the kinds of vectors previously described. Here there are several observations of each of several entities. For the purposes of this book, the focus is on these kinds of vectors—vectors that represent the same kind of measurement of several subjects. Because there are four pieces of information about each entity or subject in Table 3.3, there are four vectors in Table 3.3. Someone has collected three pieces of information and then calculated a fourth piece of information, batting average, from two of the pieces of information (dividing "hits" by "at bats"). This fourth piece of information is considered to be a new piece of information, as discussed later in the chapter.

Table 3.3

Example of Several Vectors for Each Entity (Player)

Player	Home Runs	At Bats	Hits	Batting Average
James	3	120	60	.500
Sam	5	100	30	.300
Dick	15	110	22	.200
Jim	25	100	25	.250
Juan	20	120	48	.400

Definition of a Vector

A vector is an ordered set of numbers. The number of elements in a vector is called the *dimension of a vector.* With reference to Table 3.3 and given the order of the players, it is known that James hit three home runs because James is the subject corresponding to the first element of each vector, and the numerical value of the first element of the home run vector is 3. Also, since .400 is the fifth element of the batting average vector, and Juan is the subject corresponding to the fifth element, Juan's batting average is .400.

We should emphasize here that any number can be an element of a vector. Positive numbers, zeros, and decimal values have already been used. Negative numbers and common fractions are also valid candidates as elements of vectors.

The other property of a vector, that there is some order, should not be taken lightly. The order does not have to be inherent in the data; in fact, it usually is not. Referring again to Table 3.3 and the hits vector, there seems to be no order. The numbers are not arranged from high to low nor from low to high. The reader should be well aware, though, that the order has already been defined by the first column of names.

Vectors as Pieces of Information

Some readers may more readily grasp the concept of vectors by thinking of them as pieces of information. One has a vector for each piece of information. How one decides to measure a variable, such as height, will determine how many pieces of information height will provide. If one measures height in inches as a continuous variable, then we know how tall a person is with just one piece of information. If, on the other hand, one measures height in terms of short, medium, and tall, then one would have three pieces of information.

Vector Notation

Vectors in this text are represented by capital letters, sometimes with additional letters or numbers for further clarification. When a particular element of the vector is referred to, a subscript will be used. The subscript indicates the position of the element in the vector. The hits vector of Table 3.3 could be represented as H. H_1 would then be the number corresponding to James' hits (60), H_2 would be the number corresponding to Sam's hits (30), and so on. In general terms, the vector "W" with "t" elements can be represented by the symbol W, or by the bracketed symbols, or by the bracketed numbers; where the three dots mean "and so on continuing to." These three dots are necessary because the value of t is not known; there could be 50 elements of W, or 100, or just 4.

$$W = \begin{bmatrix} W_1 \\ W_2 \\ W_3 \\ . \\ . \\ . \\ W_t \end{bmatrix} = \begin{bmatrix} 6 \\ 10 \\ 3 \\ . \\ . \\ . \\ 17 \end{bmatrix}$$

Example 3.1: Vector elements. If the vector $\begin{bmatrix} 1 \\ 5 \\ 6 \\ 13 \\ -2 \end{bmatrix}$ is

represented by X, then X_1 is equal to 1, X_2 is equal to 5, X_3 is 6, X_4 is 13, and X_5 is -2. Note that X_6, X_7, etc., are not defined for this vector.

Categorical Vectors

Many occasions will be found later to use what some call *categorical vectors.* These vectors are usually helpful in representing group membership. One number is assigned to entities that belong to the group under consideration and another number to those entities that do not belong to that group. Any two numbers could be used to designate the two groups, but it is more useful to use a one (1) if an entity is a member of the group under consideration and a zero (0) if the entity is not a member of that group. *We cannot overemphasize that the judicious use of ones and zeros for representing group membership is a crucial aspect of the GLM.* Most of the statistical literature refers to these as *dummy vectors* because of the arbitrary assignment of the 1s and 0s, but we prefer to think of the 1,0 vectors as *smart vectors,* because of their utility.

Continuous Vectors

When a vector contains more than two numbers, it is referred to as a *continuous vector.* The numbers in the vector may be any one of the following measurement metrics: (a) *ordinal,* meaning that larger numbers have more of the construct; (b) *interval,* meaning that the difference between a "1" and a "2" is the same as the difference between a "2" and a "3"; and (c) *ratio,* meaning that besides the above characteristics, the scores in the vector have a true zero point. While most statistics books make a point of requiring interval data for inferential statistics, most nevertheless provide many examples with ordinal data. As was discussed in chapter 1, we argue that the purpose of research is to find functional relationships, and therefore the measurement scale of the data is irrelevant. We present examples in chapter 9 and chapter 13 of well-known physical laws that take clearly ratio scales and make them ordinal to produce a high R^2.

Example 3.2: Group membership vectors. Suppose a researcher is interested in studying the college student population and is interested in the effect on some behavior of three variables: gender, college status, and marital status. With respect to the behavioral model introduced in chapter 1, she thinks that the behavior she is investigating depends on gender, college status, and marital status. To complete the example, assume that the behavior under investigation is the degree of political liberalism. The group membership vectors that need to be constructed for this problem are:

S1 = 1 if person is a male, 0 otherwise;
S2 = 1 if person is a female, 0 otherwise;

C1 = 1 if person is a freshman, 0 otherwise;
C2 = 1 if person is a sophomore, 0 otherwise;

C3 = 1 if person is a junior, 0 otherwise;
C4 = 1 if person is a senior, 0 otherwise;

MYES = 1 if person is married, 0 otherwise; and
MNO = 1 if person is not married, 0 otherwise.

Note that two vectors are used to represent gender, and two vectors are used to represent marital status, while four vectors are used to represent college status. Another investigator might not be satisfied with this set of group designations. An additional marital status group could well be that of "divorced." This latter group might provide relevant information and thus increase the predictability of the behavior under consideration. It is up to the researcher to define the groups that are to be included. The GLM is flexible in that it will handle as many groups as the researcher is willing to define and obtain data from. Even gender can be divided into more than two groups if one is willing to redefine the variable as "gender-role interests" and divide the continuum of responses to such a questionnaire into, say, three or four groups.

Given the following vectors, defined as previously, one knows that the first person (the first element in each vector) is a male, a freshman, and is unmarried. Inspection of the second element in each vector shows a female senior who is married.

Person	S1	S2	C1	C2	C3	C4	MYES	MNO
1	1	0	1	0	0	0	0	1
2	0	1	0	0	0	1	1	0
3	1	0	1	0	0	0	1	0
4	1	0	0	0	0	1	1	0

ELEMENTARY VECTOR ALGEBRA OPERATIONS

Addition of Two Vectors

The addition of two vectors is defined as the addition of each element in one vector to the corresponding element in the other vector. An implicit requirement is that both vectors have the same number of elements in order for addition to be possible. The addition of two vectors, A and B, to produce a third vector, C, is written symbolically as:

$$C = A + B = \begin{bmatrix} A_1 \\ A_2 \\ A_3 \\ A_4 \\ A_5 \end{bmatrix} + \begin{bmatrix} B_1 \\ B_2 \\ B_3 \\ B_4 \\ B_5 \end{bmatrix} = \begin{bmatrix} A_1 \\ A_2 \\ A_3 \\ A_4 \\ A_5 \end{bmatrix} + \begin{bmatrix} B_1 \\ B_2 \\ B_3 \\ B_4 \\ B_5 \end{bmatrix} = \begin{bmatrix} C_1 \\ C_2 \\ C_3 \\ C_4 \\ C_5 \end{bmatrix} = C$$

Example 3.3: Addition of two vectors.

$$A = \begin{bmatrix} 1 \\ 0 \\ 3 \\ -4 \end{bmatrix} \quad B = \begin{bmatrix} 2 \\ 1 \\ 5 \\ 7 \end{bmatrix} \quad D = \begin{bmatrix} 3 \\ 0 \\ 4 \end{bmatrix} \quad C = A + B$$

$$C = A + B = \begin{bmatrix} 1 \\ 0 \\ 3 \\ -4 \end{bmatrix} + \begin{bmatrix} 2 \\ 1 \\ 5 \\ 7 \end{bmatrix} = \begin{bmatrix} 1 + 2 \\ 0 + 1 \\ 3 + 5 \\ -4 + 7 \end{bmatrix} = \begin{bmatrix} 3 \\ 1 \\ 8 \\ 3 \end{bmatrix} = C$$

The addition of the vectors B and D is not possible because these two vectors do not have the same number of elements.

Subtraction of Two Vectors

The subtraction of two vectors is as straightforward as the addition of two vectors. To subtract B from A to produce the vector C, one simply subtracts the corresponding elements of B from A.

Example 3.4: Subtraction of two vectors.

$$A = \begin{bmatrix} 2 \\ 0 \\ 7 \\ -4 \end{bmatrix} \quad B = \begin{bmatrix} 1 \\ 1 \\ 5 \\ 7 \end{bmatrix} \quad A - B = C$$

$$C = A - B = \begin{bmatrix} 2 \\ 0 \\ 7 \\ -4 \end{bmatrix} - \begin{bmatrix} 1 \\ 1 \\ 5 \\ 7 \end{bmatrix} = \begin{bmatrix} 2 - 1 \\ 0 - 1 \\ 7 - 5 \\ -4 - 7 \end{bmatrix} = \begin{bmatrix} 1 \\ -1 \\ 2 \\ -11 \end{bmatrix}$$

Multiplication of a Vector by a Number

Later in this text, it is frequently necessary to multiply a vector by a number; so it is important to become familiar with this operation. The multiplication of a vector by a number is defined as the multiplication of each element of the vector by that number.

$C = k * A$ (where k is a constant; and the * indicates multiplication) is computed by multiplying each element of A by the constant k.

$$C = k * A = \begin{bmatrix} k * A_1 \\ k * A_2 \\ k * A_3 \\ \cdot \\ \cdot \\ \cdot \\ k * A_t \end{bmatrix} = C$$

Example 3.5: Multiplication of a vector by a number. Suppose that several weight observations have been made on various entities, and the data need to be changed from the original unit of pounds to the new unit of ounces. Thus, because there are 16 ounces per pound, each pound observation must be multiplied by the constant 16. This operation can be represented in vector notation as:

$$Z = 16 * P$$

where the vector P represents the original observations in terms of pounds, and the vector Z represents the observations in the new units, ounces. Suppose that the original vector looked like this:

$$P = \begin{bmatrix} 2 \\ 4 \\ 6 \\ 3 \\ 0 \end{bmatrix}$$

Then the new vector Z would be:

$$Z = 16 * P = 16 * \begin{bmatrix} 2 \\ 4 \\ 6 \\ 3 \\ 0 \end{bmatrix} = \begin{bmatrix} 16 * 2 \\ 16 * 4 \\ 16 * 6 \\ 16 * 3 \\ 16 * 0 \end{bmatrix} = \begin{bmatrix} 32 \\ 64 \\ 96 \\ 48 \\ 0 \end{bmatrix} = Z$$

The reader should note that it makes sense to multiply every element of the pounds vector by 16 (the constant) because there are 16 ounces in each pound and each of the subjects' pounds information needs to be transformed to ounce information. The fourth element of vector Z, for instance, should represent (in ounces) the number of pounds corresponding to the fourth element of P. The fourth element of Z (48 ounces) is equivalent to the fourth element of P (3 pounds).

Example 3.6: The conciseness of multiplying a vector by a number. Another example is introduced here to illustrate the simplicity and conciseness of vector algebra. Suppose one wants to change a 100-element vector of observations reported in units of feet to units in terms of inches. This operation can be represented with vectors in the following fashion. Given that F is the vector of observations in feet, I is the new vector of observations in inches, and k is the multiplication constant (12 here because there are 12 inches per foot):

$$I = k * F = 12 * F$$

Again, there is not just one multiplication implied by the above expression; every element of F is multiplied by the constant k to produce the corresponding element in I.

USEFUL PROPERTIES OF VECTORS

Combining the above knowledge of vectors with some knowledge about the properties of ordinary numbers, the following five useful properties can be expressed.

Property 1. The subtraction of a vector from itself yields the null vector.

$$X + (-1)X = X - X = 0$$

The *0* represents a vector who's every element is equal to 0. Such a vector is often called the *null vector.* The property becomes useful when one has an occasion to subtract a vector from itself. The result of an operation is a vector who's every element is equal to zero, or the null vector *0*.

$$
\begin{bmatrix} X_1 \\ X_2 \\ X_3 \\ \cdot \\ \cdot \\ \cdot \\ X_t \end{bmatrix}
+ -1
\begin{bmatrix} X_1 \\ X_2 \\ X_3 \\ \cdot \\ \cdot \\ \cdot \\ X_t \end{bmatrix}
=
\begin{bmatrix} X_1 \\ X_2 \\ X_3 \\ \cdot \\ \cdot \\ \cdot \\ X_t \end{bmatrix}
-
\begin{bmatrix} X_1 \\ X_2 \\ X_3 \\ \cdot \\ \cdot \\ \cdot \\ X_t \end{bmatrix}
=
\begin{bmatrix} X_1 - X_1 \\ X_2 - X_2 \\ X_3 - X_3 \\ \cdot \\ \cdot \\ \cdot \\ X_t - X_t \end{bmatrix}
=
\begin{bmatrix} 0 \\ 0 \\ 0 \\ \cdot \\ \cdot \\ \cdot \\ 0 \end{bmatrix}
$$

Property 2. Multiplication of the sum of two or more vectors by a constant is equivalent to the multiplication of each vector by the constant and then the addition of the resulting products, as shown below (where a is the constant):

$$ a * (X + Y) = (a * X) + (a * Y) = aX + aY $$

$$
a\left(\begin{bmatrix} X_1 \\ X_2 \\ X_3 \\ \cdot \\ \cdot \\ \cdot \\ X_t \end{bmatrix}
+
\begin{bmatrix} Y_1 \\ Y_2 \\ Y_3 \\ \cdot \\ \cdot \\ \cdot \\ Y_t \end{bmatrix}\right)
= a
\begin{bmatrix} X_1 + Y_1 \\ X_2 + Y_2 \\ X_3 + Y_3 \\ \cdot \\ \cdot \\ \cdot \\ X_t + Y_t \end{bmatrix}
=
\begin{bmatrix} a * (X_1 + Y_1) \\ a * (X_2 + Y_2) \\ a * (X_3 + Y_3) \\ \cdot \\ \cdot \\ \cdot \\ a * (X_t + Y_t) \end{bmatrix}
=
$$

$$
\begin{bmatrix} aX_1 + aY_1 \\ aX_2 + aY_2 \\ aX_3 + aY_3 \\ \cdot \\ \cdot \\ \cdot \\ aX_t + aY_t \end{bmatrix}
=
\begin{bmatrix} aX_1 \\ aX_2 \\ aX_3 \\ \cdot \\ \cdot \\ \cdot \\ aX_t \end{bmatrix}
+
\begin{bmatrix} aY_1 \\ aY_2 \\ aY_3 \\ \cdot \\ \cdot \\ \cdot \\ aY_t \end{bmatrix}
=
a\begin{bmatrix} X_1 \\ X_2 \\ X_3 \\ \cdot \\ \cdot \\ \cdot \\ X_t \end{bmatrix}
+ a
\begin{bmatrix} Y_1 \\ Y_2 \\ Y_3 \\ \cdot \\ \cdot \\ \cdot \\ Y_t \end{bmatrix}
= aX + aY
$$

Example 3.7: Simplification of two vectors multiplied by the same constant. Suppose that one wanted to multiply the total scores on the Graduate Record Exam (GRE) of five subjects by a constant of 3. As illustrated below, one could multiply both the verbal (V) and quantitative (Q) sections by 3 and add these products; or one could first sum the verbal and quantitative sections and then multiply this sum (T) by the constant 3.

$$
\text{Given:} \quad V = \begin{bmatrix} 200 \\ 250 \\ 300 \\ 400 \\ 500 \end{bmatrix}
\quad Q = \begin{bmatrix} 400 \\ 450 \\ 500 \\ 600 \\ 700 \end{bmatrix}
\quad T = (V + Q) = \begin{bmatrix} 600 \\ 700 \\ 800 \\ 1000 \\ 1200 \end{bmatrix}
$$

$$3 * \begin{bmatrix} 600 \\ 700 \\ 800 \\ 1000 \\ 1200 \end{bmatrix} = \begin{bmatrix} 1800 \\ 2100 \\ 2400 \\ 3000 \\ 3600 \end{bmatrix} = 3 * \begin{bmatrix} 200 \\ 250 \\ 300 \\ 400 \\ 500 \end{bmatrix} + 3 * \begin{bmatrix} 400 \\ 450 \\ 500 \\ 600 \\ 700 \end{bmatrix} = \begin{bmatrix} 600 \\ 750 \\ 900 \\ 1200 \\ 1500 \end{bmatrix} + \begin{bmatrix} 1200 \\ 1350 \\ 1500 \\ 1800 \\ 2100 \end{bmatrix} = \begin{bmatrix} 1800 \\ 2100 \\ 2400 \\ 3000 \\ 3600 \end{bmatrix}$$

The addition of the two vectors, before multiplying by the constant, is the most frequently used option. It involves one less mathematical operation than if one were to multiply both vectors by the constant and then add the results.

The primary reason for introducing these properties is that it will become necessary later in this chapter to reduce the number of vectors used to the simplest possible form. Although the two sides of Property 2 are numerically the same, the left-hand side of Property 2, [a * (X + Y)], is in a simpler form than the right-hand side, [(a * X) + (a * Y)].

Property 3. Multiplying a vector by the sum of two constants is equivalent to separately multiplying that vector by each constant and then summing, as below.

$$(a + b) * X = (a * X) + (b * X) = aX + bX$$

$$(a + b) * X = \begin{bmatrix} (a + b)X_1 \\ (a + b)X_2 \\ (a + b)X_3 \\ \cdot \\ \cdot \\ \cdot \\ (a + b)X_t \end{bmatrix} = \begin{bmatrix} aX_1 + bX_1 \\ aX_2 + bX_2 \\ aX_3 + bX_3 \\ \cdot \\ \cdot \\ \cdot \\ aX_t + bX_t \end{bmatrix} =$$

$$\begin{bmatrix} aX_1 \\ aX_2 \\ aX_3 \\ \cdot \\ \cdot \\ \cdot \\ aX_t \end{bmatrix} + \begin{bmatrix} bX_1 \\ bX_2 \\ bX_3 \\ \cdot \\ \cdot \\ \cdot \\ bX_t \end{bmatrix} = aX + bX$$

The above illustration shows that when a vector is to be multiplied by the sum of two constants, one can either add the two constants together and then multiply the vector by the sum of the two constants or multiply the vector by the two separate constants and then add the resultant products. Adding the two constants first is much easier and quicker because fewer operations are involved.

As with Property 2, there will be many occasions to simplify vectors and their weighting coefficients. The left-hand side of Property 3 is simpler than the right-hand side, though they are numerically equal. The left-hand side clearly shows that there is only one vector; whereas the right-hand side contains two vectors that can be reduced, or simplified, to one vector. More on these notions will be presented in the section on linear dependencies.

Property 4. Multiplication of a vector by zero yields a vector with all elements equal to zero, the null vector, as shown below.

$(0) * X = 0$

$$0 * X = 0 * \begin{bmatrix} X_1 \\ X_2 \\ X_3 \\ \cdot \\ \cdot \\ \cdot \\ X_t \end{bmatrix} = \begin{bmatrix} 0 * X_1 \\ 0 * X_2 \\ 0 * X_3 \\ \cdot \\ \cdot \\ \cdot \\ 0 * X_t \end{bmatrix} = \begin{bmatrix} 0 \\ 0 \\ 0 \\ \cdot \\ \cdot \\ \cdot \\ 0 \end{bmatrix} = 0$$

No matter what the elements of a vector, if one multiplies the vector by zero, one will end with the null vector as the product.

Property 5. Multiplication of a vector by one yields that same vector. Multiplication of any vector by the constant one (1) yields the same vector, as shown below.

$1 * X = X$

$$1 * X = 1 * \begin{bmatrix} X_1 \\ X_2 \\ X_3 \\ \cdot \\ \cdot \\ \cdot \\ X_t \end{bmatrix} = \begin{bmatrix} 1 * X_1 \\ 1 * X_2 \\ 1 * X_3 \\ \cdot \\ \cdot \\ \cdot \\ 1 * X_t \end{bmatrix} = \begin{bmatrix} X_1 \\ X_2 \\ X_3 \\ \cdot \\ \cdot \\ \cdot \\ X_t \end{bmatrix} = X$$

These last two properties may seem trivial, as indeed they are—but they are useful. The reader is urged to understand all of the above five properties before proceeding. These properties of vectors, coupled with the idea of linear combinations to be discussed in the next section, form the structure of the GLM. The more adept one becomes with the ideas presented in the present chapter, the more adequately one can handle the building and simplification of linear regression models.

LINEAR COMBINATIONS OF VECTORS

Linear Combinations of Two Vectors

There will be many situations later where vectors will be combined and where it will be necessary to figure out if certain vectors are, in fact, linear combinations of other vectors. Therefore, the important idea of linear combinations needs to be defined.

Vector X is said to be a *linear combination* of vectors Y and Z if there exist two numbers (numerical constants called *weighting coefficients*), a and b (of which at least one is not zero), such that the following relationship holds:

$X = (a * Y) + (b * Z)$

This definition may become more understandable with the following examples.

Example 3.8: Determining linear combinations.

$$\text{Given:} \quad X = \begin{bmatrix} 3 \\ 4 \\ 5 \end{bmatrix} \quad Y = \begin{bmatrix} 1 \\ 2 \\ 3 \end{bmatrix} \quad Z = \begin{bmatrix} 1 \\ 0 \\ -1 \end{bmatrix}$$

X is a linear combination of Y and Z because:

X = (a * Y) + (b * Z), when a = 2 and b = 1

$$(2 * Y) + (1 * Z) = 2 * \begin{bmatrix} 1 \\ 2 \\ 3 \end{bmatrix} + 1 * \begin{bmatrix} 1 \\ 0 \\ -1 \end{bmatrix} = \begin{bmatrix} 2 + 1 \\ 4 + 0 \\ 6 + -1 \end{bmatrix} = \begin{bmatrix} 3 \\ 4 \\ 5 \end{bmatrix} = X$$

Some Special Linear Combinations of Vectors

A total test score vector (computed by simply adding the two subtest scores) is a linear combination of the two subtest vectors. In this instance, the weighting coefficients, a and b, are both equal to 1.

Example 3.9: Total score as a linear combination of subtest scores. Consider total GRE scores, which are computed by adding the verbal and quantitative GRE subtest scores. Given the verbal vector (V) and the quantitative vector (Q):

$$V = \begin{bmatrix} 400 \\ 500 \\ 600 \\ 350 \end{bmatrix} \quad Q = \begin{bmatrix} 400 \\ 300 \\ 400 \\ 750 \end{bmatrix}$$

The total GRE vector (T) is:

T = (1 * V) + (1 * Q) = V + Q

Therefore:

$$T = (V + Q) = \begin{bmatrix} 400 \\ 500 \\ 600 \\ 350 \end{bmatrix} + \begin{bmatrix} 400 \\ 300 \\ 400 \\ 750 \end{bmatrix} = \begin{bmatrix} 800 \\ 800 \\ 1000 \\ 1100 \end{bmatrix}$$

Another special case of a linear combination occurs when a vector is multiplied by a number. The weight of the "second vector" in this instance is zero; and because it is zero, the elements of the second vector are of no consequence. That is, it does not matter what the elements of the second vector are because it is already known that multiplication of any vector by zero will yield the null vector.

Example 3.10: Linear combination of two vectors.

$$\text{Given:} \quad A = \begin{bmatrix} 4 \\ 3 \\ 0 \\ 1 \end{bmatrix} \quad B = \begin{bmatrix} B_1 \\ B_2 \\ B_3 \\ B_4 \end{bmatrix} \quad C = \begin{bmatrix} 24 \\ 18 \\ 0 \\ 6 \end{bmatrix}$$

$C = (6 * A) + (0 * B)$

$$C = 6 * \begin{bmatrix} 4 \\ 3 \\ 0 \\ 1 \end{bmatrix} + 0 * \begin{bmatrix} B_1 \\ B_2 \\ B_3 \\ B_4 \end{bmatrix} = \begin{bmatrix} 6 * 4 \\ 6 * 3 \\ 6 * 0 \\ 6 * 1 \end{bmatrix} + \begin{bmatrix} 0 * B_1 \\ 0 * B_2 \\ 0 * B_3 \\ 0 * B_4 \end{bmatrix} = \begin{bmatrix} 24 \\ 18 \\ 0 \\ 6 \end{bmatrix} + \begin{bmatrix} 0 \\ 0 \\ 0 \\ 0 \end{bmatrix} = \begin{bmatrix} 24 \\ 18 \\ 0 \\ 6 \end{bmatrix}$$

The vector C is a linear combination of the vector A. The idea of a linear combination of vectors is not restricted to just two vectors. A vector may be a linear combination of more than two vectors. The following example will help to clarify this point.

Example 3.11: Linear combination of several vectors.

$$A = \begin{bmatrix} 4 \\ 3 \\ 2 \\ 1 \end{bmatrix} \quad B = \begin{bmatrix} 1 \\ 0 \\ 0 \\ 0 \end{bmatrix} \quad C = \begin{bmatrix} 0 \\ 1 \\ 0 \\ 0 \end{bmatrix} \quad D = \begin{bmatrix} 0 \\ 0 \\ 1 \\ 0 \end{bmatrix} \quad E = \begin{bmatrix} 0 \\ 0 \\ 0 \\ 1 \end{bmatrix}$$

A is a linear combination of B, C, D, and E because:

$A = (4 * B) + (3 * C) + (2 * D) + (1 * E)$

The reader should verify the above statement by carrying out the implied multiplications. The vector B is not a linear combination of vectors C, D, and E because no weighting coefficients exist such that:

$B = (a * C) + (b * D) + (c * E)$

Mutually Exclusive Group Membership Vectors

Another special linear combination of vectors occurs when mutually exclusive group membership vectors are summed.

Example 3.12: Representation of mutually exclusive vectors. Suppose that one had occasion to deal with the variables of gender and marital status. The group membership vectors may be defined as follows:

SF = 1 if subject is female, 0 otherwise;
SM = 1 if subject is male, 0 otherwise;
MY = 1 if subject is married, 0 otherwise; and
MN = 1 if subject is not married, 0 otherwise.

	SF	SM	MY	MN
Sam	0	1	1	0
Sue	1	0	1	0
Sally	1	0	0	1
Jane	1	0	0	1
Jack	0	1	0	1
Joe	0	1	1	0

SM and SF are mutually exclusive group membership vectors; that is, all subjects belong to one or the other category of male or female. Also, the married and not married categories exhaust all the possibilities of marital status (as far as the present researcher is concerned). Other categories of marital status could have been included, but evidently the researcher's question did not require any additional categories.

One way of checking to see if the stated categories are, in fact, mutually exclusive is to compute the linear combination (using all weights equal to 1), of the vectors under consideration. If these vectors are mutually exclusive, then they will consider each subject once and only once; that is, group membership vectors are represented by ones and zeros, and the resultant sum of the mutually exclusive group membership vectors will yield a vector with all elements equal to one (1).

THE UNIT VECTOR

A vector with all its elements equal to 1 is called the *unit vector* and is symbolized as U. *Because of the frequent use of the unit vector, the symbol U is reserved for that vector.* Consider adding the two gender vectors. Here a linear combination is being computed because the weights can be thought of as equal to one (1) as in Example 3.9.

$$(1 * SF) + (1 * SM) = \begin{bmatrix} 0 \\ 1 \\ 1 \\ 1 \\ 0 \\ 0 \end{bmatrix} + \begin{bmatrix} 1 \\ 0 \\ 0 \\ 0 \\ 1 \\ 1 \end{bmatrix} = \begin{bmatrix} 0+1 \\ 1+0 \\ 1+0 \\ 1+0 \\ 0+1 \\ 0+1 \end{bmatrix} = \begin{bmatrix} 1 \\ 1 \\ 1 \\ 1 \\ 1 \\ 1 \end{bmatrix} = U$$

Thus, SF and SM are mutually exclusive because their sum is equal to the unit vector (U).

MY and MN also are mutually exclusive vectors, as shown in the following:

$$(1 * MY) + (1 * MN) = \begin{bmatrix} 1 \\ 1 \\ 0 \\ 0 \\ 0 \\ 1 \end{bmatrix} + \begin{bmatrix} 0 \\ 0 \\ 1 \\ 1 \\ 1 \\ 0 \end{bmatrix} = \begin{bmatrix} 1+0 \\ 1+0 \\ 0+1 \\ 0+1 \\ 0+1 \\ 1+0 \end{bmatrix} = \begin{bmatrix} 1 \\ 1 \\ 1 \\ 1 \\ 1 \\ 1 \end{bmatrix} = U$$

The reader should now have a good feeling for the fact that *the unit vector can be considered a linear combination of mutually exclusive group membership vectors.* In fact, the unit vector was relied upon to define mutually exclusive group membership vectors; so the above statement is simply a consequence of that definition—but a very important consequence, as will be shown in later chapters. The unit vector is assumed by computer programs to be in every regression model, so the relationship of the unit vector to the other vectors needs to be known.

PIECES OF INFORMATION

Linear Dependency

The concept of linear dependency is very important. It can be easily introduced at this point because it deals with linear combinations of vectors. A *linear dependency* occurs when one vector in a set of vectors can be expressed as a linear combination of the other vectors. Such a vector is said to be *linearly dependent* upon the other vectors. A linearly dependent vector, because it can be expressed in terms of other vectors, is redundant information—it is not a new piece of information. As such, it is not useful in terms of predicting behavior.

Example 3.13: Linearly dependent vectors. In Example 3.9, a total GRE score was expressed as the sum of the two subtest scores. The reader should verify that the verbal subtest (V) can be expressed as the total GRE (T) minus the quantitative subtest (Q); or, $V = (1 * T) + (-1 * Q)$. It is also true that the quantitative subtest is a linear combination of the total GRE and the verbal subtest; or, $Q = (1 * T) + (-1 * V)$.

Any one of the three vectors in Example 3.9 is linearly dependent on the other two because it can be expressed as a linear combination of the other two. The total GRE can be expressed neither in terms of the verbal subtest alone nor in terms of the quantitative subtest alone. That is, no weight (a) can be found such that: $T = a * V$, nor such that: $T = a * Q$. Also, no weight (a) can be found such that: $V = a * Q$.

Linear Independency

The discussion so far has centered on linearly dependent vectors. But one will generally want to figure out how many vectors in the set of vectors are linearly *independent*, which is the opposite of linearly dependent. If there is one linear dependency in a set of three vectors, then two vectors are linearly independent. In Example 3.14, the final result is a set of two vectors of information that is said to be *linearly independent.*

A vector is said to be *linearly independent* if that vector cannot be expressed as a linear combination of the other vectors in the set. If a vector can be expressed as a linear combination of the other vectors in the set, then it is redundant information and as such must be eliminated from the set of vectors when attempting to figure out the number of linearly independent vectors in a set of vectors. The following example is intended to clarify the determination of the number of linearly independent vectors.

Example 3.14: Determining the number of linearly independent vectors. Suppose that we have four pieces of information that come from a common instrument. Information that comes from the same place is often identified with the same capital letter and given a unique number after that letter, as below.

$$V1 = \begin{bmatrix} 1 \\ 2 \\ 3 \\ 4 \\ 5 \end{bmatrix} \quad V2 = \begin{bmatrix} 2 \\ 4 \\ 6 \\ 8 \\ 10 \end{bmatrix} \quad V3 = \begin{bmatrix} 4 \\ 8 \\ 12 \\ 16 \\ 20 \end{bmatrix} \quad V4 = \begin{bmatrix} 1 \\ 2 \\ 3 \\ 4 \\ 4 \end{bmatrix}$$

V2 is a linear combination of the other vectors, because

$$V2 = (2 * V1) + (0 * V3) + (0 * V4)$$

or,

$$V2 = 2 * V1$$

Therefore, V2 is eliminated from the set of vectors for determining the number of linearly independent vectors in the set. There is now a potential set of three linearly independent vectors, V1, V3, and V4. But V3 is a linear combination of the remaining vectors:

V3 = (4 * V1) + (0 * V4)

or,

V3 = 4 * V1

Therefore, V3 is eliminated from the set of vectors for determining the number of linearly independent vectors in the set.

Two vectors, V1 and V4, remain. The problem is to find a weight (a) such that: V1 = a * V4.

Note that the first four elements of V4 must be multiplied by a weight of 1, whereas the fifth element must be multiplied by a weight of 1.25 in order to equal the elements of V1. Thus, there is no single weight that will suffice. Vectors V1 and V4 are thus *linearly independent*. Therefore, there are two linearly independent vectors in the set of vectors in Example 3.14.

One could have first eliminated from the set of four vectors V1, because V1 = (.5 * V2) + (0 * V3) + (0 * V4). Then V3 could have been eliminated because it is a linear combination of V2 and V4: V3 = (2 * V2) + (0 * V4). Examine whether V2 is linearly dependent on V4. For the first four elements, the weighting coefficient would be 2, but for the last element, the weighting coefficient would have to be 2.5. Since V2 cannot be shown to be a linear combination of V4, there are two linearly independent vectors in the set of vectors in Example 3.14.

It also can be shown that V3 and V4 are two linearly independent vectors in the set of vectors in Example 3.14. It does not matter which two vectors remain in the set; the crucial point is that, in this set of four vectors, only two vectors contain new information. The other two vectors contain redundant information and thus would not increase predictability if used in a prediction equation.

Complexity

Complexity is operationally defined in the GLM as the number of linearly independent vectors. Thus, in Example 3.14, the level of complexity is two—two pieces of information exist—but in the following example, the level of complexity is five.
Example 3.15: Complexity and the number of linearly independent vectors.

S1	S2	C1	C2	C3	C4	U
1	0	1	0	0	0	1
0	1	1	0	0	0	1
0	1	0	1	0	0	1
1	0	0	1	0	0	1
0	1	0	0	0	1	1
1	0	0	0	0	1	1

The first two vectors are the two gender vectors, and the next four vectors are the four class-rank vectors. Hint: Whenever dichotomous vectors are being considered with the unit vector, it is beneficial to leave the unit vector in the set. Of the seven vectors above,

S1 = (1 * U) + (-1 * S2), and
C1 = (1 * U) + (-1 * C2) + (-1 * C3) + (-1 * C4)

Therefore, C1 and S1 are two linear dependencies in the set. No other vectors can be eliminated, and there are five pieces of information in the set of seven vectors. (One group membership vector in each mutually exclusive group can always be eliminated if the unit vector is in the initial set of vectors. Consider the unit vector as being in the set of vectors, as that is what computer programs do.) Most linear dependencies can be found by knowing the variables and how they are defined. Indeed, one would seldom want to look at the actual data and attempt to find the weighting coefficients.

Suppose the following vectors were being considered:

U = 1 for all subjects;
M = 1 if male, 0 otherwise;
F = 1 if female, 0 otherwise;
X3 = 1 if freshman, 0 otherwise;
X4 = 1 if sophomore, 0 otherwise;
X5 = 1 if junior, 0 otherwise;
X6 = 1 if senior, 0 otherwise;
MA = Math Achievement Test A;
MB = Math Achievement Test B;
IQV = Verbal IQ;
IQNV = Nonverbal IQ; and
IQTOT = Total IQ (Verbal IQ plus Nonverbal IQ).

One of the gender vectors (M or F) can be eliminated; one of the class-rank vectors (X3, X4, X5, or X6) can be eliminated; and one of the IQ vectors can be eliminated. One probably would eliminate the total IQ, so that direct measures of both verbal and nonverbal IQ would remain in the analysis. Note that although two math achievement tests are being considered, it is highly unlikely that these two tests would be providing perfectly redundant information. To do so, they would have to be perfectly correlated, which is a very unlikely state of affairs. It could be that one of the math tests, though, is empirically linearly dependent upon the whole set of vectors. Again, this is not a *likely* state of affairs but a *possible* one. It is sometimes difficult to figure out linear dependencies by inspection of the model. Indeed, the data themselves determine whether a vector is linearly dependent. Fortunately, the computer solution will verify the actual number of linearly independent vectors, although one would want to rely on that procedure as a last resort. In the prediction of a given criterion, the number of nonzero weighting coefficients will be the number of linearly independent pieces of information.

In later chapters, the prediction of a criterion variable using a set of predictor variables will be discussed. The notion of linear dependencies will be used to eliminate redundancies from the predictor set. It is the case, though, that one of the major goals of research is to find weighting coefficients such that the criterion variable is linearly dependent upon the predictor set. It is therefore desirable to have linear dependency when considering criterion variables, whereas it is not desirable to have linear dependency when considering predictor variables.

4

RESEARCH HYPOTHESES THAT EMPLOY DICHOTOMOUS PREDICTOR VARIABLES

In this chapter we discuss several general ways to test Research Hypotheses that employ dichotomous variables. Any variable in a regression model can be either a continuous or a dichotomous variable; which is used depends on how the researcher wants to consider the variable in the Research Hypothesis. A dichotomous variable is one that can obtain only two values. For several reasons (discussed below), the values 1 and 0 are usually used. Any other two values could be used, but the usage of ones and zeros simplifies several notions. A dichotomous variable can represent a real dichotomy, such as 1 if alive, 0 if dead; or the dichotomy can be artificial, such as 1 if tall, 0 if short. Most dichotomies are artificial, the dividing line being the result of some arbitrary decision. One must keep in mind that most phenomena in the real world are of a continuous nature and that imposing arbitrary boundaries probably will decrease predictability. More will be presented on this notion in later chapters.

First, applications will be considered where the predictor is dichotomous and the criterion variable is continuous. Second, Research Hypotheses that use multiple dichotomous predictors will be discussed (with the literature on orthogonal contrasts, post hoc comparisons, and a priori hypotheses). Third, the situation in which the criterion variable is also dichotomous will be discussed.

DICHOTOMOUS PREDICTORS, CONTINUOUS CRITERION

Research Hypotheses involving two treatments use dichotomous variables. Each of the treatments can be identified by a dichotomous group membership vector. We first consider the simplest and most widely used situation involving dichotomous predictors—the two-group problem.

Research Hypotheses Involving Two Groups

Two groups of subjects are randomly selected and assigned to two different treatments. The Research Hypothesis of concern might be:

For a given population, the experimental treatment is more effective than the comparison treatment with respect to the math achievement criterion under consideration.

After a reasonable amount of time treating the two groups, the criterion measure of math achievement is obtained. For the two methods to show a difference in their effectiveness, a Full Model must be developed to allow each group to have its own mean. That model would have as predictors a group membership vector for the experimental group and a group membership vector for the comparison group, two dichotomous vectors. The Full Model reflecting the Research Hypothesis would then be:

$$(\text{Model } 4.1)\ Y = a_1 G1 + a_2 G2 + E_1$$

where:

    Y = the criterion variable of math achievement

    $G1$ = 1 if criterion score is from a student in experimental method, 0 otherwise;

    $G2$ = 1 if criterion score is from a student in comparison method, 0 otherwise; and

    a_1 and a_2 are least squares weighting coefficients calculated to minimize the sum of the squared values in the error vector (E_1).

(There are two linearly independent pieces of information on the predictor side in this Full Model. We will represent the number of pieces of information in the Full Model as m1.)

After formulating the Full Model, the task then becomes finding weighting coefficients a_1 and a_2 such that the sum of the squared values in E_1 is minimized. This task is generally accomplished by a computer solution, but several values for the weighting coefficients are used in Table 4.1 to illustrate the task. Clearly, if all three students in G1 score 5 on the criterion test, the weighting coefficient a_1 ought to be 5. Note that the value of 5 produces no error in prediction ($E_1 = 0$) for any of the students in G1. The next concern is to find the numerical value of the weighting coefficient a_2. One could use the most frequently occurring score, a value of 6. Using 6, one obtains perfect prediction for two of the three students, but overpredicts by three units for the first student in G2. By using the mean score for students in G2 as the weighting coefficient, one does not perfectly predict for any subject. But the objec-

Table 4.1
Errors Produced by Various Values for Weighting Coefficients

Y	=	$a_1 G1$	+	$a_2 G2$	+	E_1	E_1	E_1
5		1		0		0	0	0
5		1		0		0	0	0
5		1		0		0	0	0
3		5 0	+	? 1		-3	-2	-1
6		0		1		0	1	2
6		0		1		0	1	2
				$\Sigma E_1^2 =$		9	6	9
						Using 6 for a_2	Using 5 for a_2	Using 4 for a_2

tive for a weighting coefficient in most statistical analyses—that of least squares—is satisfied when the group mean of 5 is used. Note that in Table 4.1 the sum of the squared error components is lowest when the group mean is used as the weighting coefficient. No other value for the weighting coefficient will produce any lower sum of squared error components.

Ones and zeros are used to represent group membership information for two reasons. First, the mean of such a vector represents the proportion of subjects in that group. The means of a set of mutually exclusive vectors represent the proportion of subjects in each group and will therefore add to one; and if this is not so, then an error has been made somewhere. Second, and more important, the sample means for each group can be easily calculated when 1,0 vectors are used.

Most computer solutions to regression models will automatically insert a unit vector into the predictor set. The unit vector may or may not be linearly dependent, depending upon what other predictor vectors are in the system. For most Research Hypotheses, this automatic inclusion of the unit vector will cause no concern as the unit vector will be desired. (Research Hypotheses not requiring the unit vector are presented in later chapters.) A complete set of dichotomous vectors reflecting all group memberships, when used with the unit vector, will contain a linear dependency. For those readers using a matrix inversion solution, such as SAS, it would be best to omit linear dependencies from the model by omitting one of the group membership vectors. An iterative solution, with the linear dependencies retained in the model or omitted from the model, will give the same R^2 value as the matrix inversion solution that cannot tolerate linear dependencies. If a linear dependency is left in the SAS model, an error message will be produced. But the Full Model R^2 will generally be correct, although other information (such as that from test statements and restriction statements) might be erroneous. Model 4.1, with the unit vector, would appear as:

$$(\text{Model 4.1a})\, Y \;=\; a_0 U + a_1 G1 + a_2 G2 + E_2$$

Models 4.1 and 4.1a produce the same E_2 vector and the same R^2 because they contain the same information.

Given that the SAS computer solution includes a unit vector, the mean for the experimental method in Model 4.1a can be found by adding the weight for the experimental vector (a_1) to the weight for the unit vector (a_0). The mean for the comparison method in Model 4.1 can be found by adding the weight for the comparison vector (a_2) to the weight for the unit vector (a_0). A quick glance below at the Restricted Model, Model 4.2, shows that only the unit vector is in the Restricted Model. Therefore, the unit vector must be in the Full Model, either explicitly or implicitly as a linear combination of other predictors. *It must always be the case that all vectors in a Restricted Model will be in the Full Model, either explicitly or implicitly as a linear combination.*

Conceptually deriving the Restricted Model. If the two treatments are not differentially effective, then they must be equally effective; thus the Statistical Hypothesis:

> For the population, the two treatments are equally effective with respect to the criterion of math achievement.

This Statistical Hypothesis implies that the two treatments could be considered to be a common treatment or that predictability is as good *without* knowledge of which treatment a student received as it is *with* knowledge. The predictor variables in the Full Model reflect knowledge about which treatment a student received; but knowing only that the student was in the total sample would be reflected by the following model:

$$(\text{Model 4.2}) \; Y \; = \; a_0 U + E_2$$

The above model, often referred to as the *unit vector model*, in which U is one (1) for all students, simply allows an overall mean to be calculated (the value for a_0). Again, no other weighting coefficient will yield a smaller sum of squared error values for this model than will the overall mean. For the data in Table 4.1, the error vector in the Full Model will have the same values as will the error vector in the Restricted Model because the overall mean is the same as each of the group means. For the data in Table 4.1, the additional knowledge about which treatment the student received did not help in predicting the criterion score. Another way of saying this is that the two treatments were not differentially effective.

Algebraically deriving the Restricted Model. The method used above for finding the Restricted Model was to reflect the Statistical Hypothesis directly. Another way of finding the Restricted Model is to make restrictions on the Full Model implied by the Statistical Hypothesis. Some students have found that making algebraic restrictions is easier than trying to conceptualize the Restricted Model directly. If the two treatments are not differentially effective as stated by the Statistical Hypothesis, then the means of the two populations will be equal. The weighting coefficients a_1 and a_2 in the Full Model are estimates of those population parameters. Hence, restricting a_1 to equal a_2 would reflect in the sample what is being hypothesized about the population. All one does then is to rewrite the Full Model with the restriction: $a_1 = a_2$.

Step 1. Recall Full Model: $Y = a_1 G1 + a_2 G2 + E_1$

Step 2. Impose restriction of $a_1 = a_2$ by replacing a_2 with a_1:

$$Y \; = \; a_1 G1 + a_1 G2 + E_2$$

(Notice that when restrictions are imposed, a potentially different error vector is expected.)

Step 3. Simplify either by collecting vectors being multiplied by like weighting coefficients, as in this case:

$$Y \; = \; a_1 (G1 + G2) + E_2$$

or by collecting weighting coefficients that multiply like vectors (using operations described in chapter 3).

Step 4. Redefine vectors into simplest form:

$$Y \; = \; a_1 (U) + E_2.$$

Note that in the Full Model and in the Restricted Model the numerical values of a_1 will likely be different. The weighting coefficients simply hold the place for numerical values. Note also that a_1 could have been replaced with a_2 in Steps 2, 3, and 4, as follows:

Step 1a. Recall Full Model: $Y = a_1 G1 + a_2 G2 + E_1$

Step 2a. $Y = a_2 G1 + a_2 G2 + E_2$

(Here a_1 has been restricted to, or replaced by, a_2.)

Step 3a. $Y = a_2(G1 + G2) + E_2$

Step 4a. $Y = a_2 U + E_2$

The point is that it does not matter which way the restriction is imposed on the data; the same Restricted Model results. The only difference in the two Restricted Models is the symbol used for the weighting coefficient for the unit vector. But these weighting coefficients simply hold the place for a numerical value, in the case of these Restricted Models the total sample mean. The general F ratio, discussed in chapter 2 and repeated here, can be applied to the above Research Hypothesis.

If the R^2 for the Full Model is *not* significantly higher than the R^2 for the Restricted Model, then the Research Hypothesis *cannot* be accepted. This does not imply that the Statistical Hypothesis is true for the population, just that enough of a discrepancy from the restriction of $a_1 = a_2$ has *not* been observed to support the hypothesis that in the population the two means are unequal.

If the R^2 for the Full Model is found to be significantly higher than the R^2 for the Restricted Model, then the Research Hypothesis can be accepted *if* the results show that the experimental method was *more* effective than the comparison method, as stated in the directional Research Hypothesis. If the unequal means are the result of the comparison method being better than the experimental method, one would surely not want to conclude that the directional Research Hypothesis is supported. If the Research Hypothesis had been nondirectional, finding a significant difference between the R^2 values for the Full and Restricted Models would allow only the conclusion that the two treatments were differentially effective.

We believe that most researchers would like to make some directional interpretations and should therefore state a directional Research Hypothesis. That is, before the data are collected, the researcher should state which treatment is expected to be better. There are often logical reasons one treatment may be expected (or at least hoped) to be better than the other treatment. Any experimental treatment should carry an expectation of whether it will yield a higher mean (or lower mean if that is the desired direction) than a comparison treatment on a particular criterion measure. The Full and Restricted Models for directional Research Hypotheses are the same as for nondirectional ones. Before interpreting "significant" results, though, one must be sure to determine whether the results were in the hypothesized direction. Thus, if one had made the following Research Hypothesis,

> The experimental treatment is better than the comparison treatment on the criterion of math achievement,

then one would have to determine whether the mean of the experimental treatment (a_1) was indeed larger than the mean of the comparison treatment (a_2). If the weight-

Generalized Research Hypothesis 4.1

Directional Research Hypothesis: For the population of interest, Group A has a higher mean than Group B on the criterion Y.

Nondirectional Research Hypothesis: For the population of interest, Group A and Group B are not equally effective on the criterion Y.

Statistical Hypothesis: For the population of interest, Group A and Group B are equally effective on the criterion Y.

Full Model: $Y = aGA + bGB + E_3$ (pieces full [m1] = 2)

Want (for directional Research Hypothesis): $a > b$; restriction: $a = b$
Want (for nondirectional Research Hypothesis): $a \neq b$; restriction: $a = b$

Restricted Model: $Y = a_0U + E_4$ (pieces restricted [m2] = 1)

Automatic unit vector analogue
Full Model: $Y = a_0U + aGA + E_3$ (pieces full [m1] = 2)
Want: (for directional Research Hypothesis) $a > 0$; restriction: $a = 0$
Want: (for nondirectional Research Hypothesis) $a \neq 0$; restriction: $a = 0$
Restricted Model: $Y = a_0U + E_4$ (pieces restricted [m2] = 1)

where:
Y = criterion;
U = 1 for all subjects;
GA = 1 if subject in Group A, 0 otherwise;
GB = 1 if subject in Group B, 0 otherwise; and
a_0, a, and b are least squares weighting coefficients calculated so as to minimize the sum of the squared values in the error vectors.

Degrees of freedom numerator = (pieces full minus pieces restricted) = (2 - 1) = 1; (or k - 1)
Degrees of freedom denominator = (N minus pieces full) = (N - 2); (or N - k)

where:
N = number of subjects;
k = number of groups or methods. (See Generalized Research Hypothesis 4.2 for more than two groups.)

Note: Remember that pieces of information is used interchangeably with linearly independent vectors.

ing coefficients are in the hypothesized direction, then the (nondirectional) probability of the resultant F ratio should rightfully be divided by two. If the means turn out opposite to the situation that was hypothesized, then the probability of the resultant F ratio must be divided by two and then subtracted from one. In this latter case, significance would not be obtained.

The previous material leads to Generalized Research Hypothesis 4.1. The two groups may be randomly assigned as discussed in our example; or the two groups may be intact, already formed groups, such as males and females, grade 6 and grade 4, and low socioeconomic status (SES) and high SES status. These intact groups were not formed by the researcher, hence they are referred to as *nonmanipulated*. Many researchers analyze such nonmanipulated predictor variables. One should note, though, that if one gender is better than the other, one does not know that gender was the sole or even partial cause. The position we take in this text is that, although it is valuable to describe in what ways certain populations differ, a more important goal is to discover why certain populations differ, and that can only be accomplished when variables are manipulated. The goals of research are discussed in more detail in chapter 13.

Many Applied Research Hypotheses are discussed in the remainder of this text as a means of providing examples of research problems and appropriate ways of testing Research Hypotheses with linear regression analysis. In addition to these Applied Research Hypotheses, we provide many associated Generalized Research Hypotheses which offer more general examples of those hypotheses with generalized Full and Restricted Models. The Applied Research Hypotheses are based on data in the data set in Appendix A. Each provides: (a) a Research Hypothesis and a Statistical Hypothesis; (b) a Full Model and a Restricted Model; (c) the information that leads to an interpretation (R_f^2, R_r^2, F, and probability); and (d) the permissible interpretation. The reader is urged to use the given hypotheses and Full and Restricted Models to get computer solutions and to check those solutions against the ones provided here.

One should turn to Applied Research Hypothesis 4.1 to apply the general notions in Generalized Research Hypothesis 4.1 to the data in Appendix A. The SAS setup for each Applied Research Hypothesis is in Appendix D. The reader will likely want to rely on these, at least initially. The first SAS setup contains only the necessary SAS statements, while later ones contain additional options.

SAS application. To use the SAS computer package, one would have to (a) collect data (data for this section are provided in Appendix A); (b) enter the data into a SAS data file; (c) develop a SAS file containing the commands that instruct the SAS program to obtain the desired statistical information; and (d) successfully run the problem. SAS examples are provided throughout the text. There are many valuable options in SAS; selected options will be introduced in each SAS problem. The necessary SAS statements are in Appendix D-4.1.

Relationship to other statistical techniques. The nondirectional Research Hypothesis in Generalized Research Hypothesis 4.1 also could have been phrased in correlational terminology: "For the population, treatment is correlated with the criterion." Analogously, the directional hypothesis stated in correlational terminology is: "For the population, the correlation between G1 and the criterion Y is positive," or alternatively, "For the population, the correlation between G2 and the criterion Y is negative." In each case, a dichotomous variable (G1 or G2) is being correlated with a continuous variable (Y). The computational formula is referred to as a *point biserial correlation.* A Pearson correlation between these two variables would yield the same

Applied Research Hypothesis 4.1

Directional Research Hypothesis: For a given population, X12 is more effective than X13 on the criterion X2.

Statistical Hypothesis: For a given population, X12 and X13 are equally effective on the criterion X2.

Full Model: $X2 = a_0U + a_{12}X12 + a_{13}X13 + E_5$

Want: $a_{12} > a_{13}$; restriction: $a_{12} = a_{13}$

Full Model without linear dependencies: $X2 = a_0U + a_{12}X12 + E_5$

Want: $a_{12} > 0$; restriction: $a_{12} = 0$

Restricted Model: $X2 = a_0U + E_6$

alpha = .05

$R_f^2 = .0003$; $R_r^2 = .00$; parameter estimate for X12 is 24.171; parameter estimate for X13 is 23.806
$F_{(1,58)} = .0150$
Computer probability = .90; directional probability = .90/2 = .45

Interpretation: Because the weighting coefficients (SAS refers to them as *parameter estimates*) are in the desired direction, the directional probability can be reported. But because it is not less than alpha, the Statistical Hypothesis cannot be rejected. Note that the sample means for X12 and X13 can be found in the Full Model output by adding the regression constant to the weight of each respective vector. The sample mean for X12 would be: $a_0 + a_{12} = 23.81 + .37 = 24.18$. The sample mean for X13 would be: $a_0 + a_{13} = 23.81 + 0 = 23.81$. (If a matrix inversion computer program, such as SAS, SPSS, or BMDP, is being used, either X12 or X13 must be omitted from the Full Model. Sample means can still be calculated as above, the only difference is that a_{12} or a_{13} would not exist, depending upon whether X12 or X13 were omitted.)

numerical value, and the square of either correlation would be equal to the R^2 of the Full Model. "Testing a correlation coefficient for significance" implies testing the Statistical Hypothesis to make sure that the population correlation is equal to zero (therefore $R^2 = 0$). Notice that the statistically hypothesized population correlation, when squared, is the R^2 value of the Restricted Model. A specific computational formula exists for testing this hypothesis; and since it is a restricted case of the general F- test formula, the general F test can be used as well.

The nondirectional Research Hypothesis in Generalized Research Hypothesis 4.1 also could have been tested with a t test for the difference between two means—

an analysis of variance approach. Indeed, most statistical advisors probably would have used the ANOVA approach rather than the correlational approach.

Most statistical texts separate correlational procedures from ANOVA procedures. The GLM approach does not make the distinction between these two procedures. In fact, the GLM approach underscores the underlying isomorphism of the two procedures and instead emphasizes the importance of the researcher asking the question that is desired in the terminology that is desired. When one is using the GLM, the emphasis is on the statistics answering the Research Hypothesis. In other statistical approaches, particularly the ANOVA approach, the design of the study is often considered inseparable from the statistics.

Geometric interpretation. The isomorphism of the *t* test for the difference between two means and the point biserial correlation coefficient also can be seen in a graphic presentation. When one is considering the difference between two means, one has just that—two means, as in Figure 4.1.

If there is no difference between the two means, then the two means would be equally high on the Y axis. In this case (Figure 4.1b), the straight line of best fit would be horizontal, and the Pearson correlation would be 0.0.

Now consider what the figure would look like in the correlational configuration. If one is correlating G1 with Y, then one would have exactly the same figure as Figure 4.1, the only difference would be that the subjects in one group would be represented by a zero and the subjects in the other group would be represented by a one. If there is a correlation between the two variables, then there is a slope, and Figure 4.1a results.

If there is no correlation between the two variables, then the line of best fit has a zero slope, or is parallel with the X axis, as in Figure 4.1b. The tests of significance of these two techniques look very different, but they are not only conceptually the same but also mathematically the same (McNeil & Beggs, 1969).

Venn diagram. Venn diagrams, which were introduced at the end of chapter 1, provide assistance in understanding various Research Hypotheses. The Venn dia-

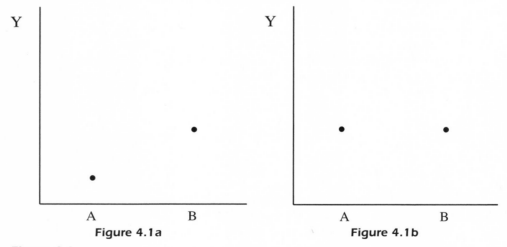

Figure 4.1a   **Figure 4.1b**

Figure 4.1.
Geometric representation of the difference between two means; 4.1a representing the Research Hypothesis that allows for mean differences, and 4.1b representing the Statistical Hypothesis that restricts the means to being equal.

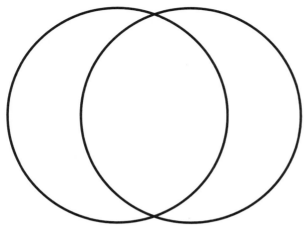

Figure 4.2.
Venn diagram of *t* test for the difference between two means and the point biserial correlation.

gram for the present technique would consist of one predictor variable (treatment versus comparison) overlapping with the criterion. That is, even though there are two groups, there is only one variable (identifying the treatment the subjects received) being used to account for the variance in Y. The amount of overlap in Figure 4.2 is the R^2 resulting from the Full Model.

Summary of the GLM Approach to Hypothesis Testing

The GLM approach to hypothesis testing was presented in the previous sections both algebraically and conceptually in the context of a two-group predictor problem. In general, the Research Hypothesis drives the entire process because it specifies all of the variables in the Full Model, as well as all of the variables in the Restricted Model. Once the models are known, the number of linearly independent vectors in the Full Model (m1) and in the Restricted Model (m2) can be determined. We let the computer determine the R_f^2 and R_r^2; and once these are known, the general F test can be calculated, as all values in the formula are known. Again, the general F test is:

$$F_{(m1 - m2, N - m1)} = \frac{(R_f^2 - R_r^2) / (m1 - m2)}{(1 - R_f^2) / (N - m1)}$$

For Applied Research Hypothesis 4.1, m1 is 2, m2 is 1, R_f^2 is .0003, and R_r^2 is .00. Plugging these values into the general F test we obtain $F = .0150$, exactly what the SAS program produces. (While the SAS program calculates the pieces of information in each model and the R^2 for each model, it only reports the R^2 for the Full Model. One would have to add another model statement to get the R^2 for the Restricted Model.) Since the Restricted Model in this example has only the unit vector, we know that the R^2 of the Restricted Model will be 0. Therefore we do not need to construct that model. Examples of Restricted Models containing more than just the unit vector will be provided in later chapters.

RESEARCH HYPOTHESES INVOLVING K GROUPS

Researchers often investigate multiple groups or treatments, instead of just two as discussed previously. (The analysis of variance analogue would be the one-factor or one-way ANOVA.) The most commonly used hypothesis for k groups is stated in Generalized Research Hypothesis 4.2. Notice that Generalized Research

---

Generalized Research Hypothesis 4.2

Directional Research Hypothesis: Not appropriate because more than one restriction (see Generalized Research Hypothesis 4.3).

Nondirectional Research Hypothesis: The k treatments are not equally effective on the criterion Y2.

Statistical Hypothesis: The k treatments are equally effective on the criterion Y2.

Full Model: $Y2 = a_1G1 + a_2G2 + \ldots + a_kGk + E_1$ (m1 = k)

Want (for nondirectional Research Hypothesis): $a_1 \neq a_2 \ldots \neq a_k$; restrictions: $a_1 = a_2 = \ldots = a_k$ (k - 1 restrictions)

Restricted Model: $Y2 = a_0U + E_2$ (m2 = 1)

    Automatic unit vector analogue
    Full Model: $Y2 = a_0U + a_1G1 + a_2G2 + \ldots + a_{k-1}Gk\text{-}1 + E_1$(m1 = [k - 1] + 1)
    Want: $a_1 \neq 0$ $a_2 \neq 0 \ldots a_{k-1} \neq 0$; restrictions: $a_1 = 0, a_2 = 0, \ldots, a_{k-1} = 0$
      (k - 1 restrictions)
    Restricted Model: $Y2 = a_0U + E_2$ (m2 = 1)

where:
    Y2 = criterion;
    U = 1 for all subjects;
    G1 = 1 if criterion from subject in Group 1, 0 otherwise;
    G2 = 1 if criterion from subject in Group 2, 0 otherwise;
    .
    .
    .
    Gk = 1 if criterion from subject in Group k, 0 otherwise; and
    $a_0, a_1, \ldots, a_{k-1}, a_k$ are least squares weighting coefficients calculated so as to minimize the sum of the squared values in the error vectors.

Degrees of freedom numerator: (m1 - m2) = (k - 1)
Degrees of freedom denominator: $(N - m1) = (N - k)$

where:
    N = number of subjects; and
    k = number of groups.

Hypothesis 4.2 is similar in format to Generalized Research Hypothesis 4.1. With two groups (Generalized Research Hypothesis 4.1) we need two dichotomous predictor vectors. With k groups (Generalized Research Hypothesis 4.2), a group membership vector needs to be constructed for each of the k groups. Notice that only a nondirectional Research Hypothesis can be tested. One way of noting this would be with the multiple restrictions being made. If significance is found, all that can be said is that the k treatments are not equally effective on the criterion Y. We know that not all restrictions are viable. Which ones (or one) are not viable is not known. One can accept the Research Hypothesis, but in reality doing so does not say much. For practical purposes, one would like to know which treatment is best. Indeed, one may have some reason to believe that a given treatment will be better than another or a combination of several others. If one has no idea at all about which treatment is best, then it is suggested that one use the means from the data to develop directional hypotheses for future verification.

Therefore, if there are four groups, there will be four group membership vectors. The weights for each of these vectors would be the mean for that group. Under the Statistical Hypothesis, there is no difference between the group means, implying the restriction that all weights in the Full Model are equal. Again, both conceptually and mathematically, we arrive at the unit vector model as the Restricted Model. If the unit vector is supplied by the computer program, then one of the group membership vectors is linearly dependent and can be eliminated. In that case, the weight for the unit vector is the mean for the group whose vector has been eliminated, and all the other weights are the difference between the mean for that group and the one whose group membership vector has been omitted. The *automatic unit vector* configuration of this design is presented in Applied Research Hypothesis 4.2. Whichever way the Full Model is represented, there will be (k - 1) restrictions (as evidenced by the number of equal signs), and the Restricted Model will be the unit vector model.

Geometric Interpretation

The k-group picture is an extension of the two-group picture in that there are k means. One possible picture for the four groups is presented in Figure 4.3. Group 2 has the highest mean and Group 3 has the lowest mean. Let us assume that these

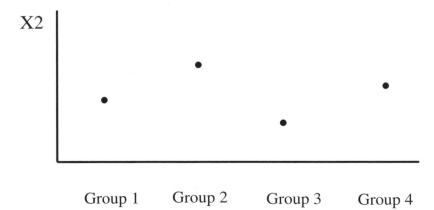

Figure 4.3.
Pictorial representation of the means for four groups.

Applied Research Hypothesis 4.2

Research Hypothesis: For a given population, the four groups (X17, X18, X19, X20) are not equally effective on the criterion X2.

Statistical Hypothesis: For a given population, the four groups (X17, X18, X19, X20) are equally effective on the criterion X2.

Full Model: $X2 = a_0U + a_{17}X17 + a_{18}X18 + a_{19}X19 + a_{20}X20 + E_5$

Want: $a_{17} \neq a_{18} \neq a_{19} \neq a_{20}$; restrictions: $a_{17} = a_{18} = a_{19} = a_{20}$

Automatic unit vector analogue
Full Model without linear dependencies: $X2 = a_0U + a_{17}X17 + a_{18}X18 + a_{19}X19 + E_5$
Want: $a_{17} \neq 0$ $a_{18} \neq 0$ $a_{19} \neq 0$; restrictions: $a_{17} = 0, a_{18} = 0, a_{19} = 0$

Restricted Model: $X2 = a_0U + E_6$

alpha = .05

$R_f^2 = .0278$; $R_r^2 = .00$

$F_{(3,56)} = .3469$.

Computer probability = .7915; directional probability is not applicable.

Interpretation: Because the probability is not less than alpha, the Statistical Hypothesis cannot be rejected.

The necessary SAS statements are in Appendix D-4.2.

results were significant, that not all four groups are equally effective on the criterion. The four groups would yield the same degree of significance if the Group 2 and Group 3 means were reversed. There are obviously many different possibilities that would satisfy the global Research Hypothesis. That is why we take the position that this type of hypothesis is not useful.

Relationship to Other Statistical Techniques

As already alluded to, Generalized Research Hypotheses 4.1 and 4.2 are both analogous to the one-way ANOVA. *Eta* is a measure of maximum curvilinearity and can be calculated by artificially grouping a continuous variable and then determining how similar (based on the dependent variable) the people are within groups, yet how different they are between groups. What might not be as obvious is that the resulting R^2 of the Full Model is equivalent to the eta (η) coefficient (McNeil, 1970b). Applied Research Hypothesis 4.3 illustrates the GLM approach to an example from Hinkle, Wiersma, and Jurs (1994, p. 525).

Applied Research Hypothesis 4.3

Research Hypothesis: For the population of interest, there is a relationship (possibly nonlinear) between anxiety and performance score.

Statistical Hypothesis: For the population of interest, there is not a relationship between anxiety and performance score.

Full Model: score $= a_1 G1 + a_2 G2 + a_3 G3 + a_4 G4 + a_5 G5 + a_6 G6 + E_5$

Want: $a_1 \neq a_2 \neq a_3 \neq a_4 \neq a_5 \neq a_6$; restrictions: $a_1 = a_2 = a_3 = a_4 = a_5 = a_6$ (5 restrictions)

Automatic unit vector analogue
Full Model: score $= a_0 U + a_1 G1 + a_2 G2 + a_3 G3 + a_4 G4 + a_5 G5 + E_5$
Want: $a_1 \neq 0 \ \ a_2 \neq 0 \ \ a_3 \neq 0 \ \ a_4 \neq 0 \ \ a_5 \neq 0$; restrictions: $a_1 = 0, a_2 = 0, a_3 = 0,$
$a_4 = 0, a_5 = 0$ (5 restrictions)

Restricted Model: score $= a_0 U + E_6$

alpha $= .05$

$R_f^2 = .818; R_r^2 = .00$

$F_{(5,10)} = 8.997$

Computer probability $= .0018$; directional probability is not appropriate because df_n greater than 1; nondirectional probability $= .0018$
Interpretation: Because the probability is less than alpha, the Statistical Hypothesis can be rejected and the Research Hypothesis can be accepted: There is a relationship (possibly nonlinear) between anxiety and performance score.

ORTHOGONAL COMPARISONS—POST HOC

Most researchers are frustrated by the permissible conclusion with the omnibus one-way F depicted in Generalized Research Hypothesis 4.2. The statement, "There is a difference between group means," invariably leads one to ask, "Which specific means are different?" So, statisticians have developed intricate systems of post hoc comparisons that test more specific questions yet protect the initially stated alpha level stipulated by the researcher. Because each orthogonal contrast is a one-degree-of-freedom question, there can be as many orthogonal questions as there are degrees of freedom in the omnibus one-way F. The orthogonal contrasts can be specified either before the omnibus one-way F is calculated (referred to as *a priori,* or *planned, comparisons*) or after the omnibus one-way F (referred to as *post hoc comparisons*). Because the post hoc questions were not specified before the data were collected and analyzed, no directional hypotheses can be tested with the post hoc comparisons. The use of orthogonal comparisons keeps the familywise alpha at the alpha stated for the omnibus F, while the effective alpha for each test is in reality the originally stated alpha divided by the number of tests.

In the following discussion, we use an example of four groups: *In all cases the hypotheses must be independent of one another, which simply means that what is found to be true with one hypothesis has no bearing at all on the other hypotheses.* This can be determined mathematically by comparing the coefficients of the groups in the various hypotheses: The coefficients must add up to zero, and the sum of the cross products must add up to zero. Even with just four groups, many sets of questions can be tested; the set chosen will be a function of how the four groups were configured and, particularly, a function of the research interests of the investigator. Specific examples follow.

Exhibit 4.1

	m1	m2	m3	m4
Research Hypothesis 1				
Nondirectional: m1 ≠ m4				
Directional: m1 > m4				
Statistical Hypothesis: m1 = m4 or				
(1 * m1) + (-1 * m4) = 0	1	0	0	-1
Research Hypothesis 2				
Nondirectional: m2 ≠ m3				
Directional: m2 > m3				
Statistical Hypothesis: m2 = m3 or				
(1 * m2) + (-1 * m3) = 0	0	1	-1	0
Research Hypothesis 3				
Nondirectional: (m1 + m4) ≠ (m2 + m3)				
Directional: (m1 + m4) > (m2 + m3)				
Statistical Hypothesis: (m1 + m4) = (m2 + m3) or				
(1 * m1) + (1 * m4) + (-1 * m2) + (-1 * m3) = 0	1	-1	-1	1

One possible set of contrast coefficients with four groups

Combination of Groups

The three Research Hypotheses in Exhibit 4.1 indicate one orthogonal set of hypotheses. What this means is that if a researcher had stated an alpha of .05 for the omnibus one-way F (as in Generalized Research Hypothesis 4.2), then the same alpha would apply to the whole set of three hypotheses. One can determine that these three hypotheses are indeed orthogonal by rearranging the Statistical Hypotheses so that zero appears by itself on one side of the equation. Then the coefficients of the four groups are inserted into the orthogonal coefficients table on the right side of Exhibit 4.1. Notice that the coefficients for Research Hypothesis 1 add up to zero, as do the coefficients for Research Hypothesis 2. Furthermore, the cross products of the coefficients for Research Hypothesis 1 and Research Hypothesis 2 add up to zero [(1 * 0) + (0 * 1) + (0 * -1) + (-1 * 0) = 0]. Therefore Research Hypotheses 1 and 2 are orthogonal contrasts and the answer to Research Hypothesis 1 does not impinge on the answer to Research Hypothesis 2. Because Research Hypothesis 1 deals only with Groups 1 and 4, while Research Hypothesis 2 deals with Groups 2 and 3, this conclusion may seem like a trivial one (it is, and it was instructionally meant to be). On the other hand, Research Hypothesis 3 deals with all four groups, yet its coefficients sum to zero, and the sum of its cross products with those of Research Hypothesis 1 sum to zero. Thus Research Hypothesis 3 is orthogonal to Research Hypothesis 1. In addition, the sum of the cross products of Research Hypotheses 3 and 2 sum to zero; thus they are orthogonal. One can conclude that the three Research Hypotheses are one set of orthogonal contrasts.

Exhibit 4.2

	m1	m2	m3	m4
Research Hypothesis 4				
Nondirectional: $(m1 + m2 + m3)/3 \neq m4$				
Directional: $(m1 + m2 + m3)/3 > m4$				
Statistical Hypothesis: $(m1 + m2 + m3)/3 = m4$ or				
$(1 * m1) + (1 * m2) + (1 * m3) + (-3 * m4) = 0$	1	1	1	-3
Research Hypothesis 5				
Nondirectional: $m2 \neq m3$				
Directional: $m2 > m3$				
Statistical Hypothesis: $m2 = m3$ or				
$(1 * m2) + (-1 * m3) = 0$	0	1	-1	0
Research Hypothesis 6				
Nondirectional: $m1 \neq (m2 + m3)/2$				
Directional: $m1 > (m2 + m3)/2$				
Statistical Hypothesis: $m1 = (m2 + m3)/2$ or				
$(2 * m1) + (-1 * m2) + (-1 * m3) = 0$	2	-1	-1	0

Another possible set of contrast coefficients with four groups

In Exhibit 4.2, Research Hypotheses 4, 5, and 6 are another set of three orthogonal contrasts. While Research Hypothesis 5 and Research Hypothesis 2 are exactly the same, Research Hypothesis 4 and Research Hypothesis 6 are different from Research Hypotheses 1 and 3. The coefficients within Research Hypotheses 4, 5, and 6 add up to zero, and the sum of the cross products adds up to zero, thus these three hypotheses constitute a different set of three orthogonal contrasts. Which set a researcher should use depends on the design of the study and the questions one has of the groups. Indeed, there are many other sets of orthogonal contrasts. As in all research, the questions should guide the analysis. With post hoc comparisons, the researcher is limited to one less question than there are groups.

An example of when Research Hypotheses 1, 2, and 3 might be of interest is when a new treatment and an existing treatment are implemented in each of two schools. In this application, consider M1 (a vector) to be the new treatment in School 1, M2 to be the existing treatment in School 1, M3 to be the existing treatment in School 2, and M4 to be the new treatment in School 2. The primary hypothesis would be: "The new treatment is better than the existing treatment," which would be tested with the contrast specified in Research Hypothesis 3. The hypothesis could be supported if there was a large difference in one school and little or no difference in the other school. If one were concerned abut this possibility, one could test Research Hypotheses 1 and 2. Research Hypothesis 1 compares the M1 and M4 groups, the new treatment in the two schools. Research Hypothesis 2 compares the existing treatment in the two schools.

An example of when Research Hypotheses 4, 5, and 6 might be of interest is when a researcher is testing the effectiveness of three different new treatments (M1, M2, and M3) and one comparison treatment (M4). Since there are four groups, three orthogonal contrast questions can be asked, and if the questions are asked before inspection of the data, directional Research Hypotheses can be tested. Research Hypothesis 4 determines if the average of the three new treatments is better than the single comparison treatment. Research Hypothesis 5 tests if the second new treatment is better than the third new treatment. Finally, Research Hypothesis 6 tests if the first new treatment is better than the average of the other new treatments. As should now be clear, the design of the research and the desired conclusion(s) determine the choice of the hypotheses and whether the hypotheses are directional or nondirectional. No one choice is always correct; the choice will depend on the research questions.

Obtaining a Better Estimate of Within-Group Variability

Recall that the F test is a ratio of two variance estimates. The numerator variance estimate is related to the restriction that is made, while the denominator variance estimate is related to the naturally occurring variable in the population. Usually the denominator variance estimate is derived from the sample data, the groups that are being compared. But one may have access to data from other subjects. If these other subjects are from the same population, then those data can be combined with the sample data to obtain an even better (more stable) estimate of the population variance. The following example illustrates this point. If a researcher had two treatment groups, the Research Hypothesis could be:

Treatment j is better than Treatment k on the criterion Y.

This is a two-group hypothesis that could be tested by the method outlined in Generalized Research Hypothesis 4.1. Now suppose that data have been collected on other groups from the same population. If the treatments applied to the other groups produce about the same variance about the group means, then the data from the other groups can be used with the two groups to obtain a better estimate of the population criterion variance than could be obtained with just the two initial groups. Notice in Generalized Research Hypothesis 4.3 that the denominator degrees of freedom is a function of the subjects in all groups, not just those in the two groups being compared. Indeed, the denominator degrees of freedom is the same as in Generalized Research Hypothesis 4.2. But because there is only one restriction being made in Generalized Research Hypothesis 4.3, the decrease in R^2 is entirely attributable to that lone restriction. Statistical significance (for the directional hypothesis in Generalized Research Hypothesis 4.3) allows one to make the definitive statement, "For the given population, Treatment j is better than Treatment k on the criterion Y, considering the variance in (i) similar treatments." (The verification that the weighting coefficient for the Treatment j vector is larger than the weighting coefficient for the Treatment k vector must be made here as in all directional hypothesis situations.)

With respect to Generalized Research Hypothesis 4.3, there are only k linearly independent vectors in the Full Model. Given that the unit vector will be in the Full Model, one of the group vectors will need to be eliminated. The unit vector in the

Generalized Research Hypothesis 4.3

Directional Research Hypothesis: For the population, Treatment j is better than Treatment k on the criterion Y, considering the variance in (i) similar treatments.

Nondirectional Research Hypothesis: For the population, Treatments j and k have different effects on the criterion Y, considering the variance in (i) similar treatments.

Statistical Hypothesis: For the population, Treatment j is as effective as Treatment k on the criterion Y, considering the variance in (i) similar treatments.

Full Model: $Y = a_1G1 + a_2G2 + \ldots + a_iGi + a_jGj + a_kGk + E_1$

Want: $a_j \neq a_k$; restriction: $a_j = a_k$ (1 restriction)

Restricted Model: $Y = a_1G1 + a_2G2 + \ldots + a_iGi + a_1Gm + E_2$

Automatic unit vector analogue
Full Model: $Y = a_0U + a_1G1 + a_2G2 + \ldots + a_iGi + a_jGj + E_1$
Want $a_j \neq 0$; restriction: $a_j = 0$ (1 restriction)
Restricted Model: $Y = a_0U + a_1G1 + a_2G2 + \ldots + a_iGi + E_2$

where:

Y = criterion;
$G1$ = 1 if criterion from subject in Treatment 1, 0 otherwise;
$G2$ = 1 if criterion from subject in Treatment 2, 0 otherwise;

.

.

.

Gi = 1 if criterion from subject in Treatment i, 0 otherwise;
Gj = 1 if criterion from subject in Treatment j, 0 otherwise;
Gk = 1 if criterion from subject in Treatment k, 0 otherwise;
Gm = 1 if the criterion is from subject in either Treatment j or k, 0 otherwise; and
$a_1, a_2, \ldots, a_i, a_j, a_k$, and a_m are least squares weighting coefficients calculated so as to minimize the sum of the squared values in the error vector.

Degrees of freedom numerator $= (m1 - m2) = [k - (k - 1)] = 1$
Degrees of freedom denominator $= (N - m1) = (N - k)$

where:

N = number of subjects; and
k = number of groups.

Restricted Model will have a weight that is equal to the mean of the subjects in Treatment j and k groups.

Trend Analysis

When the treatments are ordered on some underlying continuum, one may want to investigate the trends in the data as in Exhibit 4.3; that is, does the criterion increase linearly with an increase in the underlying continuum (as in Research Hypothesis 7), or is there a *minimum* performance (as in Research Hypothesis 8)? (By reversing all the weights in Research Hypothesis 8, one could investigate *maximum* performance.) Finally, with four groups there may be a cubic trend as in Research Hypothesis 9. Note that the coefficients for Research Hypotheses 7, 8, and 9 all add to zero and that the cross products all add to zero. Therefore, these three Research Hypotheses constitute another set of orthogonal contrasts for four groups.

Exhibit 4.3

Research Hypothesis 7 (linear)	$m1$	$m2$	$m3$	$m4$
Nondirectional: $(-3m1) + (-1m2) + (1m3) + (3m4) \neq 0$				
Directional: $(-3m1) + (-1m2) + (1m3) + (3m4) > 0$				
Statistical Hypothesis: $(-3m1) +$				
$(-1m2) + (1m3) + (3m4) = 0$				
or $(-3 * m1) +$				
$(-1 * m2) + (1 * m3) + (3 * m4) = 0$	-3	-1	1	3

Research Hypothesis 8 (quadratic)				
Nondirectional: $(1m1) + (-1m2) + (-1m3) + (1m4) \neq 0$				
Directional: $(1m1) + (-1m2) + (-1m3) + (1m4) > 0$				
Statistical Hypothesis: $(1m1) + (-1m2) +$				
$(-1m3) + (1m4) = 0$ or				
$(1 * m1) + (-1 * m2) +$				
$(-1 * m3) + (1 * m4) = 0$	1	-1	-1	1

Research Hypothesis 9 (cubic)				
Nondirectional: $(-m1) + (3m2) + (-3m3) + (m4) \neq 0$				
Directional: $(-m1) + (3m2) + (-3m3) + (m4) > 0$				
Statistical Hypothesis: $(-m1) + (3m2) +$				
$(-3m3) + (m4) = 0$ or				
$(-1 * m1) + (3 * m2) +$				
$(-3 * m3) + (1 * m4) = 0$	-1	3	-3	1

One possible set of contrast coefficients with four groups: trend analysis

Two Factors

Now suppose that the four groups differ not on just one underlying factor, as they did in the above examples, but on *two* underlying factors. Exhibit 4.4 posits the following example of two groups getting the new treatment and two groups getting the comparison treatment. Thus the first underlying factor is *treatment*: new versus comparison.

Exhibit 4.4

M1 = new treatment, AM   M2 = new treatment, PM
M3 = comparison treatment, AM   M4 = comparison treatment, PM

Research Hypothesis 10

Nondirectional: The two treatments, averaged across the two different time periods, are not equally effective.

(m1 + m2)/2 ≠ (m3 + m4)/2

Directional: The new treatment, averaged across the two different time periods, is more effective than the comparison treatment.

(m1 + m2)/2 > (m3 + m4)/2

Statistical Hypothesis: (m1 + m2)/2 = (m3 + m4)/2 or

(m1 + m2) = (m3 + m4) or

(m1 + m2) - (m3 + m4) = 0   or

	m1	m2	m3	m4
(1 * m1) + (1 * m2) + (-1 * m3) + (-1 * m4) = 0	1	1	-1	-1

Research Hypothesis 11

Nondirectional: The two time periods, averaged across the two treatments, are not equally effective.

(m1 + m3)/2 ≠ (m2 + m4)/2

Directional: The AM period, averaged across the two different treatments, is more effective than the PM period.

(m1 + m3)/2 > (m2 + m4)/2

Statistical Hypothesis: (m1 + m3)/2 = (m2 + m4)/2 or

(m1 + m3) = (m2 + m4) or

(m1 + m3) - (m2 + m4) = 0 or

	m1	m2	m3	m4
(1 * m1) + (-1 * m2) + (1 * m3) + (-1 * m4) = 0	1	-1	1	-1

Research Hypothesis 12

Nondirectional: The difference in effectiveness of the AM new treatment and the PM new treatment is different from the difference between the AM comparison treatment and the PM comparison treatment.

(m1 - m2) ≠ (m3 - m4)

Directional: The difference in effectiveness of the AM new treatment and the PM new treatment is greater than the difference between the AM comparison treatment and the PM comparison treatment.

(m1 - m2) > (m3 - m4)

Statistical Hypothesis: The difference in effectiveness of the AM new treatment and the PM new treatment is the same as the difference between the AM comparison treatment and the PM comparison treatment.

(m1 - m2) = (m3 - m4) or

(m1 - m2) - (m3 - m4) = 0 or

	m1	m2	m3	m4
(1 * m1) + (-1 * m2) + (-1 * m3) + (1 * m4) = 0	1	-1	-1	1

One possible set of contrast coefficients: two-way analysis of variance

One of the new treatment groups is in the AM and one is in the PM. One of the comparison treatment groups is in the AM and one is in the PM. Thus the second factor is *time of treatment*: AM versus PM.

What would be the Research Hypotheses of interest with this design? One probably would want to compare the new treatments to the comparison treatments, and possibly compare the AM treatments to the PM treatments. These two hypotheses are developed first, and then we turn our attention to the third orthogonal comparison.

The nondirectional Research Hypothesis for treatment would be: "The two treatments, averaged across the two different time periods, are not equally effective," resulting in the orthogonal coefficients for Research Hypothesis 10 in Exhibit 4.4. One could have stated this Research Hypothesis with a directional expectation, resulting in the same set of orthogonal coefficients. The nondirectional Research Hypothesis for time of treatment would be Research Hypothesis 11: "The two time periods, averaged across the two different treatments, are not equally effective." Again, one could have stated this hypothesis with a directional expectation. Notice that the coefficients for Research Hypothesis 11 are orthogonal to those for Research Hypothesis 10. Research Hypotheses 10 and 11 are referred to as *main-effects* hypotheses within the ANOVA framework. Unless stated directionally a priori, they are always tested in a nondirectional fashion.

Given the above two orthogonal contrasts, the third orthogonal contrast would have to be that specified in Research Hypothesis 12. The nondirectional Research Hypothesis associated with these coefficients is: "The difference between AM new treatment and PM new treatment is *different* from the difference between AM comparison treatment and PM comparison treatment." Again, one could have stated this hypothesis with a directional expectation. (For example, "The difference between AM new treatment and PM new treatment is *greater* than the difference between AM comparison treatment and PM comparison treatment.") Research Hypothesis 12 is referred to in the ANOVA literature as the *test for interaction,* which is discussed in great detail in chapter 7.

The directional Research Hypothesis could be tested with the following Full Model:

(Model 4.3) $Y = m_1 M1 + m_2 M2 + m_3 M3 + m_4 M4 + E_1$ (pieces full = 4)

Want: $(m_1 + m_2)/2 > (m_3 + m_4)/2$ or $(m_1 + m_2) > (m_3 + m_4)$
    restriction: $(m_1 + m_2) = (m_3 + m_4)$ or $(m_1 + m_2 - m_4) = m_3$

resulting in the Restricted Model:

(Model 4.4) $Y = m_1 M1 + m_2 M2 + (m_1 + m_2 - m_4)M3 + m_4 M4 + E_2$, or

$Y = m_1(M1 + M3) + m_2(M2 + M3) + m_4(M4 - M3) + E_2$ (pieces restricted = 3)

There are other ways to code group membership variables. One could use any two values, referred to as *nonsense coding* by Cohen and Cohen (1975) and by Williams (1987). One also could use the actual orthogonal weights (Williams, 1974a, 1974b). Each orthogonal comparison would be reflected by a vector. Thus the treatment main-effect vector would be: TREATMAIN = 1, if M1 or M2, and = -1, if M3 or M4. The time main-effect vector would be: TIMEMAIN = 1, if M1 or M3, and = -1, if M2 or M4. The third orthogonal contrast would be the interaction between time and treatment (discussed in chapter 7): TIMETREAT = 1, if M1 or M4, and = -1, if M2 or M3. Thus, Research Hypothesis 10 could be tested by Models 4.5 and 4.6, which are reparameterizations of Models 4.3 and 4.4:

(Model 4.5) $Y = a_0 U + bTREATMAIN + cTIMEMAIN + dTIMETREAT + E_1$
    (pieces full = 4)

Research Hypothesis 10 calls for the restriction of b = 0, resulting in the Restricted Model:

(Model 4.6) $Y = a_0 U + cTIMEMAIN + dTIMETREAT + E_2$
    (pieces restricted = 3)

The F test in comparing models 4.5 and 4.6 will yield the same degrees of freedom and same F value as the comparison of Models 4.3 and 4.4. These are two equivalent ways to test the same hypothesis.

Research Hypothesis 11 could be tested by restricting c = 0 in the Full Model, Model 4.5, resulting in the following Restricted Model:

(Model 4.7) $Y = a_0 U + bTREATMAIN + dTIMETREAT + E_3$
    (pieces restricted = 3)

Research Hypothesis 12 could be tested by restricting d = 0 in the Full Model, Model 4.5, resulting in the following Restricted Model:

(Model 4.8) $Y = a_0 U + bTREATMAIN + cTIMEMAIN + E_4$
    (pieces restricted = 3)

ORTHOGONAL COMPARISONS—A PRIORI

Orthogonal comparisons were developed to attempt to identify where the significance was after the global, nondirectional question was found to be significant. Orthogonal comparisons also are appropriate when one has a priori expectations, particularly directional expectations. As indicated in Exhibits 4.1, 4.2, 4.3, and 4.4, each of those questions *could* have been directional if there was reason to believe a priori that the results would be in one direction. The only difference between post hoc and a priori is that in a priori one has a reason (based on theory or past research) to expect or want the results to be in a particular direction. Again, stating the Research Hypothesis in a directional manner allows one to make a directional conclusion.

Stating Questions of Interest

Another aspect of orthogonal comparisons needs to be clarified: The researcher should always ask the questions that are of interest. For instance, with four groups there are three degrees of freedom and therefore three orthogonal questions that can be asked. Research Hypotheses 1, 2, and 3 comprise one possible set. Research Hypotheses 4, 5, and 6 comprise another. Research Hypotheses 7, 8, and 9 comprise another. Research Hypotheses 10, 11, and 12 comprise another. Which one is applicable depends upon the nature of the data and the question asked.

More complicated designs have been modeled with the GLM. For instance, the Solomon four-group design was developed to assess the effect of pretest sensitization. Traditional ANOVA computational formulae have not been developed for this analysis, but Williams and Newman (1982) and Newman, Benz, and Williams (1990) illustrated the GLM approach to this design.

A researcher may be interested in only one or two of the questions in an orthogonal set. One does not have to test all the possible one-degree-of-freedom questions. If a researcher wishes to make a directional conclusion, then the contrasts must be stated before the data are collected. We would argue that, in this situation, the omnibus one-way F does not need to be calculated, and the analysis can proceed directly to the stated questions. (That is, the omnibus one-way F was not a stated question of interest and therefore should not be tested.)

A More Complicated Design

Perhaps the research design includes three slightly different comparison groups and two slightly different discussion treatment groups, along with four other kinds of treatment groups. A Research Hypothesis of interest might be:

> The average effect, on the criterion, of the two discussion treatments is better than the average effect of the three comparison groups, considering the variance of the four other kinds of treatment groups.

The antithesis of the Research Hypothesis would be the following Statistical Hypothesis:

> The average effect, on the criterion, of the two discussion treatments is equal to the average of the three comparison groups, considering the variance of the four other kinds of treatment groups.

The Full Model reflecting the Research Hypothesis contains a dichotomous vector for each group, and is:

(Model 4.9) $Y2 = a_0U + c_1C1 + c_2C2 + c_3C3 + t_1T1 + t_2T2 + t_3T3 + t_4T4 + d_1D1 + d_2D2 + E_3$

where:

$Y2$ = the criterion variable;
U = the unit vector containing a 1 for each subject;
$C1$ = 1 if criterion from Comparison Group 1, 0 otherwise;
$C2$ = 1 if criterion from Comparison Group 2, 0 otherwise;
$C3$ = 1 if criterion from Comparison Group 3, 0 otherwise;
$T1$ = 1 if criterion from Treatment Group 1, 0 otherwise;
$T2$ = 1 if criterion from Treatment Group 2, 0 otherwise;
$T3$ = 1 if criterion from Treatment Group 3, 0 otherwise;
$T4$ = 1 if criterion from Treatment Group 4, 0 otherwise;
$D1$ = 1 if criterion from Discussion Group 1, 0 otherwise;
$D2$ = 1 if criterion from Discussion Group 2, 0 otherwise; and
$a_0, c_1, c_2, c_3, t_1, t_2, t_3, t_4, d_1,$ and d_2 are least squares weighting coefficients calculated so as to minimize the sum of the squared values in the error vector, E_3.

If the cells have the same number of subjects in each cell or are proportional, then the restrictions implied by the Research and Statistical Hypotheses are:

$$\frac{c_1 + c_2 + c_3}{3} = \frac{d_1 + d_2}{2}$$

If the number of subjects is not proportional, then the number of subjects in each cell needs to be considered, as shown in a later chapter. The restrictions can be solved for any of the five weighting coefficients—we solve for c_1 here. Multiplying both sides by 3 yields:

$$c_1 + c_2 + c_3 = 3(d_1 + d_2)/2$$

Subtracting c_2 and c_3 from both sides yields:

$$c_1 = (3/2 * d_1) + (3/2 * d_2) - c_2 - c_3$$

Now inserting the right-hand expression for c_1 into the Full Model yields:

$$Y2 = a_0U + [(3/2 * d_1) + (3/2 * d_2) - c_2 - c_3]C1 + c_2C2 + c_3C3 + t_1T1 + t_2T2 + t_3T3 + t_4T4 + d_1D1 + d_2D2 + E_4$$

$$Y2 = a_0U + (3/2 * d_1)C1 + (3/2 * d_2)C1 - c_2C1 - c_3C1 + c_2C2 + c_3C3 + t_1T1 + t_2T2 + t_3T3 + t_4T4 + d_1D1 + d_2D2 + E_4$$

Expanding and collecting terms results in the following Restricted Model:

$$(\text{Model } 4.10)\ Y2 = a_0U + c_2(C2 - C1) + c_3(C3 - C1) + t_1T1 + t_2T2 + t_3T3 + t_4T4 + d_1[D1 + (3/2 * C1)] + d_2[D2 + (3/2 * C1)] + E_4$$

There are ($m1 = 9$) linearly independent pieces of information in the Full Model (Model 4.9) because one of the nine group vectors is linearly dependent upon the other eight group vectors plus the unit vector. There are ($m2 = 8$) good pieces of information in the Restricted Model (Model 4.10). Therefore, $df_n = (9 - 8) = 1$ and $df_d = N - 9$. If the F ratio in comparing the Full and Restricted Models produced a probability value lower than the predetermined alpha, one could accept the Research Hypothesis, *if* the average of the two discussion weighting coefficients was greater than the average of the three comparison weighting coefficients. Some of the vectors appearing in the Restricted Model look strange but could easily be generated by the computer (see Appendix D4.4). The fortunate researcher having SAS available could make the following test statement:

$$\text{TEST } (C1 + C2 + C3)/3 - (D1 + D2)/2 = 0;$$

Advantages to Conceptualizing Post Hoc Comparisons in the GLM

The statistically sophisticated reader may see the similarities between what has just been presented and the myriad techniques referred to as *multiple comparisons* or *post hoc comparisons*.

The above procedure is similar to the *planned comparison* techniques. We see three differences between what has just been presented and the body of traditional literature referred to as *post hoc comparisons*.

First, in the GLM approach as presented above, the researcher is forced to state the Research Hypothesis. The Research Hypothesis should be directional in nature, allowing the researcher to make a conclusive statement. (The "planned comparison" literature does not usually consider the notion of directional hypothesis testing.)

Second, because the Research Hypothesis is stated by the researcher and is a well-thought-out analysis of the design, no adjustment to the resultant probability is necessary as is the case in post hoc comparisons. It should be noted that a priori (as contrasted to post hoc) comparisons do not differ from the regression presentation on this point, nor on the next point.

Third, the researcher does not initially have to compare the Full Model to the "unit vector model." (The post hoc comparisons require the one-way F to be significant before any specific comparisons can be made.) The "one-way F hypothesis" does not have to be run for two reasons—the one-way F question is not of interest to the researcher, and global statistical significance might not be obtained, even though some specific questions of interest might produce significance. This could occur, for instance, when nine comparison groups and only one experimental group are examined. The 10 groups might not be statistically differentially effective, even though the one experimental group might well be better than the average of the nine comparison groups.

Recommendations Concerning Directional Hypothesis Testing

We strongly believe that past findings and theoretical expectations should dictate the Research Hypothesis. If one is groping in the dark regarding the field of investigation, then the nondirectional hypothesis might yield a set of variables that may be relevant to understanding a particular criterion. Following hunches is certainly one of the inductive phases of inquiry, and directionality may not be called for. Indeed, if one is searching for relevant variables, inferential statistics in themselves may not be needed. Bivariate correlations and sample means are fine descriptors for expressing hunches.

On the other hand, if prior knowledge is available in a researcher's domain, we firmly believe that the directional hypothesis is not only to be preferred but that it is the only hypothesis to test. The continuing emphasis upon directional Research Hypotheses will appear throughout the remainder of this text. It should be noted that the regression approach can be applied to either directional or nondirectional Research Hypotheses. The choice of either directional or nondirectional Research Hypotheses is not a *statistical* choice; it depends on theory, past research, and what the researcher wants to conclude from the analysis.

As indicated in Exhibits 4.1, 4.2, 4.3, and 4.4, any of the previously discussed sets of orthogonal comparisons could have been specified before the data were collected, and one could have tested a stated a priori directional Research Hypothesis. Present statistical developments limit the directional Research Hypothesis to a single restriction on a Full Model. Thus, there will only be one degree of freedom for the numerator of the F ratio. The F ratio is equal to t^2 when there is only one degree of freedom in the numerator. Therefore, the directional Research Hypotheses are sometimes referred to as *one-degree-of-freedom questions.* If the degrees of freedom in the numerator of the F ratio is greater than one, then a directional Research Hypothesis has not been tested.

DICHOTOMOUS PREDICTOR, DICHOTOMOUS CRITERION

Although the statistical and computer solutions to dichotomous variables are the same as with continuous variables, researchers may formulate their Research Hypotheses differently depending on whether their variables are continuous or dichotomous. Statisticians have developed alternate computational formulae, depending on whether the variables are continuous or dichotomous. Fortunately, the computer neither knows nor cares whether the variables are dichotomous or continuous. The following sections demonstrate how the GLM can be used to answer questions dealing with dichotomous variables in both the predictor and the criterion sides. We begin first with a simple dichotomy in both sets; later examples deal with more than two designations in the predictor set and then with more than two in both sets.

Research Hypotheses Involving One Dichotomous Predictor and One Dichotomous Criterion

The Research Hypothesis for a dichotomous predictor and a dichotomous criterion could be generalized as:

For the population, there is a relationship between A and B.

Alternatively:

For the population, given knowledge of A, B can be predicted.

And finally, in terms of proportions:

For the population, the proportion of B1 in A1 is different from the proportion of B1 in A2.

While these hypotheses are nondirectional, the analogous directional hypotheses should be clear:

For the population, A is positively related to B.

For the population, people in A1 tend to be in B1.

For the population, the proportion of B1 people in A1 is higher than the proportion of B1 people in A2.

Both predictor and criterion variables are dichotomous and, if the data are as in Table 4.2, the vectors would look like Model 4.11.

Table 4.2
Sample Data for a Single Dichotomous
Predictor and a Single Dichotomous Criterion

	A1	A2	Total
B1	3	1	4
B2	2	4	6
Total	5	5	10

$$(\text{Model 4.11}) \quad B2 \ = \ a_1A1 + a_2A2 + E_3$$

1	1	0
1	1	0
0	1	0
0	1	0
0	1	0
1	0	1
1	0	1
1	0	1
1	0	1
0	0	1

where:
>   B2 $=$ 1 if in B2, 0 if in B1;
>   A1 $=$ 1 if in A1, 0 otherwise; and
>   A2 $=$ 1 if in A2, 0 otherwise.

The weight for A1, a_1, will again be the mean of the criterion scores; but because the criterion scores are ones and zeros, the mean will be the proportion of the A1 subjects who are in B2, in this case, .40. Likewise, the numerical value for a_2 will be the proportion of A2 subjects who are in B2, or .80.

The antithesis of the Research Hypothesis is:

For the population, there is no relationship between A and B.

The restriction on the Full Model thus would be: $a_1 = a_2$, resulting in the unit vector model:

$$(\text{Model 4.12}) \quad B2 \ = \ a_0U + E_4$$

Again the general F can be applied to the above Full and Restricted Models:

$$F_{(m1-m2,\,N-m1)} \ = \ \frac{(R_f^2 - R_r^2) \,/\, (m1 - m2)}{(1 - R_f^2) \,/\, (N - m1)}$$

$$F_{(2-1, N-2)} = \frac{(R_f^2 - 0) / (2 - 1)}{(1 - R_f^2) / (N - 2)}$$

These results can be obtained by using the SAS program as in Appendix D-4.7. Note that the probability for the overall model and the probability associated with the specific restriction are exactly the same. Since the overall test of the Full Model tests the Full Model against the unit vector model, the two tests are exactly the same *in this application*, which will not always be the case, illustrating that the test for the overall model will not always be of interest.

Relationship to other statistical techniques. If the same Research Hypothesis had been tested with the contingency coefficient, the resulting phi (ϕ) coefficient would have been .40. The probability associated with the chi-square (χ^2), though, would have been slightly larger than that associated with the general F test. This is because the chi-square test is not sensitive to the number of subjects. (See Appendix D-4.5 for the SAS verification of this statement.) Persons influenced by correlational thinking would compute a phi coefficient on this kind of hypothesis, while ANOVA-trained persons would compute a chi-square on this contingency table. *Neither of these two special terms is needed, as they are both computational simplifications of the general least squares procedure.* A Research Hypothesis of the nature of Generalized Research Hypothesis 4.4 will yield a probability value exactly the same as the test for the phi coefficient, and more exact than that for the chi-square analysis (Leitner, 1979; McNeil, 1974). These references extend these notions to more than two levels on one predictor variable. The GLM approach is limited to having a criterion of two categories. If the researcher has more than two categories on the criterion but only two on the predictor, then the GLM can be used if the researcher temporarily interchanges the concepts of predictor and criterion. If there are more than two categories on both variables, the discriminant function must be employed.

The R^2 of the Full Model in Generalized Research Hypothesis 4.4 is equal to ϕ^2 and is also equal to χ^2/N. The GLM approach to hypothesis testing is a better method for at least four reasons:

1. The GLM approach forces the researcher to state the Research Hypothesis. Unfortunately, this is not always done in chi-square analyses and in phi coefficient analyses.

2. The GLM approach is easily generalized to all other least squares hypotheses. Separate computing formulae and different rules for calculating degrees of freedom are not necessary.

3. The stating and testing of directional hypotheses is encouraged by the GLM approach, whereas directional chi-square analyses and directional phi coefficient analyses are at best mentioned only briefly in statistics books.

4. The GLM approach considers the number of subjects and the number of categories in the predictor variable in the calculation of the denominator degrees of freedom. Because the chi-square test of significance assumes many subjects, the probability statement is more exact when it is calculated from the F test that results from the multiple linear regression approach.

Caveats

Directionality. As indicated above, Generalized Research Hypothesis 4.4 also could be directional, depending on whether the researcher had reason to believe that

---

Generalized Research Hypothesis 4.4

Directional Research Hypothesis: The dichotomous variable D2 is positively [or negatively] correlated with the dichotomous variable D1.

Nondirectional Research Hypothesis: The dichotomous variable D2 is correlated with the dichotomous variable D1.

Statistical Hypothesis: The dichotomous variable D2 and the dichotomous variable D1 are uncorrelated.

Full Model: $D1 = a_0 U + a_1 D2 + E_1$

Want (for directional Research Hypothesis): $a_1 > 0$; restriction: $a_1 = 0$
Want (for nondirectional Research Hypothesis): $a_1 \neq 0$; restriction $a_1 = 0$

Restricted Model: $D1 = a_0 U + E_2$

where:

$\quad\quad$ D1 $\;=\;$ dichotomous criterion; and
$\quad\quad$ D2 $\;=\;$ a dichotomous variable.

Degrees of freedom numerator $= (m1 - m2) = (2 - 1) = 1$
Degrees of freedom denominator $= (N - m1) = N - 2$

where:

$\quad\quad N \quad = \;$ number of subjects.

---

the proportion of D2 would be larger in one population than in the other and depending on whether the researcher wanted to conclude that the proportion was larger in one population than in the other. As always, sample results must be checked to see if they are in the hypothesized direction; if they are, the computed probability must be divided by 2. If the results are opposite to the hypothesized direction, the reported probability must be (1 - prob/2).

Power. The GLM approach is more powerful than the chi-square or contingency table approach. It is more powerful in the statistical sense in that one is more likely to find significance if in fact there is significance to be found. The GLM approach is also more flexible in that the same conceptual approach, the same general F test, and the same computer procedure can be used.

R^2. While some researchers who use the contingency table approach report the phi coefficient, few transform that value to the squared value and talk about the proportion of variance accounted for. As always, significance can be obtained with little variance accounted for. The GLM approach provides that index of accounted-for variance.

Research Hypotheses Involving One Multichotomous Predictor and One Dichotomous Criterion

When the predictor variable has more than two categories, the GLM can still be employed. The approach is analogous to the one-way ANOVA; the only difference is that the criterion is dichotomous. The solution, though, would fit the mold of Generalized Research Hypothesis 4.2. Because more than one restriction is being made on the Full Model, a global nondirectional question is being asked. The researcher who is trying to predict a dichotomous criterion from a multichotomous predictor could use a 2-by-k chi-square analysis. The advantage with the GLM approach is that the F test takes into account the number of subjects. In addition, one should readily see that a k-by-2 chi-square problem also can be handled by the GLM by making the dichotomous variable the criterion. The R^2 and significance level are all that are of concern, so it really does not matter which variable is treated as the criterion. As long as no causal interpretation is made, it does not matter which variable is the criterion and which is the predictor. When there are more than two categories on each of the predictor and criterion sides, then the GLM cannot be used, and one must turn to a multivariate procedure, such as discriminant analysis (with multichotomous criteria) or multivariate analysis of variance (with continuous criteria).

ADVANTAGE OF THE GLM APPROACH OVER CORRELATION AND ANOVA

If one had asked a bivariate correlational question, relating only one predictor variable (and the unit vector) to the criterion, one could find the correlational value in the correlation matrix automatically calculated and printed by most computerized regression programs. The advantage of constructing a regression model is in finding the weighting coefficients (to find group means or to plot the lines of best fit) and in being able to ascertain the probability of that high a correlation, or one higher, occurring by chance alone. Most statistics books have tables of necessary correlational values for *selected* alpha levels. But it just might be that an investigator has an alpha level that is not one of the "selected" ones. Furthermore, our position in this text is consistent with the emerging notion of reporting the probability of the sample data occurring under the Statistical Hypothesis rather than imposing one's (arbitrarily chosen perhaps) alpha level on one's readers. This position does not support the selection of alpha after the results are computed—instead it treats research readers as thinkers rather than as blind followers. A given journal author may adopt an alpha of .05 and find that the results are significant for the author. If the actual probability is not reported, a more conservative reader cannot determine whether the results are significant at .01 or not. However, if the author reports the actual probability of .007, the conservative reader knows that the Research Hypothesis is tenable for the reader as well as for the author. On the other hand, if the actual probability is .03 and is reported, then the Research Hypothesis is *not* tenable for the conservative reader, though the reader realizes that the Research Hypothesis is tenable for the author.

Many researchers are aware that statistical significance can be obtained, yet practical significance may really not be obtained. Statistical significance is indeed a necessary but not sufficient condition for practical significance. Statistical significance simply indicates that something other than chance is operating, but statistical significance does not indicate to what degree that nonchance phenomenon is operating. Research studies using extremely large sample sizes often do find statistical significance, but little practical significance is obtained. The larger the sample size, the smaller the difference between two means needs to be before statistical significance is obtained. One treatment mean could be 14.0001 and the other 14.0000, and if enough subjects were in the sample, statistical significance could be obtained. In most practical applications, a mean difference of .0001 would not be of any importance.

When the Statistical Hypothesis is not true, increasing the sample size lowers the probability that a result of a given magnitude occurs by chance alone. The R^2 value, though, is not artificially inflated by the increase in sample size. Increasing sample size merely produces a more stable and closer estimate of the population R^2 value. This is one of the reasons that some researchers have been reporting R^2 values with their indexes of statistical significance. Persons who are writing a literature review should be encouraged to compute and include the R^2 value of each finding reviewed. When an F value is reported in the literature (when R^2 of the Restricted Model is .00), the R^2 value of the Full Model is:

$$R^2 \quad = \quad \frac{df_n * F}{df_d + (df_n * F)} \qquad (4.1)$$

When a t is reported, the R^2 value is:

$$R \quad = \quad \frac{t^2}{df + t^2} \qquad (4.2)$$

For instance, a t of 2.4 with 116 subjects would be "highly significant" ($p <$ .01); yet using Equation 4.2, the R^2 is found to be .05, showing that the researcher is accounting for only 5% of the phenomenon. One might interpret this as knowing something for sure about very little. In the past, probability values have been used to support theories. We hope that, in the future, R^2 values instead of probability values will be used to indicate the satisfaction one has with one's theoretical framework.

5

RESEARCH HYPOTHESES THAT EMPLOY CONTINUOUS PREDICTOR VARIABLES

In the previous chapter, we discussed the use of the GLM with categorical predictor variables. The flexible GLM also can be used with continuous predictor variables. All the procedures are the same, but the way the Research Hypothesis is stated is often different because the data being analyzed are continuous instead of categorical.

The simple linear model and its components are presented. A simple data set is given to provide concrete meaning for these components. The calculation of linear components in complex models is an easy process with existing computer programs. Thus, procedures for calculating weighting coefficients are discussed only briefly. The emphasis of this chapter is on stating the Research Hypothesis and on constructing linear models designed to answer that Research Hypothesis. Actual problems, with data and answers, are provided.

RESEARCH HYPOTHESES REQUIRING A SINGLE STRAIGHT LINE OF BEST FIT

Given a criterion behavior (Y) exhibited by a group of individuals under study, one may wish to know if the variance in these criterion scores can be accounted for. The information at hand to account for criterion behavior may take many forms (gender of the subject, previous test scores, knowledge of which treatment the subject received, etc.).

For illustrative purposes, suppose one is interested in establishing the following Research Hypothesis: "There is a relationship between ability and performance." Now suppose that scores on the performance behavior (Y) and the ability (X) hypothesized to be relevant to that performance behavior are obtained for five subjects. Implied by the Research Hypothesis just given is a supposition that there *is* a systematic relationship between the X and Y variables. This relationship could take many forms, but since the particular form is not specified, the default assumption is that the relationship being investigated is a linear one. (Chapter 7 deals with nonlinear relationships, which we feel should more often be investigated.)

Figure 5.1 is the graph of the observed performance and ability scores for the five individuals (A, B, C, D, and E). The description of this graph of scores involves using the formula for a straight line:

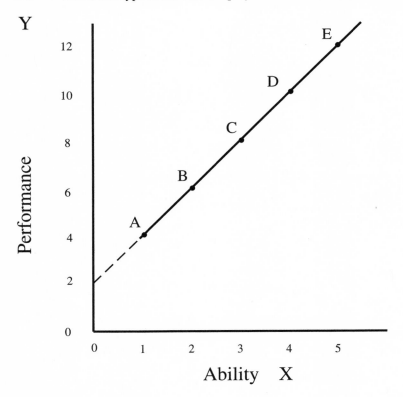

Figure 5.1.
The observed X and Y scores for five individuals.

$$Y = a + bX \qquad\qquad (5.1)$$

where:

Y = the scores on the Y-axis variable (performance scores);
X = the scores on the X-axis variable (ability scores);
a = the point at which the straight line crosses the Y axis (called the *Y-intercept*); and
b = the increase in the Y-axis variable for every one-unit increase in the X-axis scores (called the *slope of the line*).

In Figure 5.1, notice the line that goes through each of the five points. This line can be specified by finding values for a and b in Equation 5.1. To get a feel for the value of the slope, b, notice that Person A scored 1 on ability (and 4 on performance) while Person B scored 2 on ability (and 6 on performance). Thus, for a one-unit increase in ability (from 1 to 2), performance increased two units (from 4 to 6). For every one-unit increase on X, there is a two-unit increase on Y. Therefore, b (in Equation 5.1) must equal 2. Putting this value into the equation, the value for the intercept, a, also can be determined.

So far, Y = a + 2X has been obtained, where $\hat{Y}$ = predicted Y, and b = 2. To complete the equation, assume for the moment that a = 0. In vector form, the equation would be:

$$\hat{Y} \quad = a * \quad U \quad + 2 * \quad X$$

Person A	2		1		1
Person B	4		1		2
Person C	6	= 0 *	1	+ 2 *	3
Person D	8		1		4
Person E	10		1		5

Note that a can be represented as a * U. Indeed, the value of a is added to each subject. Also note that the first element in vector X is 1 and must be the score on X for Person A. Also, the X score of 5 must represent Person E's score on X. The vector U provides a one for every person and, when multiplied by a, adjusts each score by a constant amount (therefore the weighting coefficient for the unit vector is often referred to as the *regression constant*).

When each element in X is multiplied by 2 (the value of b) and added to the value of "0 times the unit vector," the result is the vector $\hat{Y}$. In finding values for a and b, one is trying to make $\hat{Y}$ equal to Y. If a really equals zero, as has been assumed above, then $\hat{Y}$ should equal Y. Note, however, that with a = 0, $\hat{Y}$ and Y are not equal.

	Y	$\hat{Y}$	Y - $\hat{Y}$
Person A	4	2	2
Person B	6	4	2
Person C	8	6	2
Person D	10	8	2
Person E	12	10	2

To make $\hat{Y}$ equal to Y, a value of 2 must be added to every score in $\hat{Y}$. This means that a = 2 (not 0). Now with a = 2 and b = 2, a line of perfect fit is obtained (see Figure 5.2).

$$\hat{Y} \quad = 2 * \quad U \quad + 2 * \quad X \quad (Y - \hat{Y})$$

4		1		1		0
6		1		2		0
8	= 2 *	1	+ 2 *	3		0
10		1		4		0
12		1		5		0

None of the observed ability scores was zero, but if there were an X score of zero, the value of Y would be 2, and the line of best fit would cross the Y axis at the Y value of 2, which is therefore the intercept for this data. The original Research Hypothesis for which these data were obtained was: "There is a [linear] relationship between ability and performance." Figure 5.2 shows that there is a linear relationship between X and Y; without exception, as X increases, so does Y. And since the increase has a constant value (i.e., 2), Y is a linear function of X.

However, the observed relationship may be due to sampling error, as discussed in chapter 2. If one wished to generalize beyond these five subjects, one would want to ask, "How likely is it that the observed linear relationship between X and Y in the

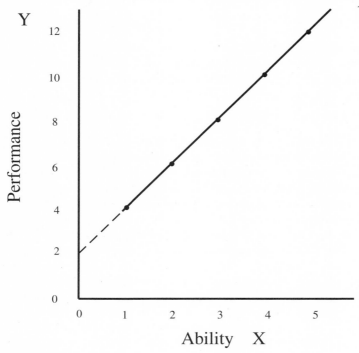

Figure 5.2.
The vector representation of $\hat{Y} = 2 + 2X$ and the line of best fit.

sample is due to sampling variation?" This can be determined by the F statistic. In the data that were collected, there was only one person at each of the five X scores. If there had been more subjects and therefore more observations at each point on X, it would be unlikely that all subjects with a particular score on X would have had the same criterion (Y) score. Each point on the X variable has a population of people with that score. In any particular study, the individuals at the scale point X = 2 are a sample of the population of people who have an X score of 2—a statement that holds true for all scale points on X. When a line is fit to the data, then all observed squared deviations on Y from that line for all scale points on X can be viewed as a within-group sum of squares. Knowledge of X cannot explain these deviations. The difference between this within-group sum of squares and the total sum of squares yields the variation explained by knowledge of the estimated linear relationship between X and Y.

For the data in Figure 5.2, there are no data points that are not on the line. Thus, there is perfect prediction for these data, a very unlikely situation. Figure 5.3 more clearly shows the notion of variability about the line of best fit for each value of the predictor variable X. Note that in Figure 5.3 two persons have an ability score of 6, but one is at a performance level of 5, whereas the other is at a performance level of only 3. Figure 5.3 also portrays an unlikely situation in that the criterion means for each ability level fall right on the line of best fit. Further discussion of that situation is presented later.

To test a Research Hypothesis, one must compare the accuracy of two linear models in fitting the same data, one model reflecting the state of affairs supposed by the Research Hypothesis, the other model reflecting the state of affairs supposed by the Statistical Hypothesis. The two models will provide two estimates of the population

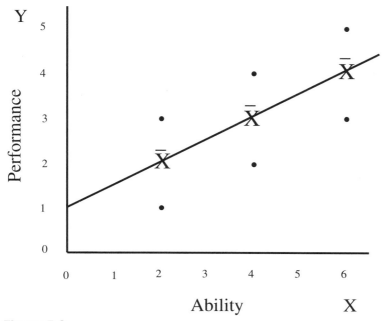

Figure 5.3.
Criterion variability for subjects having the same X values.

criterion variance accounted for by the predictor variables. An *F* test can then be calculated from these two variance estimates. Because the present Research Hypothesis specifies a systematic straight-line relationship, one regression model would depict a single straight line:

$$(\text{Model 5.1}) \ Y = aU + bX + E_1$$

where:

- Y = performance scores;
- U = the unit vector;
- X = ability (predictor) scores;
- E_1 = a vector of deviations from the line of best fit (when squared and summed, yields the within sum of squares);
- a = regression constant; and
- b = the slope of the line.

The model reflecting the Research Hypothesis will always contain more information than the model reflecting the Statistical Hypothesis, so the model reflecting the Research Hypothesis is referred to as the *Full Model*.

Dividing the sum of squared error components (ESS_f) by the number of data points free to vary, as in Equation 5.2, yields a within estimate of the criterion variance ($\hat{v}_w$) (some readers may be more familiar with MS_w/df):

$$\hat{v}_w = \frac{ESS_f}{df_w} \tag{5.2}$$

For the example, $\hat{v}_w = (\Sigma E_1^2) / (N - 2)$. The degrees of freedom denominator for the single straight-line model is $N - 2$ because two weighting coefficients had to be calculated to find the straight line of best fit (i.e., a and b).

The Statistical Hypothesis states that there is no relationship between ability and performance. This hypothesis implies that the best prediction is the mean of all the criterion scores or, in other words, that the line of best fit is horizontal—implying that the slope is equal to zero. The slope in Model 5.1 can be set equal to zero, which effectively eliminates X from the model. A simplification of Model 5.1, with the restriction that b = 0, results in the following (restricted) model:

$$(\text{Model 5.2}) \; Y \; = \; aU + E_2$$

This model has one piece of predictor information (the unit vector). The weighting coefficient for the unit vector will likely be different in the two models; the coefficient a simply holds the place for a numerical value. Indeed, if the weighting coefficient for X is other than zero in the Full Model, the numerical value for the unit vector will be different in the two models. Weighting coefficients will not always be subscripted differently in the Full and Restricted Models in the remaining sections of this text. The reader must be aware that weighting coefficients simply hold the place for some numerical value. The weighting coefficients in Model 5.1 and Model 5.2 are referred to as *raw-score weights*, or *b weights*. They are applicable for raw data. If all variables are standardized so that they have a mean of 0.0 and a standard deviation of 1.0, then the weighting coefficients are referred to as *standard weights*, or *standard partial regression weights*, or *beta weights*. Seldom is there any value in standardizing the data, so raw-score weights are used throughout the text.

The error vector is identified differently in the Full Model than in the Restricted Model. This is done to imply that the Full Model (Model 5.1) is allowed to have a different degree of predictability (and therefore error) than is the Restricted Model (Model 5.2). The criterion vector is the same in both models since the two models are reflections of the same Research Hypothesis. That is, the same data are being "modeled" in two different ways. The Full Model simply uses more predictor information than does the Restricted Model in accounting for the variance in a given criterion.

The error in prediction (the sum of the squared error, ESS or ΣE^2) probably will be lower in Model 5.1 (which incorporated the X information) than in Model 5.2. The question again is, "How likely is it that the decrease in error in prediction (or increase in predictability) is due to randomness?" The increase in predictability would be the difference between the two sums of squared errors ($\Sigma E_2^2 - \Sigma E_1^2$). Dividing this difference by the number of pieces of information that account for the difference (the number of pieces of information in the Full Model minus the number of pieces of information in the Restricted Model) yields another variance estimate. (The $\hat{v}_p$ is a variance estimate from a particular piece of information.)

$$\hat{v}_p = \frac{ESS_r - ESS_f}{df_p} \tag{5.3}$$

For the example, $\hat{v}_p = [\Sigma E_2^2 - \Sigma E_1^2] / 1$

Now that two variance estimates are available, the F test can be performed. Because it considers all the predictor information that appears in the Full Model, $\hat{v}_w$ (again, MS_w/df) is the best estimate of the criterion variance. And, $\hat{v}_p$ is a variance estimate influenced by the "degree of worth" or gain in accuracy achieved by the predictor (namely, X) included in the Full Model that is absent from the Restricted Model (traditionally called the *variance between groups*).

The F test for deciding if a statistically significant gain has been produced by use of the Full Model, as compared to the Restricted Model, is Equation 5.4:

$$F_{(df_p, df_w)} = \frac{\hat{v}_p}{\hat{v}_w} = \frac{(ESS_r - ESS_f)/df_p}{(ESS_f)/df_w} \tag{5.4}$$

At the end of chapter 2 the F test was expressed in terms of R^2. Since that formula is more easily generalizable to other Research Hypotheses, the F test stated in Equation 5.4 is translated to R^2 terms in Equation 5.5 (the demonstration of the equivalency of the two formulae is in Appendix F):

$$F_{(df_p, df_w)} = \frac{(R_f^2 - R_r^2) / df_p}{(1 - R_f^2) / df_w} \tag{5.5}$$

It is the case that $(R_f^2 - R_r^2)$ yields the gain in accuracy due to the additional information in the Full Model. This gain is compared to the error variation shown by the denominator in Equation 5.4 as ESS_f and in Equation 5.5 as $(1 - R_f^2)$. Thus, the amount of gain is compared against the measure of random error variation. If this gain is large in comparison to the measure of random variation, as determined by the magnitude of F and its associated chance probability distribution, then the Statistical Hypothesis (or null hypothesis) may be rejected (with the realization that such a decision will be wrong a proportion of the time equal to alpha). In general, the degrees of freedom for the numerator and denominator will be referred to as df_n and df_d, respectively. Therefore, the F test used throughout the remainder of this text is the same as first introduced in Equation 2.4 (and used in chapter 4 with dichotomous predictors) and is reproduced here for use with continuous predictors:

$$F_{(df_n, df_d)} = \frac{(R_f^2 - R_r^2) / (df_n)}{(1 - R_f^2) / (df_d)} \tag{5.6}$$

The R^2 values for the Full Model and Restricted Model are now calculated. Model 5.1 was the Full Model, and its vector representation is:

$$Y \quad = a_1 * \quad U \quad + b_1 * \quad X \quad + \quad E_1 \quad (E_1)^2$$

$$\begin{bmatrix} 4 \\ 6 \\ 8 \\ 10 \\ 12 \end{bmatrix} = 2 * \begin{bmatrix} 1 \\ 1 \\ 1 \\ 1 \\ 1 \end{bmatrix} + 2 * \begin{bmatrix} 1 \\ 2 \\ 3 \\ 4 \\ 5 \end{bmatrix} + \begin{bmatrix} 0 \\ 0 \\ 0 \\ 0 \\ 0 \end{bmatrix} \begin{bmatrix} 0 \\ 0 \\ 0 \\ 0 \\ 0 \end{bmatrix}$$

$$\Sigma(E_1)^2 = 0 = ESS_f$$

If each element in E_1 is squared and then summed, the error sum of squares associated with the predictors in that model is obtained. The total sum of squares can be calculated by subtracting the criterion mean from each criterion score, squaring this deviation, and then summing all the squared deviations. The total sum of squares for the data under consideration is 40. Therefore,

$$R_f^2 = \frac{(SS_t - ESS_f)}{SS_t} = \frac{(40 - 0)}{40} = 1.00$$

Model 5.2 was the Restricted Model, and its vector representation is:

$$Y \quad = a_0 * \quad U \quad + \quad E_2 \quad (E_2)^2$$

$$\begin{bmatrix} 4 \\ 6 \\ 8 \\ 10 \\ 12 \end{bmatrix} = 8 * \begin{bmatrix} 1 \\ 1 \\ 1 \\ 1 \\ 1 \end{bmatrix} + \begin{bmatrix} -4 \\ -2 \\ 0 \\ +2 \\ +4 \end{bmatrix} \begin{bmatrix} 16 \\ 4 \\ 0 \\ 4 \\ 16 \end{bmatrix}$$

$$\Sigma(E_2)^2 = 40 = ESS_r$$

To minimize the sum of the squared elements in E_2, the weight a will in this case be the mean of the scores in the criterion vector (Y). The mean of Y is 8; therefore, a = 8. Each element of E_2 is the difference of the individual score from the criterion mean. The first element in vector U is multiplied by the weighting coefficient of 8, and the resultant value is 8. The first element in Y is 4; thus (4 - 8) = -4, which is the first element in E_2. Vector E_2 contains the deviation of each individual's score from the criterion mean. When these elements are squared and summed, the error sum of squares for the Restricted Model (for these data, 40) is obtained.

No other value than 8 (the criterion mean) for a will give a smaller error sum of squares for the Restricted Model. The reader may wish to try other values for a to see what happens to E_2, and, more important, what happens to $\Sigma(E_2)^2$.

The proportion of criterion variance accounted for by the Restricted Model is:

$$R_r^2 = \frac{(SS_t - ESS_r)}{SS_t} = \frac{(40 - 40)}{40} = .00$$

We are ready to discuss the calculation of degrees of freedom. The concept of degrees of freedom was presented in chapter 2 in terms of how many data points are free to vary. Degrees of freedom also can be conceptualized in terms of the number of linearly independent predictor vectors in the Full and Restricted Models and the number of subjects. The notion of linearly independent vectors was introduced at the end of chapter 3.

The degrees of freedom in the numerator (df_n) in Equation 5.6 is the difference between the number of linearly independent vectors in the Full Model (symbolized as m1 in this text) and the number of linearly independent vectors in the Restricted Model (symbolized as m2 in this text). One way to conceptualize the degrees of freedom for the numerator of the F test that is being constructed is to ask, "How many linearly independent predictors were deleted from the predictors in the Full Model to form the Restricted Model?"

The number of linearly independent vectors in a model would include neither the criterion vector nor the error vector, only the predictor information. Weighting coefficients for the predictor vectors are mathematically determined to make the criterion vector linearly dependent upon the weighted combination of the predictor vectors; therefore, the criterion vector is not a linearly independent vector. A perfect fit will usually not be possible; therefore, an error vector is included. But note that the values in the error vector are determined only after the other weighting coefficients are determined; therefore, the error vector is not a predictor vector.

The Full Model ($Y = aU + bX + E_1$) has two linearly independent vectors (U and X). The Restricted Model ($Y = aU + E_2$) has only one linearly independent vector (U). Therefore, $df_n = (2 - 1) = 1$. The Restricted Model has one less piece of information than the Full Model, and that piece of information accounts for the difference between R_f^2 and R_r^2.

The R_r^2 is not always equal to .00. Often, several predictors are under consideration, and the researcher wishes to restrict one of the predictors from the Full Model to test the proportion of unique variance that that one piece of information adds to the others when used with them. (Such problems are presented in the next chapter.) The Restricted Model will then contain information in addition to the unit vector.

The F ratio for the comparison of Models 5.1 and 5.2 will now be calculated. The general F equation has as the denominator: $(1 - R_f^2)/df_d$. The 1 represents the maximum proportion of criterion variance that could be accounted for. When the proportion of accounted-for variance in the Full Model (R_f^2) is subtracted from 1, the proportion of unaccounted-for variance, or *error variance*, is obtained. The number of observations (N) minus the number of linearly independent vectors in the Full Model will always be df_d (there are two linearly independent vectors in the Full Model and $N = 5$; therefore $df_d = 5 - 2 = 3$). Another way to conceptualize degrees of freedom is that for every weighting coefficient calculated in the Full Model, one less criterion score is free to vary. This is analogous to saying that, once the group mean is calculated, one of the observed values in the group is fixed.

In the problem of the relationship of X to Y tested by Models 5.1 and 5.2, the following data are obtained:

$$R_f^2 = 1.00 \qquad R_r^2 = .00$$

$$df_n = 2 - 1 \qquad df_d = N - 2$$

$$F_{(df_n, df_d)} = \frac{(R_f^2 - R_r^2) / (df_n)}{(1 - R_f^2) / (df_d)} = \frac{(1.00 - .00) / (2 - 1)}{(1 - 1.00)/(5 - 2)} = \frac{100/1}{0/3} =$$

infinity.

F is equal to infinity because of the zero in the denominator. This difficulty is encountered because we wanted to facilitate understanding of the slope and Y-intercept concepts, and so data that yield an R_f^2 of 1.00 for the Full Model were employed. To overcome the ill effects of the unusual data we chose for this illustration, a little error can be introduced. Let $R_f^2 = .94$ and thus $1 - R_f^2 = .06$, and the resultant F is:

$$F_{(1,3)} = \frac{(.94 - .00)/1}{(1 - .94)/3} = \frac{.94/1}{.06/3} = \frac{.94}{.02} = 47.0$$

By looking in an F table, one will note that, with 1 and 3 degrees of freedom, an observed F of 34 or greater is found fewer than one time in a hundred ($p < .01$) due to sampling variation. The observed F (with a little error thrown in) was 47.0, which is larger than 34.

The next step is to find out how tenable is the Research Hypothesis: "There is a linear relationship between ability and performance." It is known that the observed F is a very rare occurrence where there is no systematic relationship and that knowledge of ability explains 94% of the criterion variance. Knowing that the five subjects are a random sample of a specified population, most researchers would reject the Statistical Hypothesis and accept the Research Hypothesis; that is, they would accept that "There is a linear relationship between ability and performance."

Yet it must be remembered that the observed data may be that 1-time-in-100 chance finding. It is highly unlikely, but possible. Whether one rejects the Statistical Hypothesis depends on the importance of the outcome. If someone's life could be lost by incorrectly accepting the Research Hypothesis, one may want several more samples—if the decision involves only a few dollars, one probably would accept the Research Hypothesis based upon the one sample.

We must stress that statistics is only an aid to decision making. The a priori establishment of a level of probability and percentage of error variance one can tolerate are *not* statistical decisions. Logical forethought, including an analysis of the costs of the various decisions, must be accomplished by the researcher *before* the collection and analysis of the data.

SUMMARY OF F AND R^2

The general F formula using R^2 is:

$$F_{(df_n, df_d)} = \frac{(R_f^2 - R_r^2) / (df_n)}{(1 - R_f^2) / (df_d)}$$

The R_f^2 is the proportion of observed criterion variance accounted for by the Full Model. The R_r^2 is the proportion of observed criterion variance that the Restricted Model explains. Some information in the Full Model is restricted (e.g., if b = 0 is hypothesized, then the variable associated with b [X] does not appear in the Restricted Model).

In the numerator (R_f^2 - R_r^2) is the proportion of unique variance that the deleted variable(s) explains. The degrees-of-freedom term in the numerator (df_n) is the number of linearly independent vectors used to account for the proportion of variance difference between R_f^2 and R_r^2. The difference between the number of linearly independent vectors in the Full Model and the number of linearly independent vectors in the Restricted Model is df_n. One minus R_f^2, (i.e., 1 - R_f^2) is the proportion of variance unexplained by the Full Model (error variance), and df_d equals the number (N) of observations minus the number of linearly independent vectors in the Full Model. In essence, df_d is the number of observations that are free to vary after weights for each of the linearly independent vectors in the Full Model have been calculated.

A SECOND ILLUSTRATIVE EXAMPLE OF A SINGLE STRAIGHT LINE OF BEST FIT

The second example contains data that are not as systematic as those in the first example and are closer, therefore, to the real-world state of affairs. The Research Hypothesis under consideration is the same as for the first example and can be stated in several ways:

1. There is a linear relationship between ability and performance.
2. For every unit of increase in ability, there is a constant change in performance.
3. The correlation between ability and performance is other than zero.
4. Ability is linearly predictive of performance.

The rival or Statistical Hypotheses for these Research Hypotheses would be, respectively:

1. There is no linear relationship between ability and performance.
2. For every unit increase in ability, there is no change in performance.
3. The correlation between ability and performance is zero.
4. Ability is not linearly predictive of performance.

The observed data points for six subjects' ability and performance are given in Figure 5.4. The six data points reflected in this scattergram seem to follow a trend; however, a single straight line cannot be cast that would go through all the data points. Two people had an ability score of 2, but their performance scores differed: one had a performance score of 3, and the other had a performance score of 1. A similar type of discrepancy is noted for ability levels 4 and 6. There is not a perfect linear relationship because a single straight line does not fit all the data points. A line of best fit, though, can be cast that will minimize the sum of the squared distances from that line (the sum of the squared elements in the error vector).

The Full Model required by the Research Hypothesis is the same single straight-line model as for the first example. We know that this is the model by answering three questions:

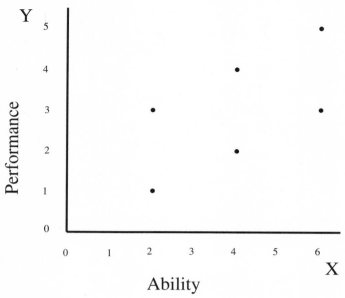

Figure 5.4.
Observed performance scores (Y) and ability scores (X) for six individuals.

1. What criterion are we trying to understand? (Answer: Y, the performance score.)
2. What information about the subjects are we going to use to predict the criterion? (Answer: X, knowledge of their ability.)
3. What functional relationship are we going to investigate? (Answer: linear, because no other functional relationship was specified.)

These answers lead to the following Full Model:

$$(\text{Model } 5.3)\ Y\ =\ aU + bX + E_3$$

The task is to find values for a and b so as to minimize the sum of the squared elements in E_3. First, a and b will be solved for intuitively, and then the weights will be derived formally. The vector representation of the Full Model is:

$$
\begin{array}{cccccccc}
Y & = a & U & + b & X & + & E_3 \\
\begin{bmatrix} 1 \\ 3 \\ 2 \\ 4 \\ 3 \\ 5 \end{bmatrix}
& = a &
\begin{bmatrix} 1 \\ 1 \\ 1 \\ 1 \\ 1 \\ 1 \end{bmatrix}
& + b &
\begin{bmatrix} 2 \\ 2 \\ 4 \\ 4 \\ 6 \\ 6 \end{bmatrix}
& + &
\begin{bmatrix} ? \\ ? \\ ? \\ ? \\ ? \\ ? \end{bmatrix}
\end{array}
$$

Look at the scattergram presented in Figure 5.4, and place a point halfway between the two scores for the individuals who scored 2 on the ability measure (the mean score is 2). That point will be the mean criterion score for these two individuals. Likewise, if one obtains the mean criterion scores for those who had an ability score of 4 and those who had an ability score of 6 and places them between the

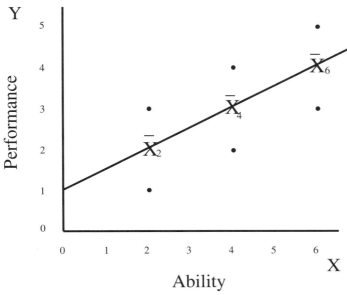

Figure 5.5.
An intuitive selection of the line of best fit.

respective observed points, one can easily cast a straight line (Figure 5.5) that is the line of best fit; that is, the sum of the squared elements in the error vector (E_3) is minimized.

The line in Figure 5.5 is straight and passes through the mean data points; thus the line of best fit has been obtained. What is the value for a? The line intersects the Y axis at 1; therefore, a = 1. Also, a two-unit increase on X from 0 to 2 gives a one-unit increase on Y. Since the slope is the same throughout, 2 to 4 and 4 to 6, b should be .5. One can check this value by noting that the value on the line at X = 2 is 2, and at X = 4 is 3. So for each unit increase on X, a .5 increase on Y is observed. The next step is to calculate E_3 with these weights and then square the elements of E_3, as illustrated below.

$$
\begin{array}{cccccccc}
Y & = 1 * & U & + .5 * & X & + & E_3 & E_3{}^2 \\
\begin{bmatrix} 1 \\ 3 \\ 2 \\ 4 \\ 3 \\ 5 \end{bmatrix} & = 1 * & \begin{bmatrix} 1 \\ 1 \\ 1 \\ 1 \\ 1 \\ 1 \end{bmatrix} & + .5 * & \begin{bmatrix} 2 \\ 2 \\ 4 \\ 4 \\ 6 \\ 6 \end{bmatrix} & + & \begin{bmatrix} -1 \\ +1 \\ -1 \\ +1 \\ -1 \\ +1 \end{bmatrix} & \begin{bmatrix} 1 \\ 1 \\ 1 \\ 1 \\ 1 \\ 1 \end{bmatrix}
\end{array}
$$

$$\Sigma(E_3)^2 = 6$$

The first element in vector E_3 is -1. This value was obtained by multiplying a times U (1 * 1) and adding that to b times the first element of X (.5 * 2), which gives a predicted score of 2. Solving for E_3, the observed score (1) minus the predicted score (2) equals -1. This procedure is followed to obtain each element in vector E_3. Squaring each element in the error vector E_3 and summing yields the error sum of squares found with the Full Model. No values for a and b other than 1 and .5, respectively, yield a smaller error sum of squares.

The Statistical Hypothesis sets b = 0, reflecting the notion that there is no linear relationship between ability and performance. The sample data indicate that b is not 0, but that b = .5. However, is this value of .5 merely a chance sampling discrepancy from a true population value of 0? To answer this question, the F test between a Full Model and a Restricted Model can be calculated and evaluated according to a predetermined alpha level. Suppose the alpha level is set at .05 (that is, if the observed F can be expected fewer than 5 times in 100, the Statistical Hypothesis will be rejected, and the Research Hypothesis will be accepted).

The Full Model was Model 5.3 ($Y = aU + bX + E_3$). To test the Statistical Hypothesis, b is restricted to 0—effectively deleting X from the Full Model. The Restricted Model is then:

$$(\text{Model } 5.4)\ Y = cU + E_4$$

Note that the weighting coefficient for the unit vector is designated as c (rather than as a), just to point out that this weight will in general *not* be the same in the Restricted Model as in the Full Model; indeed, in this case it is not. The value for c that will satisfy the minimal sum of squared elements in E_4 is the criterion mean of Y, $\Sigma X/N$, or 18/6 = 3. Therefore, c = 3. Below is the vector representation of the Restricted Model with E_4 and E_4^2.

Y	= 3 *	U	+	E_4	E_4^2
1		1		-2	4
3		1		0	1
2	= 3 *	1	+	-1	1
4		1		+1	1
3		1		0	0
5		1		+2	4

$$\Sigma(E_4)^2 = 10$$

As in the case of the first example, when only the criterion mean is used to predict the criterion, the sum of the squared elements in the error vector is numerically equal to the total sum of squares. For the data under consideration, $SS_t = 10$. To obtain the R_f^2, one must place the sum of squares explained by that model over the total sum of squares.

The sum of squares in vector E_3 was 6 and is the sum of squares *not* accounted for by knowledge of ability. The total sum of squares is 10. The proportion of variance accounted for in the Full Model is the difference between the total sum of squares (or SS_t) and $\Sigma(E_3)^2$ (or ESS_f) divided by the total sum of squares. Equation 5.7 presents this notion symbolically and finds the value of R_f^2 for the example data.

$$R_f^2 = \frac{(SS_t - ESS_f)}{(SS_t)} = \frac{(10 - 6)}{10} = \frac{4}{10} = .40 \qquad (5.7)$$

The proportion of variance accounted for in the Restricted Model is the total sum of squares (SS_t) minus the sum of the squared elements in E_4 (or ESS_r) over the total sum of squares. In this case, the sum of the squared elements in E_4 is equal to the total sum of squares. This will always be the case when the Restricted Model contains only the unit vector, although not all Restricted Models contain just the unit vector. Therefore, the R^2 for the Restricted Model can be found from Equation 5.8:

$$R_r^2 = \frac{(SS_t - ESS_r)}{SS_t} = \frac{(10 - 10)}{10} = .00 \tag{5.8}$$

where ESS_r is equal to the sum of the squared elements in the error vector associated with the Restricted Model. Please do not memorize these formulae: they are presented here only to help those who have overlearned their ANOVA.

Substituting the sample information into the general F formula, Equation 5.5, the F value for comparing Model 5.3 with Model 5.4 can be found:

$$F_{(df_n, df_d)} = \frac{(R_f^2 - R_r^2) / (df_n)}{(1 - R_f^2) / (df_d)}$$

$$F_{(1,4)} = \frac{(.40 - .00)/(2 - 1)}{(1 - .40)/(6 - 2)} = \frac{.40/1}{.60/4} = \frac{.40}{.15} = 2.667$$

Here, df_n equals the difference between the number of linearly independent vectors in the Full Model and the Restricted Model; therefore, $df_n = 1$ because the Full Model has two linearly independent vectors, and the Restricted Model has one linearly independent vector; and df_d equals N minus the number of linearly independent vectors in the Full Model. $N = 6$, so $df_d = (6 - 2) = 4$.

With 1 and 4 degrees of freedom, an F value of 7.71 or greater is observed 5 times in 100 when the population value of b really is zero. (The critical values for F can be found in most statistics texts. Because the computer program will print the actual probability of the observed F occurring by chance, we can avoid the F and directly compare the computer-provided probability to the chosen alpha.) An F value of 2.667 is observed more than 5 times in 100 due to sampling variation; therefore, because an alpha of .05 had been adopted, the Statistical Hypothesis that b = 0 cannot be rejected, and it is concluded that the observed apparent linear relationship between ability and performance may only reflect a chance event.

A straight line will not often go through all the data points as it did in the first example, nor will it often go through all the mean data points as in the second example. Therefore, mathematical formulae are usually required to find the weighting coefficients for the line of best fit, as discussed in the following section.

MATHEMATICAL CALCULATION OF THE SINGLE STRAIGHT LINE OF BEST FIT

The data used up to this point were constructed to provide intuitive solutions to obtaining the line of best fit. In a bivariate case where the line of best fit is not intuitively obvious, the following equations, Equations 5.9 and 5.10, can be used to obtain the slope weight (b) and the Y-intercept weight (a) that will minimize the sum of the squared elements in the error vector:

$$b = \frac{\Sigma XY - \dfrac{(\Sigma X)(\Sigma Y)}{N}}{\Sigma X^2 - \dfrac{[(\Sigma X)2]}{N}}$$

(5.9)

$$a = (\text{mean of Y}) - (b * \text{mean of X})$$

(5.10)

To find the values for a and b in the example just given, Equation 5.9 can be used to derive the weight b, Equation 5.10 to solve for a. The calculations are given below. The weights (b = .5 and a = 1) are the same values previously obtained using the intuitive approach.

Y	X	X^2	XY
1	2	4	2
3	2	4	6
2	4	16	8
4	4	16	16
3	6	36	18
5	6	36	30
$\Sigma Y = 18$	$\Sigma X = 24$	$\Sigma X^2 = 112$	$\Sigma XY = 80$

$\bar{Y} = (\Sigma Y / N) = (18 / 6) = 3$
$\bar{X} = (\Sigma X / N) = (24 / 6) = 4$

$$b = \frac{80 - \dfrac{(24)(18)}{6}}{112 - \dfrac{(24)^2}{6}} = \frac{80 - 72}{112 - 96} = \frac{8}{16} = .5$$

$a = [3 - .5 (4)] = (3 - 2) = 1$

It should be apparent that the real world seldom yields data that can be solved intuitively, and indeed real-world problems extend beyond the bivariate case such that many predictors are usually needed to account for the criterion variance. However, as complex models are investigated, the reader should be aware that the complex models break down into subsets of the basic linear model that has the following form: $Y = aU + bX + E$. The mathematical solution of weights for the multiple predictor set becomes involved and has been treated extensively elsewhere (Maxwell & Delaney, 1990). The intent of this book is to explicate conceptual research problems and the linear models required to answer the resulting hypotheses. We let the computer do its work in figuring out the numbers.

VENN DIAGRAM

When one continuous predictor is used, one is attempting to account for the criterion variance with one variable, hence the Venn diagram is the same as in the previous chapter. The proportion of overlap of the two circles is equal to the R^2 of the Full Model.

GEOMETRIC INTERPRETATION

As discussed in this chapter, the single continuous predictor allows a single straight line to fit the data. If the data do not conform to the single line, or to a straight line, then the single straight-line model will not fit it well, and the R^2 will be substantially lower than 1.00. If it appears that there is a systematic relationship in the data that goes beyond a single straight line, then another model of the data needs to be tested.

RELATIONSHIP TO OTHER STATISTICAL TECHNIQUES

The single straight-line model and the Pearson product moment correlation are isomorphic. The R^2 of the Full Model is equivalent to the square of the Pearson product moment correlation. That the Pearson correlation only measures the linear relationship between two variables is thus made clear in the regression formulation.

THE USE OF APPLIED RESEARCH HYPOTHESES: SINGLE STRAIGHT-LINE COMPUTER PROBLEM

Applied Research Hypothesis 5.1 is designed to fit the single straight-line notions just presented. The vectors in the problem are the same as those in the data set in Appendix A. At this point, we strongly recommend that time be spent using a linear regression program to obtain the information (R_f^2, R_r^2, F, and probability) provided in Applied Research Hypothesis 5.1.

This computer run is somewhat simple but illustrates one widespread application of the GLM. The Research Hypothesis stated in Applied Research Hypothesis 5.1 requires a single straight line of best fit. Most regression programs automatically provide the Pearson product moment correlation between each variable and every other variable in the analysis. Note that the correlation between the predictor variable and the criterion variable is provided on the computer output; if one were to square this correlation, one would obtain the R_f^2 listed on the output. Thus, the single straight-line model provides the same information as does the correlation coefficient. (The letters used to designate predictors and criteria are arbitrarily chosen.

Applied Research Hypothesis 5.1

Research Hypothesis: For a given population, X3 is linearly predictive of X2.

Statistical Hypothesis: For a given population, X3 is not linearly predictive of X2.

Full Model: $X2 = a_0U + a_3X3 + E_1$ (pieces full = m1 = 2)

Want: $a_3 \neq 0$; restriction: $a_3 = 0$

Restricted Model: $X2 = a_0U + E_2$ (pieces restricted = m2 = 1)

alpha = .001
$R_f^2 = .92$; $R_r^2 = .00$
$F_{(1,58)} = 747$
Computer probability = .0001; nondirectional probability = .0001

Interpretation: Because the calculated probability is less than the predetermined alpha, the following conclusion can be made: X3 is linearly predictive of X2.

Often in this text, both predictors and criteria are referred to as X variables [i.e., X2, X3, X27, etc.]. This corresponds to the use of X's in the Applied and Generalized Research Hypotheses and helps to show that any variable may serve as a predictor in one hypothesis but as a criterion in another hypothesis. The general application of the single straight-line model is in Generalized Research Hypothesis 5.1.)

A NOTE ABOUT THE UNIT VECTOR IN COMPUTER SOLUTIONS

Both the Full Model and the Restricted Model require the unit vector when testing the single straight-line hypothesis. Most computerized regression programs provide the unit vector automatically because of its widespread use in linear models. Because the unit vector is automatically provided, the researcher does not have to enter it. The unit vector must, though, be considered in the determination of linearly independent vectors. Because each element of the unit vector is a 1 and therefore the weighting coefficient of the unit vector will be added to every individual's predicted score, the weight of the unit vector is often referred to as the *regression constant*. The SAS computer program refers to the unit vector weight as INTERCEP; SPSS refers to the unit vector weight as CONSTANT; and BMDP refers to the unit vector weight as Y-INTERCEPT.

Generalized Research Hypothesis 5.1

Directional Research Hypothesis: For some population, X is positively related with Y1.

Nondirectional Research Hypothesis: For some population, X is related with Y1.

Statistical Hypothesis: For some population, X is not related with Y1.

Full Model: $Y1 = a_0 U + bX + E_1$ (pieces full $= m1 = 2$)

Want (for directional Research Hypothesis): $b > 0$; restriction: $b = 0$
Want (for nondirectional Research Hypothesis): $b \neq 0$; restriction: $b = 0$

Restricted Model: $Y1 = a_0 U + E_2$ (pieces restricted $= m2 = 1$)

where:
- $Y1$ = criterion;
- U = 1 for all subjects;
- X = predictor score for subject; and
- a_0 and b are least squares weighting coefficients calculated so as to minimize the sum of the squared values in the error vectors.

Degrees of freedom numerator = (pieces full minus pieces restricted) = $(m1 - m2) = (2 - 1) = 1$
Degrees of freedom denominator = (N minus pieces full) = $(N - m1) = (N - 2)$

where:
- N = number of subjects.

Note: SAS makes the restrictions by restricting the variable names, in this case X = 0. Although this is definitely more user-friendly, it is not algebraically correct. Because the user does not define the weighting coefficients in SAS, this is the only option for SAS.

6

MULTIPLE CONTINUOUS PREDICTORS

In the previous chapter, we presented a limited case of researching behavior. Although a single continuous variable is definitely a parsimonious way of researching behavior, most behavior is probably a function of more than one variable. Therefore, there is a need for statistical models that can include more than one continuous variable. In this chapter, we present the more general case of multiple continuous predictor variables.

We begin with the simplest case of more than one continuous predictor—the case of two continuous predictors. General questions related to multiple continuous predictors are discussed. First, the overall values of the variables in the model are tested; second, the value of a particular variable in the model is tested; and third, the full intent of chapter 1 is fulfilled by using Full and Restricted Models to compare two competing models of behavior.

TWO CONTINUOUS PREDICTORS

Rationale

Using an example with two continuous predictor variables allows the extension of several concepts and comes a little closer to the intent presented in chapter 1. The situations discussed in the previous two chapters occur in approximately 90% of the empirical research studies on human behavior. The careful reader should realize, though, the limiting aspects of the models discussed in those chapters. Therefore this section allows us to approach more closely the multivariable stance taken in chapter 1.

Geometric Representation

While the single continuous predictor variable model can be represented by a single straight line in a two-dimensional space, the two continuous predictor variable model requires another dimension; it can be represented by a single flat plane in a three-dimensional space.

The two-dimensional ellipse of data points becomes a three-dimensional football of data points, with a plane of best fit as in Figure 6.1. The weighting coeffi-

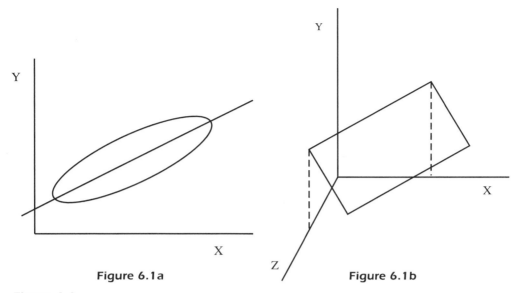

Figure 6.1a

Figure 6.1b

Figure 6.1.
Geometric representation of a one-variable model (Figure 6.1a) and a two-variable model (Figure 6.1b).

cients are determined so that the plane fits the data as well as possible, minimizing the squared distance of the real data points from the predicted data points that are on the plane of best fit. Figure 6.2 provides three different sets of data, each using two continuous predictors.

If one of the predictor variables is in actuality not needed in the model, then the plane of best fit is perpendicular to that axis, as in Figure 6.3a. This situation reduces to a two-dimensional ellipse of data, requiring only a single line of best fit, as in Figure 6.3b.

Research Hypothesis, Full Model, and Restricted Model

When two continuous predictor variables are used to predict a criterion, a researcher could ask the following Research Hypothesis:

The two predictors (X and Z) account for variance in the criterion (Y).

An alternate way of stating the Research Hypothesis would be:

The two predictors (X and Z) predict the criterion (Y).

For either hypothesis, the Full Model implied by the Research Hypothesis contains the two continuous predictors predicting the criterion:

Full Model: $Y = a_0U + a_1X + a_2Z + E_1$

Want: $a_1 \neq 0$; $a_2 \neq 0$; restrictions: $a_1 = 0$; and $a_2 = 0$; or $a_1 = a_2 = 0$

Restricted Model: $Y = a_0U + E_2$

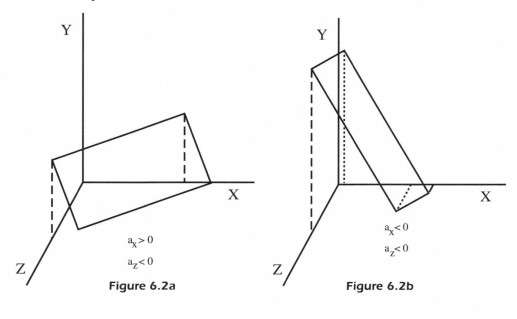

$a_x > 0$

$a_z < 0$

Figure 6.2a

$a_x < 0$

$a_z < 0$

Figure 6.2b

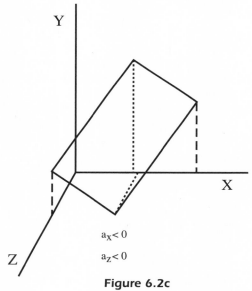

$a_x < 0$

$a_z < 0$

Figure 6.2c

Figure 6.2.
Three different geometric representations of planes of best fit.

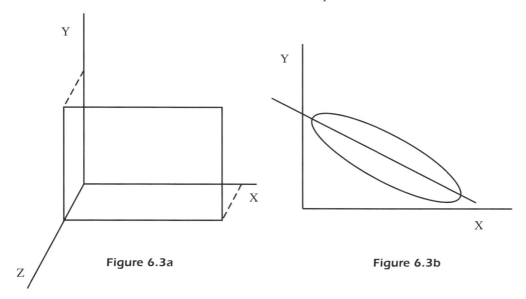

Figure 6.3a

Figure 6.3b

Figure 6.3.
Geometric interpretation when variable Z accounts for no additional variance and therefore is not needed.

Although the unit vector is not specified overtly in the Research Hypothesis, to include the unit vector in each regression model has become an automatic default. This is because the unit vector acts like a scale adjuster, making the mean of the predicted criterion equal to the mean of the actual criterion. That this makes sense relies on the nature of how variables are scaled in the behavioral sciences. In most situations, one would not be able to assume that the value of zero on X will predict the value of zero on Y. Therefore, the need for the scale adjuster. If one did not use both U and X, the line would be forced to go through the origin (0,0), or with the Z predictor, the plane would be forced to go through the origin (0,0,0).

The above Research Hypothesis is vague because there are two restrictions being made. Thus, while the Research Hypothesis can indeed be tested, it does not provide definitive information for the researcher. That is, if there is a significant difference between the two models, we know that the two variables together account for a significant amount of variance, but we do not know if the significance is a result of one variable, the other variable, or both variables. All that we know is that the two variables together produce the increment in R^2 from that of the Restricted Model ($R_r^2 = 0$) to that of the Full Model (R_f^2). Whether both variables are useful is not answered, because that was not the question asked.

Relationship to Other Statistical Techniques

Because multiple continuous predictor variables are used in the Full Model, the statistical test has become referred to as *multiple correlation*. The Research Hypothesis tested above can be thought of as testing to see if a multiple correlation is significantly different from zero. When there are two predictors, the multiple correlation can be determined from the bivariate Pearson correlations, using Equation 6.1:

$$R^2_{Y.XZ} = \frac{r^2_{YX} + r^2_{YZ} - (2\, r_{YZ}\, r_{YX}\, r_{ZX})}{1 - r^2_{ZX}}$$

(6.1)

The general F test, although applicable, is usually presented in statistical texts in a "simpler" form, as in Equation 6.2 from Hinkle et al. (1994, p. 461):

$$F = \frac{R^2 / k}{(1 - R^2) / (N - k - 1)}$$

(6.2)

where

k   = number of predictor variables.

The 1 in $N - k - 1$ is for the unit vector. This version of the general F formula does not recognize that the unit vector itself can be tested. This form also assumes that the unit vector will always be in the Full Model. While this will almost always be the case, the formula perpetuates the myth that the unit vector will always be in both the Full and Restricted Models. The GLM approach used throughout this text forces the researcher to specify whether the unit vector is in either model.

The R^2 is often referred to as the *coefficient of determination* because the R^2 shows how much variance is determined, or accounted for. The degree of unpredictability ($1 - R^2$, or the amount of error variance) is referred to as the *coefficient of nondetermination*. In some ways, this terminology is better than "error variance," because the only error in this variance is from either the error of measurement in the predictors and criterion or the misspecification of the predictor variables in the model. A more accurately specified model would transfer some of that "nondetermined" or error variance into "determined" variance.

Venn Diagram

The introduction of a second continuous predictor variable also can be understood with Venn diagrams. In the orthogonal contrast designs discussed in chapter 4, each of the independent (predictor) variables is required to be uncorrelated. For example, in Exhibit 4.4, not only are the main effects uncorrelated, but the interaction effect is also uncorrelated. Thus, each source of variance accounts for a totally separate amount of criterion variance. Figure 6.4 repeats the Venn diagrams presented in chapter 4. Note that, with the introduction of the uncorrelated predictor variable B, the amount of criterion variance accounted for by the A variable remains the same because the B and A * B interaction account for totally separate amounts of variance. Note, though, that the error variance is reduced by the additive effects of A, B, and A * B.

However, with two continuous predictor variables that are allowed to be correlated (a more realistic representation of the real world), the Venn diagram becomes a little more complicated. Nevertheless, we shall proceed. Suppose that we have the situation presented in Exhibit 6.1.

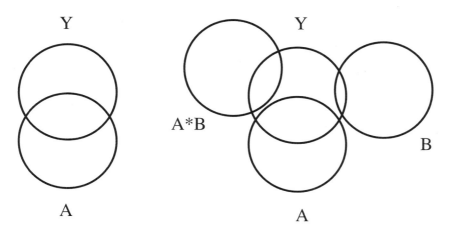

One-Way ANOVA   Two-Way ANOVA

Figure 6.4a   **Figure 6.4b**

Figure 6.4.
Venn diagram for *t* test (one-way ANOVA; Figure 6.4a) and orthogonal contrasts with four groups (two-way ANOVA; Figure 6.4b).

Exhibit 6.1

r_{XY} = .50 r_{ZX} = .50 r_{ZY} = .50

Example correlations between two predictors (X and Z) and a criterion (Y)

Now we know that there is a 25% overlap of X and Y and a 25% overlap of Z and Y. But these two percentages of variance cannot simply be added because X and Z are themselves correlated. If we did not know the correlation between X and Z, we could have any of the situations occurring in Figure 6.5, with Figure 6.5a showing the minimum amount of variance the two could account for, Figure 6.5b showing the maximum amount, and Figure 6.5c a more likely situation between those two extremes.

The multitude of possibilities identified in Figure 6.5 should reinforce the value of finding the correlation between the predictors and of letting the computer find the answers to the weighting coefficients and to the R^2.

Another way to represent the Research Hypothesis tested in this section is to reflect a Venn diagram as in Figure 6.6. Because the unit vector has the same "score"—one (1) for each subject—it provides no differential predictability and therefore does not appear in the Venn diagram.

The thoughtful reader might suggest that the Full Model be compared to a different Restricted Model, one that contains only one of the predictor variables. Although the same Full Model would be used in the two situations, the two different restrictions would result in two different Restricted Models, implying that two different questions were being tested and two different answers were possibly resulting. Indeed, that is the focus of the third section of this chapter.

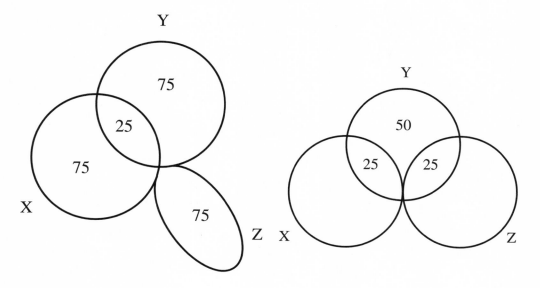

Figure 6.5a   Figure 6.5b

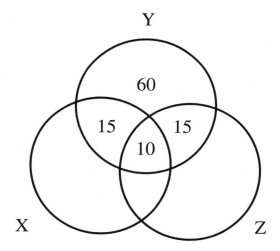

Figure 6.5c

Figure 6.5.
Three possible Venn diagrams resulting from the use of two continuous predictors.

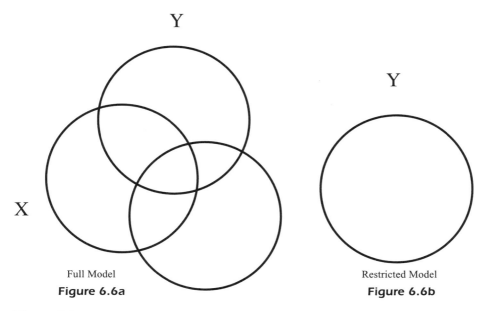

Full Model   Restricted Model
Figure 6.6a   **Figure 6.6b**

Figure 6.6.
Venn diagrams of Full and Restricted Models with two continuous predictors, X and Z.

TESTING THE OVERALL PREDICTABILITY OF THE MODEL

Rationale

The discussion in chapter 1 implied that a researcher probably would want to include multiple predictors when attempting to understand a particular phenomenon. While statistical models have already been discussed for one- and two-predictor situations, the more general case of multiple predictors offers the researcher the chance to "model" the phenomenon as complexly as it is really determined. We have never understood why researchers would want to reflect what they acknowledge to be a complexly determined phenomenon with an overly simplified one- or two-variable model.

Geometric Representation

One may think of a three-predictor situation as it was presented in Figure 6.1: a football spinning through a fourth dimension of time. Three snapshots of the plane of best fit can be seen in Figure 6.7.

The reader's mind (as well as each author's) is obviously being challenged by the four-dimension case. Geometric representation of more than three predictor variables (four dimensions) are beyond our ability. It is now time to turn to the computer to identify the weighting coefficients and the resulting R^2.

Venn Diagram

As the geometric representation becomes more involved with multiple variables, so does the Venn diagram representation. But it is instructive to develop a hypothetical Venn diagram representation when variables are added one at a time.

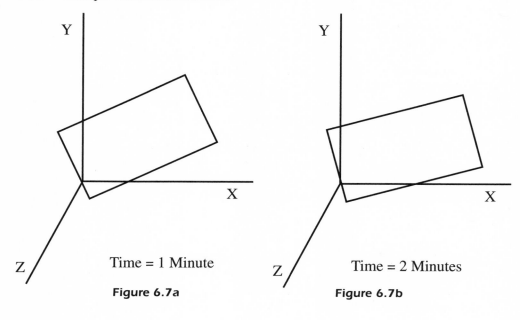

Time = 1 Minute

Figure 6.7a

Time = 2 Minutes

Figure 6.7b

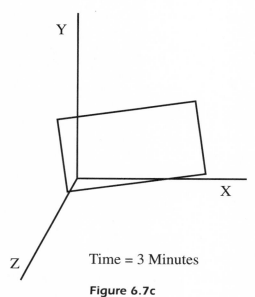

Time = 3 Minutes

Figure 6.7c

Figure 6.7.
Geometric representation of four dimensions, the fourth dimension being time.

Suppose that the model in Figure 6.6a (repeated in Figure 6.8a) has another variable added to it, with the result being shown in Figure 6.8b.

Predictor variables X and Z accounted for unique variance in Y and had a common, shared amount (the shaded area in Figure 6.8a.). Likewise, when the third variable, P, is added to the model, P accounts for some unique variance (the right-leaning diagonal stripe in Figure 6.8b), as well as for some shared variance between P and X (dotted area) and some shared variance between P and Z (horizontal lines) and some shared variance between P, X, and Z (area of blocks). Predictor variable Q, added in Figure 6.8c, although highly correlated with the criterion (determined by the large amount of overlap between Q and Y) also overlaps with the other predictor variables. When the Q variable is added to the model containing P, X, and Z, the increase in R^2 will be only the unique variance accounted for by Q (the dotted area in Figure 6.8c).

As indicated in chapter 1, inclusion of several variables from a particular area (such as within person) may lead the researcher to a point of diminishing returns. Some authors suggest that five or six variables will be the upper limit of useful variables (Hinkle et al., 1994). We strongly feel that all behavior is predictable and that researchers simply have to find the right variables (or the right functional relationships—interactions and nonlinear relationships, which will be discussed in later chapters).

In addition, the researcher may have to look for predictor variables in areas other than the within-person area. We do not argue that once a researcher has used five within-person predictors additional within-person predictors may not increase the prediction. What we do argue is that the researcher should look for predictors in other "focal stimuli" and "contextual" areas. The search for predictor variables that will increase the R^2 should continue until an R^2 close to 1.00 is obtained.

Research Hypothesis, Full Model, and Restricted Model

When testing the overall predictability of the Full Model, one is formally testing the following Research Hypothesis:

The predictor variables X1, X2, . . ., Xk predict the criterion (Y).

Implied by this Research Hypothesis is the following Full Model:

$$(\text{Model 6.1})\ Y \ = \ a_0 U + a_1 X1 + a_2 X2 + \ldots + a_k Xk + E_3$$

The antithesis of the Research Hypothesis is that the set of predictor variables does not predict the criterion Y, implying the following restrictions: $a_1 = 0$; $a_2 = 0$; . . .; $a_k = 0$, resulting in the following Restricted Model:

$$(\text{Model 6.2})\ Y \ = \ a_0 U + E_4$$

The general F test is applicable to this Research Hypothesis as the same criterion is in both models, the unit vector is in both models, and the Restricted Model is a restricted subset of the Full Model. In testing to see if the multiple correlation is significant, a researcher is testing the Full Model against one with zero predictability. The combined predictive power of the predictors is being questioned. Because more than one restriction is being made, a directional Research Hypothesis cannot be tested. The hypothesis presented as Applied Research Hypothesis 6.1 illustrates

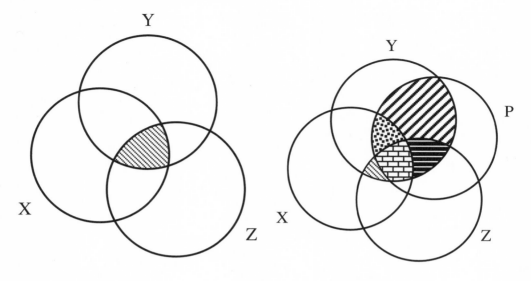

Figure 6.8a   Figure 6.8b

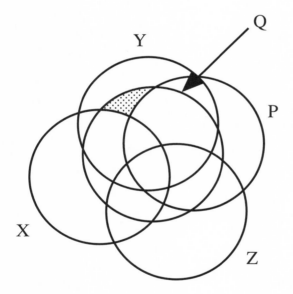

Figure 6.8c

Figure 6.8.
Figure 6.6a repeated in Figure 6.8a, with another variable added to it in Figure 6.8b. Finally, a fourth variable, Q, is added in Figure 6.8c.

Applied Research Hypothesis 6.1

Suppose that a researcher suspects that variables X4, X5, and X6 predict the criterion X2 in the data set in Appendix A. This leads the researcher to the formal statement of the hypothesis:

Research Hypothesis: The set of variables X4, X5, and X6 predict the criterion X2.

The antithesis of this Research Hypothesis is:

Statistical Hypothesis: The set of variables X4, X5, and X6 do not predict the criterion X2.

The regression models implied by these hypotheses are:

Full Model: $X2 = a_0U + a_4X4 + a_5X5 + a_6X6 + E_3$ (pieces full = m1 = 4)

Want: $a_4 \neq 0$, $a_5 \neq 0$, $a_6 \neq 0$; restrictions: $a_4 = 0$, $a_5 = 0$, $a_6 = 0$

Restricted Model: $X2 = a_0U + E_4$ (pieces restricted = m2 = 1)

alpha = .05
$F = 26.179$
Computer probability = .0001; nondirectional probability = .0001

the Research Hypothesis being tested (see the SAS setup in Appendix D-6.1). We do not recommend testing this hypothesis as it is not definitive.

Notice in Applied Research Hypothesis 6.1 that the F test value on the SAS output has an associated probability that is lower than an alpha of .05. Thus the researcher can reject the Statistical Hypothesis and accept the Research Hypothesis. Notice also that this F value and its associated probability value are exactly the same as those shown at the top of the SAS printout. We have therefore finally learned what Research Hypothesis this F test actually tests. While that question is of interest in this situation, it usually will not be of interest. Because many users in the past asked that question, most computer programs automatically include it. The point is that every time an F is reported, a question has been tested, and that question should have been stated a priori. If the user did not have that question, then it is entirely inappropriate to look at that particular F value.

Because this was a nondirectional Research Hypothesis, the conclusion, "The set of three variables predicts the criterion," does not allow the determination of which variables are or are not "valuable" in this prediction. Nor does the analysis allow for a directional interpretation, because df_n is not equal to 1, and the Research Hypothesis was not directional. This nondirectional Research Hypothesis is analogous to the omnibus F test discussed in chapter 4. The information gained from such a Research Hypothesis is of very limited, if any, value. In the following section, we discuss how one might ask a definitive question.

TESTING ONE VARIABLE'S CONTRIBUTION

Rationale

In response to the limitation identified in the previous section, one might be interested in focusing on the "value-added benefit" of a particular variable. If a variable does not add a significant amount of accounted-for variance to the variance accounted for by the variables already in the model, then that variable can be omitted from further consideration. On the other hand, it may be valuable to know how a predictor affects the criterion within a set of other predictors.

Geometric Representation

How a predictor affects the criterion within a set of other predictors can be phrased in geometric terms as, "Is a k-dimensional space needed, or can one less dimension be just as effectively used to reflect the data?" If the F test is significant, then the larger space is needed. If the F test is *not* significant, then the smaller space is all that is needed.

Venn Diagram

The Research Hypothesis here focuses on the unique variance accounted for by the targeted variable. The Full Model R^2 is not of interest—only as the denominator in the general F test. Thus the Research Hypothesis focuses on the dotted area in Figure 6.9.

Research Hypothesis, Full Model, and Restricted Model

The formal statement of this question about variance would be the following Research Hypothesis:

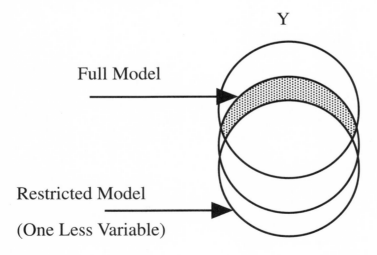

Figure 6.9.
Unique variance accounted for by an additional variable—the value-added model.

Xk accounts for additional variance in the criterion, over and above X1, X2, . . ., Xk-1.

Or:

Xk predicts the criterion Y, over and above X1, X2, . . ., Xk-1.

The Research Hypothesis implies the following Full Model:

$$(\text{Model 6.3}) \; Y = a_0 U + a_1 X1 + a_2 X2 + \ldots + a_k Xk + E_5$$

The restriction implied by the Research Hypothesis is $a_k = 0$, which leads to the Restricted Model:

$$(\text{Model 6.4}) \; Y = a_0 U + a_1 X1 + a_2 X2 + \ldots + a_{k-1} Xk\text{-}1 + E_6$$

As always, the general F test can be used to test these two models, and it is instructive to repeat here the general F test:

$$F_{(m1 - m2, N - m1)} = \frac{(R_f^2 - R_r^2) / (m1 - m2)}{(1 - R_f^2) / (N - m1)}$$

There are k continuous predictors in the Full Model, along with the unit vector, therefore there are k + 1 pieces of information in the Full Model. With the one restriction, there are [(k + 1) - 1] pieces of information in the Restricted Model. There was one restriction, so the numerator degrees of freedom is one, and the general F test will be:

$$F_{(1, N - k - 1)} = \frac{(R_f^2 - R_r^2) / (1)}{(1 - R_f^2) / (N - k - 1)}$$

The reduction in the R^2 from the Full Model to the Restricted Model is a function of the one variable that is restricted from the Full Model. Therefore, any difference in variance accounted for is directly attributable to that one variable.

SAS Caveat

One could test each of the three variable's contribution to the Applied Research Hypothesis 6.1 model in SAS by using the following test statements:

TEST X4 = 0;
TEST X5 = 0;
TEST X6 = 0;

One should note that the p values of these three tests of significance are the same as the p values on the same lines for these variables. Furthermore, since these are one-degree-of-freedom questions, t^2 will equal F, and that is exactly the case for variable X4 ($t = 1.934$, $t^2 = 3.740$, $F = 3.740$) and for variable X5 ($t = 8.553$, $t^2 = 73.15$, $F = 73.15$).

Again, researchers have so often chosen to ask this type of question that computerized statistical packages now tend to provide this information automatically. This information should be attended to only if one has stated a priori the value-added Research Hypothesis.

Relationship to Other Statistical Techniques

Testing the "value added" of a particular variable has become routine by way of a procedure referred to as *stepwise regression analysis*. In this procedure, the computer completely takes over and takes the researcher on a gigantic fishing expedition. In Step 1, a model is identified that consists of the variable most highly related to the criterion—the one that accounts for the most variance in the criterion.

Step 2 consists of testing the value added of *each* of the other variables. Thus, with, say, nine predictor variables, once the first step is concluded and the one "best" variable is identified, then each of the other eight are tested with the format discussed in the previous pages of this section. Eight tests of significance are performed in Step 2, and Step 2 is finished when the computer identifies the best two-variable model. Step 3 starts with this two-variable model and performs the statistical test discussed in this section seven times, once for each of the remaining seven variables, with the result being the best three-variable model. Step 4 starts with this three-variable model and performs the statistical test discussed in this section six times, once for each of the remaining six variables, with the result being the best four-variable model.

This process continues until one of several criteria imposed by the researcher is met: (a) a minimum R^2 value is obtained; (b) a prespecified p value is not met; or (c) all nine variables are entered into the model. If the four-variable model turns out to be the "best" model, a total of 30 (9 in Step 1, 8 in Step 2, 7 in Step 3, and 6 in Step 4) F tests have been calculated. Many users of stepwise regression analysis either forget this or were never aware of it in the first place. This procedure is one of the most successful fishing expeditions a researcher can go on because a "best" model will always be identified. But the technique should be viewed for what it is—a fishing expedition for *identifying* Research Hypotheses for future verification; it is *not* for verifying hypotheses (since none were asked) on the data at hand.

Other versions of stepwise regression analysis can be found in the SAS manual. In particular, backward stepwise analysis works in just the opposite direction, starting with the best nine-variable model and then identifying the variable that can be eliminated from the nine-variable set in order to result in the best eight-variable model. If one has already saved a fortune to go on a fishing expedition, then we recommend that one use the backward stepwise method because the resulting model will never have a smaller R^2 than the forward stepwise method did: *This is because in the multivariable world variables often work together, and their contribution may well exceed the sum of the parts.*

There are several problems inherent in the stepwise regression procedure: (a) the multitude of tests of significance; (b) different answers from different stepwise computer programs; (c) different answers depending upon the probability values chosen for inclusion and for exclusion; and (d) different answers for forward stepwise analysis and for backward stepwise analysis. We discuss stepwise regression in more detail in chapter 11. (See Appendix D-11.3 for stepwise SAS setup.)

MULTIPLE CONTINUOUS PREDICTORS: COMPARING TWO MODELS OF BEHAVIOR

Rationale

The Full and Restricted Models can be envisioned as two competing ways of modeling or explaining behavior. One model has fewer variables and is thus more parsimonious. The other model, because it has more predictor variables, will yield a higher R^2. The cost of the additional pieces of information, as compared to the benefit of a higher R^2, can be arbitrated by the general F test. Alternatively, one could arbitrarily require that each piece of information result in, say, an R^2 increase of 10%. The general F test simply assures the researcher that the significant variables do better than what random variables would do.

Geometric Interpretation

As in the previous section of this chapter, the Full and Restricted Models identify two different dimensional spaces in which the data can be described or modeled. The only difference here, in this section, is that instead of a difference of one variable (and therefore one dimension), the present section may have a difference of any number of variables (and therefore any number of dimensions). We will therefore rely on the computer to provide the answer.

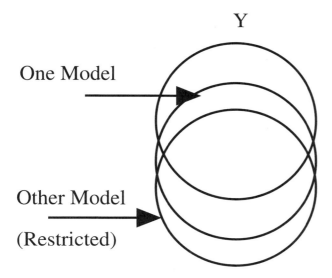

Figure 6.10.
Venn diagram comparing two different models.

Venn Diagram

Because variables can "work together," the Venn diagrams can best be understood in terms of the total set of variables (rather than in terms of each of the separate variables), as pictured in Figure 6.10.

Research Hypothesis, Full Model, and Restricted Model

The comparison of two models of behavior requires that the same behavior be considered and that one model be a restricted case of the other model. Looking at the situation the other way, one model is an extension of the other model. In either case, one model is likely to be the "accepted" or "traditional" model of behavior. The researcher either has identified a simpler way to explain the behavior *or* has identified additional mediating or necessary variables. Since most R^2 in the behavioral sciences are pathetically low, the researcher has probably identified some additional variables that:

• were not specified in the original theory,
• have recently become feasible to measure,
• have surfaced as a necessary variable,
• have surfaced as a mediating variable,
• interact with variables already in the model, or
• specify an untapped nonlinear relationship.

These last three cases will be dealt with extensively in later chapters. The first three cases can be tested as follows:

Research Hypothesis: Dr. Smith's six-variable model of behavior can be improved with the addition of variables X7 and X8.

Statistical Hypothesis: The six variables in Dr. Smith's model cannot be improved with the addition of variables X7 and X8.

Full Model: $Y = a_0U + a_1X1 + a_2X2 + \ldots + a_6X6 + a_7X7 + a_8X8 + E_7$

Want: $a_7 \neq 0$; $a_8 \neq 0$

Restrictions: $a_7 = 0$; $a_8 = 0$
Restricted Model: $Y = a_0U + a_1X1 + a_2X2 + \ldots + a_6X6 + E_8$

One should note that the Restricted Model is Dr. Smith's model and that the Full Model is the proposed improved model. The general F test can again be used to arbitrate the argument:

$$F_{(m1 - m2, N - m1)} = \frac{(R_f^2 - R_r^2) / (m1 - m2)}{(1 - R_f^2) / (N - m1)}$$

Because there are 9 pieces of information in the Full Model and 7 pieces in the Restricted Model, the above equation reduces to:

$$F_{(m1-m2, N-m1)} = \frac{(R_f^2 - R_r^2) / (2)}{(1 - R_f^2) / (N - 9)}$$

Whatever increase there is in R^2 from the Restricted Model to the Full Model can be attributed to the two variables that have been added to the Restricted Model. If one variable is "valuable," the value of the other variable (the increase in R^2) will be averaged over both variables. (This is shown in the numerator of the F test.) In such a case, significance may not be obtained, whereas if only one "valuable" variable had been tested, significance might have been obtained. Whenever a researcher tests more than one variable simultaneously such a risk is taken. Only when a one-degree-of-freedom question is tested do we know exactly what piece of information is responsible for the decrease in R^2 (within the context of all the variables remaining in the Restricted Model—the *over and above* concept).

7

INTERACTION

Behavior is often a result of more than the effects of one or more variables alone; behavior is usually a result of variables in combination. Previous chapters have illustrated how researchers can consider dichotomous predictors together as "main effects" and continuous predictors together as "linear effects."

In this chapter we illustrate how one can use the GLM to model behavior when the effect of one predictor "depends upon" another predictor. Statisticians call this phenomenon interaction and represent it mathematically with multiplication of the variables. See also Lewis-Beck (1980) for a discussion of interaction as indicated by "it depends."

When you think about it, most behavior is not a function of just one variable by itself but depends on the presence or absence (dichotomous variable) or numerical value (continuous variable) of another variable. Interaction has traditionally been viewed as a difficult concept and as one that researchers would rather not have around because its presence clouds the interpretation of the main effects. We take a strong stance in treating interaction as a phenomenon to be expected, as another predictor variable, and as a clarifying phenomenon. If interaction does exist, then we need to include it in our model. We need to be able to test for it rather than casually assume that it does not exist. Most important, we need to be able to interpret interaction if it does exist in our data. Based on our experience, interaction will likely exist. Few causal statements are simply "X causes Y," but "X causes Y, depending on the value of Z."

We discuss interaction from three vantage points, following the structure provided in the previous three chapters. The first section illustrates how the GLM can reflect interaction between two dichotomous predictors. The second section illustrates how the GLM can reflect interaction between a dichotomous predictor and a continuous predictor. The third section illustrates how the GLM can reflect interaction between two continuous predictors. In the final section, we discuss the relative advantages in employing dichotomous and continuous predictors.

INTERACTION BETWEEN DICHOTOMOUS PREDICTORS

We first discuss interaction between dichotomous predictors when each predictor has only two levels. An example with more than two levels is presented at the end of the section.

Exhibit 7.1

M1 = new treatment, AM   M2 = new treatment, PM
M3 = comparison treatment, AM   M4 = comparison treatment, PM

Research Hypothesis 10
Nondirectional: The two treatments, averaged across the two different time periods, are not equally effective.
$(m1 + m2)/2 \neq (m3 + m4)/2$
Directional: The new treatment, averaged across the two different time periods, is more effective than the comparison treatment.
$(m1 + m2)/2 > (m3 + m4)/2$
Statistical Hypothesis: $(m1 + m2)/2 = (m3 + m4)/2$ or
$(m1 + m2) = (m3 + m4)$ or
$(m1 + m2) - (m3 + m4) = 0$ or

	m1	m2	m3	m4
$(1 * m1) + (1 * m2) +$				
$(-1 * m3) + (-1 * m4) = 0$	1	1	-1	-1

Research Hypothesis 11
Nondirectional: The two time periods, averaged across the two treatments, are not equally effective.
$(m1 + m3)/2 \neq (m2 + m4)/2$
Directional: The AM period, averaged across the two different treatments, is more effective than the PM period.
$(m1 + m3)/2 > (m2 + m4)/2$
Statistical Hypothesis: $(m1 + m3)/2 = (m2 + m4)/2$ or
$(m1 + m3) = (m2 + m4)$ or

	m1	m2	m3	m4
$(m1 + m3) - (m2 + m4) = 0$ or				
$(1 * m1) + (-1 * m2) +$				
$(1 * m3) + (-1 * m4) = 0$	1	-1	1	-1

Research Hypothesis 12
Nondirectional: The difference in effectiveness of the AM new treatment and the PM new treatment is different from the difference between the AM comparison treatment and the PM comparison treatment.
$(m1 - m2) \neq (m3 - m4)$
Directional: The difference in effectiveness of the AM new treatment and the PM new treatment is greater than the difference between the AM comparison treatment and the PM comparison treatment.
$(m1 - m2) > (m3 - m4)$
Statistical Hypothesis: The difference in effectiveness of the AM new treatment and the PM new treatment is the same as the difference between the AM comparison treatment and the PM comparison treatment.
$(m1 - m2) = (m3 - m4)$ or
$(m1 - m2) - (m3 - m4) = 0$ or

	m1	m2	m3	m4
$(1 * m1) + (-1 * m2) +$				
$(-1 * m3) + (1 * m4) = 0$	1	-1	-1	1

One possible set of contrast coefficients: two-way analysis of variance (repeat of Exhibit 4.4)

Rationale

We repeat Exhibit 4.4 in the form of Exhibit 7.1, with the purpose of now discussing Research Hypothesis 12, the interaction hypothesis. In chapter 4, we discussed the two main-effect hypotheses, Research Hypothesis 10 and Research Hypothesis 11. Remember that with four groups these three hypotheses comprised one set of orthogonal hypotheses. Note that the Research Hypothesis 11 nondirectional Research Hypothesis states that "The two time periods, averaged across the two treatments, are not equally effective." Note that the Research Hypothesis 12 nondirectional Research Hypothesis qualifies that difference by not averaging across the two different treatments. Instead, Research Hypothesis 12 actually specifies that the difference between the two time levels will be different in the two treatments. Research Hypothesis 12 states that whatever difference exists at one level, that difference will not be the same at the other level. The differences will be different. And if the differences are different, then the main-effect statement of Research Hypothesis 11, even though statistically significant, is not equally true at each level. It may even be the case, as demonstrated in the next section, that the overall main effect may be a result of a large difference at one level and no difference at the other level (or even a small opposite difference at the other level).

Alternatively, the Research Hypothesis 12 nondirectional Research Hypothesis could have been stated as "The difference in effectiveness of the AM new treatment and the AM comparison treatment is different from the difference between the PM new treatment and the PM comparison treatment." This Research Hypothesis results in the following symbols: $(m1 - m3) \neq (m2 - m4)$ and contrast coefficients of $+1, -1, -1,$ and $+1$, which are the same as for Research Hypothesis 12. Thus, this alternative way of stating the Research Hypothesis does actually test the same question, but in this formulation it emphasizes the differences between treatments rather than the differences between time. The point is that it is one and the same interaction hypothesis and, if found to be significant, it qualifies not only Research Hypothesis 11, the time main effect, but also Research Hypothesis 10, the treatment main effect; that is, finding a significant interaction between A and B qualifies the interpretation of both the A main effect and the B main effect.

Geometric Interpretation

Figure 7.1 illustrates three possible situations of interaction in a two-by-two design. Each graph in the figure represents the same amount of interaction, although the impact on the main-effects interpretation for treatment is very different. In Figure 7.1a the new treatment is superior to the comparison treatment, particularly for the AM. In Figure 7.1b the new treatment is superior to the comparison treatment, but the difference is essentially nil for the PM. It really does not matter which treatment is implemented in the PM. If only one treatment can be implemented in both the AM and the PM, then the new treatment ought to be the one. Figure 7.1c presents a very different state of affairs. Here, the relative superiority of the new treatment over the comparison treatment at the AM level is completely opposite to the superiority of the comparison treatment over the new treatment at the PM level, and here, which (new or comparison) treatment is implemented is crucial. One would hope that one had the capability of implementing either treatment, depending upon time of day. That is, to maximize outcome effects, one would implement the new treatment in the AM and the comparison treatment in the PM.

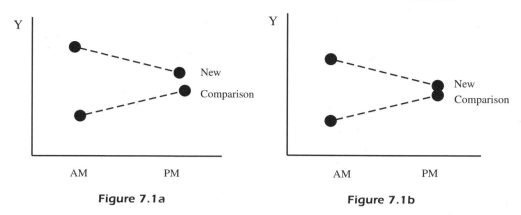

Figure 7.1a   Figure 7.1b

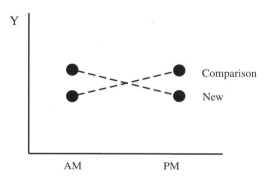

Figure 7.1c

Figure 7.1.
Three possible interaction results.

Venn Diagram

When all four groups have an equal number of subjects (or are proportional) the effects will be orthogonal, which means that they do not overlap. In other words, the significance of one does not impinge on the significance of any other. Figure 7.2 shows three possible states of affairs. In Figure 7.2a the two main effects are significant, whereas the interaction is not. Figure 7.2b shows all three sources as being significant. Finally, Figure 7.2c shows neither main effect as being significant but the A * B interaction to be significant.

Full Model, Restricted Model, and *F* Test

Suppose that one wanted to test the directional hypothesis in Research Hypothesis 12. For each group to have its own mean, the Full Model must allow for all four group means:

$$(\text{Model 7.1}) \; Y = a_1 M1 + a_2 M2 + a_3 M3 + a_4 M4 + E_1$$

where:

Y = the criterion;

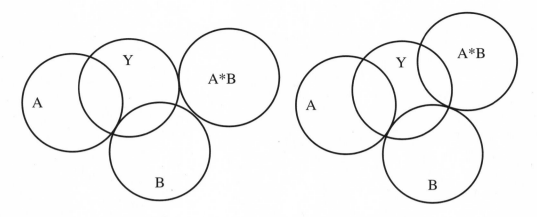

Figure 7.2a   Figure 7.2b

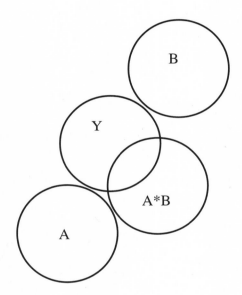

Figure 7.2c

Figure 7.2.
Possible two-way ANOVA results in Venn diagram.

M1 = 1 if subject in new treatment AM, 0 otherwise;
M2 = 1 if subject in new treatment PM, 0 otherwise;
M3 = 1 if subject in comparison treatment AM, 0 otherwise; and
M4 = 1 if subject in comparison treatment PM, 0 otherwise.

The restriction implied by Research Hypothesis 12 is:

$(a_1 - a_2) = (a_3 - a_4)$; or $a_1 = (a_3 - a_4 + a_2)$

Forcing these restrictions into Model 7.1 yields:

$$(\text{Model } 7.2a)\ Y = (a_3 - a_4 + a_2)M1 + a_2M2 + a_3M3 + a_4M4 + E_2$$

or:

$$(\text{Model } 7.2b)\ Y = a_3M1 + (-a_4M1) + a_2M1 + a_2M2 + a_3M3 + a_4M4 + E_2$$

or:

$$(\text{Model } 7.2c)\ Y = a_2(M2 + M1) + a_3(M1 + M3) + a_4(M4 - M1) + E_2$$

There are three rather strange looking vectors in the above model. That there are three vectors makes sense because we made one restriction on four pieces of information. Appendix B-3 shows algebraically how Model 7.2c is equivalent to the following alternative:

$$(\text{Model } 7.2d)\ Y = b_0U + b_1\text{New} + b_2\text{Comparison} + b_3\text{AM} + b_4\text{PM} + E_3$$

Since (New + Comparison) = U , and (AM + PM) = U , the above model can be reduced to:

$$\text{Model } 7.3{:}\ Y = b_0U + b_1\text{New} + b_3\text{AM} + E_3$$

The above Full and Restricted Models make sense conceptually because in order to allow for interaction we need to know each of the four group means, which are available in the Full Model. If there is no interaction, then all we need to know is the AM, PM, new, and comparison means, which are available in the Restricted Model.

One restriction was made on a model with four pieces of information; thus the degrees of freedom are (4 - 3) and (N - 4). In general, the Full Model will have a mean for each "cell" and the Restricted Model will only have main-effects information. The degrees of freedom for the interaction hypothesis will be a function of the number of levels on each variable, where A equals the number of levels on variable A, and B equals the number of levels on variable B:

([A - 1] * [B - 1]) and (N - [A * B]).

SAS Caveat

As usual, the SAS statistical package user must remember that SAS will automatically provide the unit vector. Therefore, one "cell" vector needs to be eliminated, otherwise SAS will produce an error message. If one has dichotomous predictors, PROC GLM (the SAS procedure for the GLM) is much easier to use than PROC REG (the SAS procedure for regression analysis) if one has the traditional ANOVA Research Hypotheses. On the other hand, if one has specific Research Hypotheses, then PROC REG must be used. Appendix D-7.1 contains the necessary PROC GLM and PROC REG statements for testing Applied Research Hypothesis 7.1.

Applied Research Hypothesis 7.1

Directional Research Hypothesis: For a given population, the relative effectiveness of Method A (X10) as compared to Method B (X11) on the criterion of interest (X9) will be greater for Group A (X12) than for Group B (X13).

Statistical Hypothesis: For a given population, the relative effectiveness of Method A (X10) as compared to Method B (X11) on the criterion of interest (X9) will be the same for Group A (X12) as for Group B (X13).

Full Model: $X9 = a_0U + b(X10 * X13) + c(X11 * X12) + d(X11 * X13) + E_1$ (pieces full = m1 = 4)

Want: $c > (b - d)$; restriction: $c = (b - d)$
Restricted Model: $X9 = a_0U + eX10 + fX12 + E_2$ (pieces restricted = m2 = 3)
alpha = .10
$R_f^2 = .0497$; $R_r^2 = .0152$
$F_{(1,56)} = 2.0319$
Computer probability = .1596; directional probability = .0798

Interpretation: Since the weighting coefficient c is numerically larger than (b - d), the directional probability is appropriate and the following conclusion can be made: "For a given population, the relative effectiveness of Method A (X10) as compared to Method B (X11) on the criterion of interest (X9) will be greater for Group A (X12) than for Group B (X13)."

Relationship to Other Statistical Techniques

Consider a two-way design with four levels of IQ and three drugs as in Table 7.1. The interaction question is only one of the infinite number of possible Research Hypotheses relevant to this design. There are many sets of multiple comparisons possible with this design, and traditional interaction is a part of some of those comparisons. The Research Hypothesis for interaction could be stated as:

The differences between the four IQ levels are not constant across the three drug treatments. [More explicitly]: The differences between the four IQ levels given Drug 1, and the differences between the four IQ levels given Drug 2, and the differences between the four IQ levels given Drug 3 are not all equal.)

The Statistical Hypothesis could then be:

The differences between the four IQ levels are constant across the three drug treatments.

The regression formulation for testing the significance of the interaction has been well documented (Bottenberg & Ward, 1963; Jennings, 1967; Kelly, Beggs, McNeil, Eichelberger, & Lyon, 1969). A Full Model, allowing for interaction, is needed to let the 12 cell means manifest themselves. Then a Restricted Model, allowing only the row (IQ) and column (Drug) means to manifest themselves, can be compared to the

Table 7.1
Design, Source Table, and Comparison of Pooling and Not Pooling for a Four-by-Three Design

Source of Variation	Case 1: Not Pooling		Case 2: Pooling	
	df	F	df	F
IQ	3	$\dfrac{\text{MS IQ}}{\text{MS within}}$	3	$\dfrac{\text{MS IQ}}{\text{MS within + Interaction}}$
Drug	2	$\dfrac{\text{MS Drug}}{\text{MS within}}$	2	$\dfrac{\text{MS Drug}}{\text{MS within + Interaction}}$
IQ * Drug	6	$\dfrac{\text{IQ * Drug}}{\text{MS within}}$		
Within	N - 12			
Total	N - 1		N - 1	

	Drug 1	Drug 2	Drug 3
IQ Level 1	a_1	a_2	a_3
IQ Level 2	a_4	a_5	a_6
IQ Level 3	a_7	a_8	a_9
IQ Level 4	a_{10}	a_{11}	a_{12}

Full Model via the generalized F ratio. The Full Model allowing the 12 means to manifest themselves would have 12 linearly independent vectors, or 12 pieces of information:

(Model 7.4a) $Y1 = a_1 C1 + a_2 C2 + a_3 C3 + a_4 C4 + a_5 C5 + a_6 C6 + a_7 C7 + a_8 C8 + a_9 C9 + a_{10} C10 + a_{11} C11 + a_{12} C12 + E_4$

where:
- $Y1$ = the criterion score;
- $C1$ = 1 if IQ Level 1 and Drug 1, 0 otherwise;
- $C2$ = 1 if IQ Level 1 and Drug 2, 0 otherwise;
- $C3$ = 1 if IQ Level 1 and Drug 3, 0 otherwise;
- $C4$ = 1 if IQ Level 2 and Drug 1, 0 otherwise;
- $C5$ = 1 if IQ Level 2 and Drug 2, 0 otherwise;
- $C6$ = 1 if IQ Level 2 and Drug 3, 0 otherwise;
- $C7$ = 1 if IQ Level 3 and Drug 1, 0 otherwise;
- $C8$ = 1 if IQ Level 3 and Drug 2, 0 otherwise;
- $C9$ = 1 if IQ Level 3 and Drug 3, 0 otherwise;
- $C10$ = 1 if IQ Level 4 and Drug 1, 0 otherwise;

C11 = 1 if IQ Level 4 and Drug 2, 0 otherwise;
C12 = 1 if IQ Level 4 and Drug 3, 0 otherwise; and
$a_1, a_2, a_3, \ldots, a_{12}$ are least squares weighting coefficients calculated to minimize the sum of the squared values in the error vector, E_4.

Model 7.4a allows each of the 12 means to manifest itself and is called by some the *cell means model*. Model 7.4a could be written another, equivalent way, which will have more value for later discussion. We feel that the following model (7.4b) more clearly shows the allowance for interaction between IQ and drugs.

$$\begin{aligned}\text{(Model 7.4b) } Y1 = \ &a_1(Q1 * D1) + a_2(Q1 * D2) + a_3(Q1 * D3) + \\ &a_4(Q2 * D1) + a_5(Q2 * D2) + a_6(Q2 * D3) + \\ &a_7(Q3 * D1) + a_8(Q3 * D2) + a_9(Q3 * D3) + \\ &a_{10}(Q4 * D1) + a_{11}(Q4 * D2) + a_{12}(Q4 * D3) + E_4\end{aligned}$$

The multiplication signs in this model indicate the allowance for interaction effects, as they will whenever they appear in subsequent models. Since Model 7.4a and Model 7.4b are equivalent models, forcing the restrictions of no interaction on either model will result in the same Restricted Model. The restrictions specifying no interaction are as follows:

$(a_1 - a_2) = (a_4 - a_5) = (a_7 - a_8) = (a_{10} - a_{11})$
$(a_2 - a_3) = (a_5 - a_6) = (a_8 - a_9) = (a_{11} - a_{12})$

Forcing these restrictions onto either Full Model would result in really weird looking models. But, using the information provided in the previous section and in Appendix D-7.1, they are equivalent to a Restricted Model that has only information about the main effects:

$$\text{(Model 7.5) } Y1 = q_1Q1 + q_2Q2 + q_3Q3 + q_4Q4 + d_1D1 + d_2D2 + d_3D3 + E_5$$

where:
Y1 = the criterion score;
Q1 = 1 if IQ Level 1, 0 otherwise;
Q2 = 1 if IQ Level 2, 0 otherwise;
Q3 = 1 if IQ Level 3, 0 otherwise;
Q4 = 1 if IQ Level 4, 0 otherwise;
D1 = 1 if Drug 1, 0 otherwise;
D2 = 1 if Drug 2, 0 otherwise; and
D3 = 1 if Drug 3, 0 otherwise.

Again, most computer programs will automatically provide the unit vector. In most applications, the unit vector is desired, but when dealing with categorical data it can cause trouble unless appropriately considered. By this we mean the consideration of linearly dependent vectors. In Models 7.4a and 7.4b, if all interaction vectors are added, the unit vector is the result. Therefore, one of those 13 vectors (the 12 interaction vectors plus the unit vector) is not a new piece of information. For heuristic purposes, one might want to ignore the unit vector. But for computer solution and for being less likely to make an error in determining the number of linearly independent vectors, the unit vector should be retained. Therefore, one of the interaction vectors is not a new piece of information; and therefore, there are 12 pieces of information in both versions of Model 7.4. In Model 7.5, the unit vector also needs to be considered. Here, the four IQ vectors added yield the unit vector, so if the unit vector is considered a piece of information, then one of the IQ vectors must be considered redundant. The same argument holds for the three drug vectors. If the three drug vectors are added together, the unit vector results. Therefore, one of the drug vectors must be considered redundant. Which IQ vector and which drug vector are considered redundant is of no concern, indeed all three drug vectors, the four IQ vectors, and the unit vector may remain in the analysis. The important point to remember is that there are six pieces of information in Model 7.5: the unit vector, three IQ vectors, and two drug vectors. Model 7.5 thus has six linearly independent pieces of information.

Another way to conceptualize the number of pieces of information in the Full Model and in the Restricted Model is to think of the design and of the number of cells in the "inside"; 12 here. The Restricted Model that does not have information about the inside cells but about the "outside" information will have six pieces of information. The degrees of freedom can be easily determined, with the degrees of freedom numerator here being (m1 - m2), or (12 - 6 = 6). The degrees of freedom denominator are $(N - m1)$, here $(N - 12)$.

If the interaction is significant, most statistical authors indicate that the main-effects question is not appropriate and that simple effects should be investigated. Each simple effects is simply a one-way ANOVA at any one level. The regression formulation for the simple effect of IQ Level 1 would require a Full Model:

$$(\text{Model 7.6}) \; Y1 \; = \; c_1(Q1 * D1) + c_2(Q1 * D2) + c_3(Q1 * D3) + E_6$$

The restriction implied by the simple effects is $c_1 = c_2 = c_3$. Forcing these restrictions into Model 7.6 yields the Restricted Model:

$$(\text{Model 7.7}) \; Y1 \; = \; a_0 U + E_7$$

Because there are three pieces of information in the Full Model and one in the Restricted Model, the degrees of freedom would be $(3 - 1)$ and $(N - 3)$. The Research Hypothesis implied by Models 7.6 and 7.7 deals only with subjects at IQ Level 1; therefore, only those subjects at IQ Level 1 would be analyzed. N here then is the number of subjects in IQ Level 1.

A more stable estimate of the within-group variance is available by considering all the cells, not just those three directly considered in the Research Hypothesis (similar to Generalized Research Hypothesis 4.3). The Full Model would be either version of Model 7.4; and the restrictions on that model would be:

$$a_1 = a_2 = a_3$$

resulting in a Restricted Model with two fewer predictor variables. Just as in Model 7.7, the Restricted Model would not have information about IQ Level 1. N here is all the subjects in the study.

Not Pooling the Interaction Term

Even when there is significant interaction, we would argue that one might still want to look at the main effects, and this would be appropriate if the interaction were ordinal. Ordinal interaction occurs when the rank order of the categories of one variable (based on their criterion scores) is the same within each category of the second predictor variable. When significant ordinal interaction exists in the data, the magnitude of the main effects is not indicative of the effects at each level, but it is the case that one level is uniformly superior, as in Figure 7.1a and 7.1b. Figure 7.1c, on the other hand, indicates a disordinal interaction in which the dashed lines are crossed. (A note should be made here that a significant main effect indicates a nondirectional difference and does not allow one to make a directional interpretation.)

Testing for main effects can be done in two ways: one in which the within term is used as the best estimate of expected variance and the other in which the within source is pooled with the nonsignificant interaction and used as a new estimate of expected variance. These notions are more fully delineated by Jennings (1967). Jennings (1967) presents the regression models appropriate for the case in which the sources are not pooled. If the main effect for IQ were to be tested, the following restriction would be made on Model 7.4a or 7.4b, if the cell numbers (N_1, N_2, etc.) were not proportional:

$$\frac{N_1 a_1 + N_2 a_2 + N_3 a_3}{N_1 + N_2 + N_3} = \frac{N_4 a_4 + N_5 a_5 + N_6 a_6}{N_4 + N_5 + N_6} =$$

$$\frac{N_7 a_7 + N_8 a_8 + N_9 a_9}{N_7 + N_8 + N_9} = \frac{N_{10} a_{10} + N_{11} a_{11} + N_{12} a_{12}}{N_{10} + N_{11} + N_{12}}$$

resulting in an extremely complicated Restricted Model, which has little conceptual value. The pooling analogue is much more conceptually pleasing and is discussed below.

Pooling the Interaction Term

When the interaction source of variance is not significant, one can consider this variance to be essentially random and thus pool that variance with the variance from the within source. The argument for this procedure (when there is little sample interaction) is that the resulting variance estimate is a more stable estimate of the expected chance variance. Winer (1971) developed a rule of thumb that states that, if the significance of the interaction source had a p value greater than .30 (e.g., $p = .35$), then one could pool the interaction and within sources of variance.

Model 7.5 depicted a state of affairs in which interaction was not allowed. If Model 7.5 is not significantly less predictable than Model 7.4b, then Model 7.5 can be treated as the more meaningful model. Testing for the IQ main effect on this model can be accomplished by the restrictions on Model 7.5 ($q_1 = q_2 = q_3 = q_4$), resulting in the following Restricted Model:

$$(\text{Model 7.8}) \ Y1 \ = \ d_1 D1 + d_2 D2 + d_3 D3 + E_8$$

Because there are six pieces of information in the Full Model (Model 7.5) and three pieces of information in the Restricted Model (Model 7.8), then the test of significance would be an F with (6 - 3) and (N - 6) degrees of freedom, similar to

Case 2 in Table 7.1. One can think of the IQ main-effect question here as having information about IQ and drugs and simply giving up information about IQ. Alternatively, one can ask, "Does knowledge of IQ account for criterion variance, over and above knowledge of drugs?"

The drug main-effects question would be tested by forcing the restrictions $d_1 = d_2 = d_3$ on Model 7.5, resulting in the following Restricted Model:

$$(\text{Model 7.9}) \ Y1 \ = \ q_1 Q1 + q_2 Q2 + q_3 Q3 + q_4 Q4 + E_9$$

resulting in an F with 2 and N - 6 degrees of freedom.

INTERACTION BETWEEN A DICHOTOMOUS PREDICTOR AND A CONTINUOUS PREDICTOR

Ambiguous results regarding the effects of specific treatments are often found in research domains. For example, in the field of physical conditioning and strength improvement, some studies report that isometric procedures are more effective in increasing strength than isotonic procedures, while others report the opposite outcome. Isometrics are exercises in which the individual tenses the muscle to be strengthened against a static object; isotonic exercises use the muscle in such activities as weight lifting and push-ups. (The material dealing with isotonic and isometric conditioning is adapted from an article by Bender, Kelly, Pierson, & Kaplan, 1968). When studies consistently provide contradictory results, one may suspect that another, undetected variable is operating to cause one method to work best for one type of individual and another method to work best for another type of individual. (It is assumed that the researchers are competent and conscientious; thus, one hopes experimenter bias is eliminated and the data are good.) The subjects that have been previously researched may represent different populations with respect to this undetected variable.

As an example, Figures 7.3a and 7.3b may represent the data regarding two such contradictory studies on improving arm strength. Study 1 in Figure 7.3a shows that isotonic exercises resulted in a higher mean posttest arm strength, while Study 2 in Figure 7.3b shows that isometric exercises resulted in a higher mean posttest arm strength.

The search for the underlying variable that can clarify the issue would most likely start with an examination of the details of the contradictory studies to determine if they represent different populations. Suppose that in the literature review it was found that most of the studies that indicated superiority for isometric procedures used skilled athletes, who were relatively strong initially, and that most of the studies supporting the superiority of isotonic procedures used subjects with average pretest strength levels. Some studies may have reported the initial arm strength of the subjects. A subsequent researcher could incorporate that information and modify Figures 7.3a and 7.3b to result in Figures 7.4a and 7.4b. Note that Figure 7.4a shows that isotonic procedures result in higher posttest arm strength for all pretest scores studied except a few at the higher end; whereas Figure 7.4b shows that isometric procedures result in higher posttest arm strength for all pretest scores studied except a few at the lower end.

The hypothetical data represented in Figures 7.4a and 7.4b reveal several important considerations. The first thing one should note is that Study 1 (Figure 7.4a)

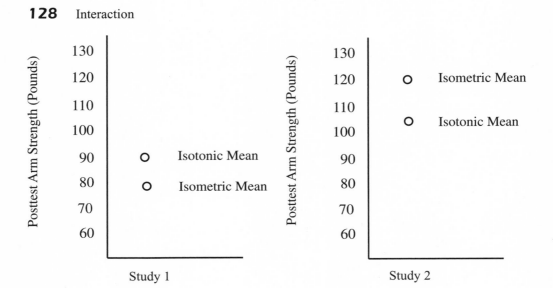

Figure 7.3a.
Study 1 mean differences on posttest arm strength for isotonic and isometric procedures, favoring isotonic.

Figure 7.3b.
Study 2 mean differences on posttest arm strength for isotonic and isometric procedures, favoring isometric.

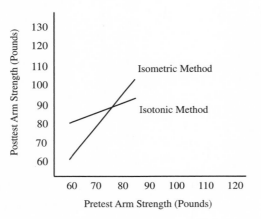

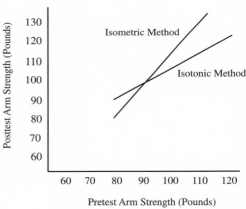

Figure 7.4a.
Study 1: Lines of best fit for data from isotonic and isometric procedures, when viewed across levels of pretest arm strength.

Figure 7.4b.
Study 2: Lines of best fit for data from isotonic and isometric procedures, when viewed across levels of pretest arm strength.

included subjects with a range of pretest scores from 60 pounds to 85 pounds, whereas the pretest strength range for subjects in Study 2 (Figure 7.4b) was from 80 pounds to 120 pounds. It appears, then, that these two studies included subjects from different populations with respect to pretest arm strength.

In addition, one might note another consideration by looking at the lines for the isotonic method in both Figures 7.4a and 7.4b. In Figure 7.4a, the posttest score is higher for the isotonic method at all levels of observed pretest scores, indicating that the individuals' strength improved. But in Figure 7.4b, the line for the isotonic method shows less and less difference between pretest and posttest scores until at a pretest strength of 120 pounds the posttest score is also 120 pounds, indicating no improvement. It seems that if a person is originally relatively weak, isotonic procedures are better than isometric for improvement; and if a person is already relatively strong, isometric procedures are more effective than isotonic. Suppose one reviews the literature and the observed trend is found repeatedly. It appears that the variable that explains the contradictory results reported in the literature has been isolated. These review findings lead to the following Research Hypothesis:

> Isometric exercises are superior to isotonic exercises in terms of posttest arm strength for men who are high on the pretest strength continuum, and isotonic exercises are superior to isometric exercises for men who are initially low on the pretest strength continuum.

The rival, Statistical Hypothesis is:

> The differential effects (if any) of isotonic and isometric methods are the same across the range of observed pretest scores.

Another way of stating the Statistical Hypothesis is:

> The slopes of the lines for the two treatments are the same across pretest scores, although the Y-intercepts are allowed to be different.

Because very weak men (below a pretest score of 80) are not included in Study 2, and very strong men (above 85) are not included in Study 1, a new study is required to test the Research Hypothesis against the Statistical Hypothesis to determine if the two studies differed only in pretest arm strength.

Suppose the researcher has the resources to select and measure 60 men who range in pretest arm strength from 60 pounds through 110 pounds. He wants to select two groups, one to be given isometric treatment and the other isotonic treatment. Random selection is needed, but he also wants to be sure that all levels of strength are included in both treatments. He may take strength intervals of 10 pounds (e.g., from 60 to 70 pounds) and randomly assign half of the individuals within that range to each treatment group. He can do this for all 10-pound intervals to be sure the two treatment groups represent the population.

Testing for Interaction

The Research Hypothesis requires that the data for both treatments be compared over the same range of interest on pretest arm strength. It would not be appropriate to find and look at a line of best fit only for the isotonic procedure or to look at a line of best fit only for the isometric procedure. Two lines of best fit (one for each method) must be found simultaneously, and then their slopes must be compared. Figure 7.5 represents the data obtained from the 60 subjects.

An inspection of Figure 7.5 suggests that the Research Hypothesis is supported because weak men improve more with isotonic exercise, and strong men improve

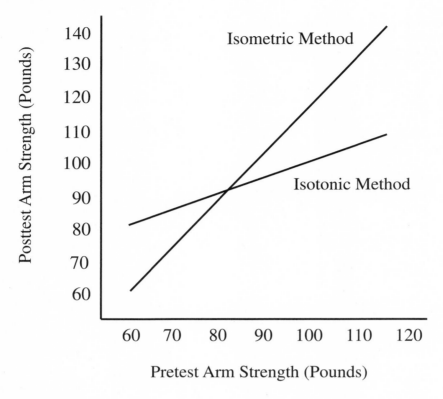

Figure 7.5.
The two lines of best fit between pretest and posttest arm strength measures for isometric and isotonic exercise groups.

more with isometric exercise. But how likely is it that the observed interaction is due to sampling error? To answer the Research Hypothesis, the researcher must analyze all the data simultaneously.

One linear model must be constructed to reflect the two lines with differing slopes in the Research Hypothesis. That Full Model is:

$$\text{(Model 7.10)} \; Y1 \; = \; a_1 U1 + b_1 X1 + a_2 U2 + b_2 X2 + E_{10}$$

where:

$Y1$ = posttest arm strength scores;

$U1$ = 1 if the score on the criterion is from a man given isotonic exercises, 0 otherwise;

$X1$ = pretest arm strength if the score on the criterion is from a man given isotonic exercises, 0 otherwise;

$U2$ = 1 if the score on the criterion is from a man given isometric exercise, 0 otherwise;

$X2$ = pretest arm strength if the score on the criterion is from a man given isometric exercise, 0 otherwise; and

$a_1, a_2, b_1,$ and b_2 are least squares weights calculated to minimize the sum of the squared elements in E_{10}.

Model 7.10 yields the following vectors:

$$\text{Individual}\quad Y1 \;=\; a_1 U1 \;+\; b_1 X1 \;+\; a_2 U2 \;+\; b_2 X2 \;+\; E_{10}$$

Individual	Y1		$a_1 U1$		$b_1 X1$		$a_2 U2$		$b_2 X2$		E_{10}
1	78		1		60		0		0		?
2	99		1		100		0		0		?
3	91		1		80		0		0		?
.	.		1		.		0		0		?
.	.		1		.		0		0		?
.	.	$= a_1$	1	$+ b_1$	.	$+ a_2$	0	$+ b_2$	0	$+$	?
30	120		1		120		0		0		?
31	84		0		0		1		80		?
.	.		0		0		.		.		?
.	.		0		0		.		.		?
.	.		0		0		.		.		?
60	120		0		0		1		122		?

Subject identification numbers are at the left in the vector display for each of the 60 subjects. Individual 1 has a criterion score of 78 (the first element in vector Y1). His corresponding value in U1 is 1 because he participated in the isotonic exercise group. His pretest score (X1) was 60 and his values for U2 and X2 are 0 because he was not in the isometric group.

Individual 31 has a Y1 score of 84; and he has a 0 in vectors U1 and X1, which means he was not given isotonic exercise. The value for person 31 in U2 is 1 and in X2 is 80, thus he was given isometric exercise and his pretest score was 80 pounds.

Model 7.10 has four linearly independent predictor vectors. These vectors allow for two lines because they allow for two slopes (b_1 and b_2) and for two Y-intercepts (a_1 and a_2). If only the top halves of the vectors (individuals 1 through 30) in the vector display are considered, the vectors U2 and X2 would contain only zeros, so the model reduces to ($Y1 = a_1 U1 + b_1 X1 + E_{10}$), a single straight line for the isotonic group. If only the bottom halves of the vectors are considered, the model reduces to ($Y1 = a_2 U2 + b_2 X2 + E_{10}$), a single straight line for the isometric group.

To support the Research Hypothesis under study, the plot of the two lines of best fit must show that the isometric treatment yields higher performance than the isotonic treatment for the strong, and the isotonic treatment yields higher performance than the isometric treatment for the weak. Figure 7.5 shows that this condition is met. The algebraically equivalent way to express this hypothesis is to expect that the slope b_2 is greater than the slope b_1.

Because Model 7.10 fits the Research Hypothesis (by allowing the lines to have separate slopes and intercepts), it is the Full Model. (This also would be the Full Model if the hypothesis were that b_1 is greater than b_2. An inspection of the actual slopes is necessary to verify that the difference [if any] in slopes is in the direction hypothesized.)

Solving for weights a_1, a_2, b_1, and b_2, one obtains an R_f^2 value that specifies the proportion of criterion variance the four predictor pieces of information account for. The hypothesis that b_2 is greater than b_1 can be tested by casting a regression model that forces $b_1 = b_2$. This restriction implies two straight lines that have a single common slope for both treatments—reflected by the vector X3, which contains pretest scores for all subjects, despite which treatment they received.

$$(\text{Model 7.11a}) \; Y1 \; = \; a_1 U1 + a_2 U2 + b_3 X3 + E_{11}$$

Model 7.11a gives up one piece of information from the Full Model, which is the knowledge of the treatment group associated with pretest scores.

When considering only the first 30 subjects (the isotonic subjects), Model 7.11a reduces to $(Y1 = a_1 U1 + b_3 X3 + E_{11})$. When considering only the isometric subjects, Model 7.11a reduces to $(Y1 = a_2 U2 + b_3 X3 + E_{11})$. The Restricted Model (Model 7.11a) allows for two lines, with possibly different intercepts (a_1 and a_2), but the same slope (b_3). It can be thought of as the *two-parallel-line model.*

The algebraic restriction on Model 7.10 is to set the two slopes equal ($b_1 = b_2$) or both equal to a common slope (b_3). Model 7.10, rewritten with the restriction, is:

$$(\text{Model 7.11b}) \; Y1 \; = \; a_1 U1 + a_2 U2 + b_3 X1 + b_3 X2 + E_{11}$$

From the properties in chapter 3, the above can be simplified to:

$$(\text{Model 7.11c}) \; Y1 \; = \; a_1 U1 + a_2 U2 + b_3 (X1 + X2) + E_{11}$$

But $(X1 + X2)$ is a new vector of pretest scores and, if represented by X3, results in Model 7.11a.

If there is a difference in slopes, then the sum of the squared values in the error vector E_{11} will be larger than in E_{10} and the R_r^2 obtained using only three pieces of information will be smaller than R_f^2. The *F* test indicates how likely it is that the difference between R_f^2 (using Model 7.10) and R_r^2 (using Model 7.11a) is due to sampling error. Again, the same general *F* test can be used to answer this Research Hypothesis.

$$F_{(m1 - m2, N - m1)} \; = \; \frac{(R_f^2 - R_r^2) / (m1 - m2)}{(1 - R_f^2) / (N - m1)}$$

where:

$m1$ = the number of linearly independent vectors in the Full Model [which is four: (U1, U2, X1, and X2)]; and

$m2$ = the number of linearly independent vectors in the restricted model [three: (U1, U2, and X3)].

It should be recalled that $[(R_f^2 - R_r^2)/df_n]$ is a proportional estimate of the population variance and in this application of the general *F* test, rather than being an among-group estimate, it is an estimate of the population variance using interaction. Likewise, $[(1 - R_f^2)/df_d]$ is the best proportional estimate of the population variance using all the information in the Full Model.

Recall that 60 subjects were studied, and assume alpha is .01. Therefore, the degrees of freedom are 1 and 56. With 1 and 56 degrees of freedom, an *F* of 7.12 or larger is observed fewer than one time in 100. Therefore, if the resultant *F* is 7.12 or larger, one would reject the Statistical Hypothesis that $b_1 = b_2$. The Research Hypothesis would be accepted *only if* b_2 is indeed greater than b_1. A large *F* can be obtained if b_1 is greater than b_2, but this result would contradict the Research Hypothesis. This is the same issue regarding probability levels associated with directional and non-directional Research Hypotheses that has been dealt with extensively.

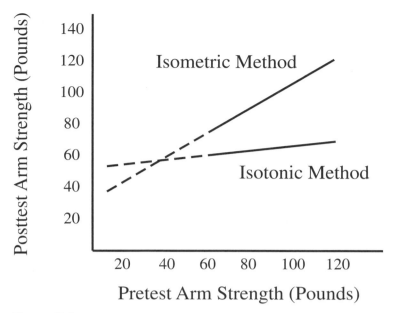

Figure 7.6.
A treatment-by-pretest interaction, where the lines cross beyond the range of interest.

Even though a large F (and low probability) is obtained, the plotting of the lines of best fit as illustrated in Figure 7.5 must be inspected closely before the Research Hypothesis can be accepted. When $b_2 > b_1$, the directional interaction exists; however, before the Research Hypothesis that was specified can be accepted, the lines of best fit must cross within the range of observed pretest scores. The SAS setup in Appendix D-7.2 indicates how one can use SAS to plot the two lines. These lines can be plotted by hand by relying on the weighting coefficients (SAS labels them *parameter estimates*).

Looking back to Figure 7.5 one notes that the lines cross at about 80 pounds on the pretest, thus the isotonic treatment seems superior in producing strength increase for men who initially have an arm strength of less than 80 pounds, and the isometric treatment is superior for those who initially have an arm strength of 80 pounds or more.

Inspection of the lines of best fit is crucial because several data configurations can generate a large F value. Figure 7.6 is one such situation. It is possible that the data could have turned out this way. The Research Hypothesis indicated that, among other things, initially weak men will benefit more from isotonic exercise. Yet an inspection of Figure 7.6 shows that the lines cross at 40 pounds on the pretest. The weakest men in the study had a pretest strength of 60 pounds. If results such as in Figure 7.6 occur, then the Research Hypothesis must be rejected. The results in Figure 7.6 show that isometric exercise is best for all pretest strength levels sampled; however, the stronger the man, the relatively more effective is the isometric treatment. Such a finding provides data that leave the original problem unresolved; the relevant variable that accounts for the previously reported contradictory findings would not yet have been found.

If the data are like those in Figure 7.5 (the lines crossing in the range of interest), the Research Hypothesis can be accepted, and the data are worth reporting because it looks like at least one source of "contamination" contributing to contradictory results has been found. If the researcher had done a good job in analyzing past research and the Research Hypothesis was based upon past findings, it would be surprising if the results did not conform to the situation as shown in Figure 7.5.

There is a procedure available (the Johnson-Neyman technique) that allows one to determine the regions where the differences between groups are significant. Pohlmann (1993) discussed how to use this technique with SAS. The approach taken in this text is to pay attention to the whole range. We acknowledge that, with respect to Figure 7.5, the isometric procedure may not be significantly better than the isotonic procedure for some subjects with pretest arm strength greater than 80. But if the two procedures are equally costly and one has the option of giving either treatment, then one would assign the isometric treatment to subjects who have a pretest arm strength greater than 80, whereas the isotonic treatment would be assigned to those whose pretest arm strength was less than 80.

Review of the Directional Interaction Example

Contradictory results were reported in the literature regarding the effectiveness of two treatment methods. The examination of the literature led to the expectation that isometric treatment would yield a steeper slope across pretest arm strength levels when compared with the slope associated with men given isotonic exercise. Furthermore, the researcher was led to expect that the starting point (the Y-intercept) would be higher for the isotonic treatment because previous data indicated that weaker men performed better under isotonic exercise treatment.

1. A Research Hypothesis was stated that expressed the expectation that two straight lines with differing slopes would be needed to reflect the data. From this Research Hypothesis, a Full Model was constructed that allowed for two Y-intercepts and two slopes.
2. A Statistical Hypothesis was stated that expressed the expectation that two straight lines with a single common slope would reflect the data. From the Statistical Hypothesis a Restricted Model was constructed that allowed for two Y-intercepts and only one common slope.
3. Upon observing that it was unlikely that the two slopes were the same, the lines of best fit from the Full Model were plotted to determine whether the lines crossed within the range of interest. Given the data expressed in Figure 7.5, along with the statistical rejection of the rival hypothesis, the Research Hypothesis can then be accepted.

Generalizing the Differential Effect of Two Treatments over the Range of a Continuous Predictor

Generalized Research Hypothesis 7.1 summarizes the regression approach to solving an interaction kind of Research Hypothesis and presents a way of testing interaction between one of two treatments individuals receive and some initial score on a single continuous variable. The notion of interaction can easily be extended to more than two groups and to more than one continuous predictor variable.

While chapter 8 goes into more detail on this subject, we would like to point out here that if more than two groups are investigated, a directional hypothesis cannot be tested. For instance, if a third treatment were considered with the isotonic-isometric example, we would have a three-line model for the Full Model and three

Generalized Research Hypothesis 7.1

Directional Research Hypothesis: For a given population, as X1 increases, the relative superiority of Method A over Method B on X4 will linearly increase.

Nondirectional Research Hypothesis: For a given population, as X1 increases, the relative superiority of Method A over Method B on X4 will linearly change.

Statistical Hypothesis: For a given population, as X1 increases, the difference between Method A and Method B on X4 will remain the same.

Full Model: $X4 = a_1U1 + b_1X2 + a_2U2 + b_2X3 + E_7$

Want (for directional Research Hypothesis): $b_1 > b_2$; restriction: $b_1 = b_2$

Want (for nondirectional Research Hypothesis): $b_1 \neq b_2$; restriction: $b_1 = b_2$

Restricted Model: $X4 = a_1U1 + a_2U2 + b_3X1 + E_8$

Automatic unit vector analogue
Full Model: $X4 = a_0U + a_1U1 + b_1X2 + b_2X3 + E_7$ (pieces full = 4)
Want: (for directional Research Hypothesis): $b_1 > b_2$; restriction: $b_1 = b_2$
Want: (for nondirectional Research Hypothesis): $b_1 \neq b_2$; restriction: $b_1 = b_2$
Restricted Model: $X4 = a_0U + a_2U2 + b_3X1 + E_8$ (pieces restricted = 3)

where:
X4 = the criterion;
U1 = 1 if the score on the criterion is from a subject in Method A, 0 otherwise;
X1 = the continuous predictor variable;
X2 = (U1 * X1) = the continuous predictor variable if the criterion is from a subject in Method A, 0 otherwise;
U2 = 1 if the score on the criterion is from a subject in Method B, 0 otherwise;
X3 = (U2 * X1) = the continuous predictor variable if the criterion is from a subject in Method B, 0 otherwise; and
a_1, a_2, b_1, b_2, and b_3 are least squares weighting coefficients calculated so as to minimize the sum of the squared values in the error vectors.

Degrees of freedom numerator = (m1 - m2) = (4 - 3) = 1

Degrees of freedom denominator = $(N - m1) = (N - 4)$

where:
N = number of subjects;
m1 = pieces full; and
m2 = pieces restricted.

parallel lines for the Restricted Model. The restriction in going from the Full Model to the Restricted Model would be to force the three lines to have a common slope, say, $b_1 = b_2 = b_3$. Since there are two equal signs, there are two restrictions, and thus a nondirectional Research Hypothesis. Once again, we see that *just because a question can be tested does not mean that it is an interesting question.*

Partial Interaction

We suggest here that one should consider models that allow interaction to occur only in those parts of the design where interaction is suspected of occurring. Often a researcher will suspect that interaction is occurring in only one segment of the design and, rather than test this specific question, will unfortunately test the overall, omnibus interaction question. Andrews, Morgan, and Sonquist (1967) presented some notions that are supportive of investigating interaction in specific aspects of the design.

Figure 7.7 depicts an extreme case in which the distance between the two levels of one independent variable at point Q4 is different from that of points Q1, Q2, and Q3. What is happening at Q4 is clearly different from what is happening at Q1, Q2, and Q3. If the omnibus interaction question had been calculated, the exact source of significance would not have been found out, nor could a directional interpretation have been made. But if interaction is tested by equating the difference at Q4 with the average of the differences at Q1, Q2, and Q3, then, because only one restriction is being made, an unambiguous interpretation can be made on a significant *F*. (Again, it will be unambiguous only if the direction of the interaction had been specified before analysis.) The careful reader will note that this is very similar to the notion presented in the previous section—a specific hypothesis and the testing of one interaction variable. Fraas and Drushal (1987) provide an interesting application of the partial interaction question.

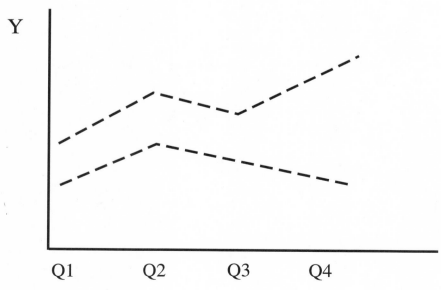

Figure 7.7.
Interaction occurring at only one level of IQ.

Relationship to Other Statistical Techniques

The reader familiar with ANOVA may realize that testing for parallel slopes is equivalent to testing the assumption of homogeneity of regression coefficients underlying the analysis of covariance F test. Also, if the pretreatment strength were represented qualitatively with group membership vectors, the approach would be equivalent to the usual test of no interaction in a two-factor design.

Directional and Nondirectional Hypotheses Miscellany

Directional and nondirectional Research Hypotheses. In most research fields, the investigator goes to much effort to review the literature and to develop a strong rationale for her expectation that a particular state exists, which logically requires the use of a directional hypothesis. Numerous examples exist in educational research of new methods being laboriously developed to ensure that the new method is better than the existing method; yet when the Research Hypothesis is cast, it states: "The two methods are different." The hypothetical expectations are *directional*, yet the test is cast in *nondirectional* terms!

There are at least two harmful outcomes as a consequence of using *nondirectional* tests when the real question is directional. First, some researchers who use nondirectional hypotheses find that the F is larger than the tabled F for their stated alpha level and cite that the findings are "significant but in the opposite direction than hypothesized." This is a serious error because the original Research Hypothesis did not state directionality. Indeed, the findings may be contrary to the theoretical rationale, but they are not contrary to the Research Hypothesis as stated. If the question were directional, then the only finding that can support the hypothesis is a significant F value with the weighting coefficient in the hypothesized direction. If the F is large but the weight is not in the direction hypothesized, then the Research Hypothesis is untenable. However, one cannot say the results are "significant but in the opposite direction," because that question was not under investigation. Such a finding may be very important because it may suggest something is seriously wrong with the theoretical background of the methods of treatment. These data can be used for reconsideration of the theoretical position under investigation, but no inferential statement should be made because these are findings after the fact rather than findings that were theoretically sensible and consequently predicted.

The second harmful outcome from inappropriately using the nondirectional hypothesis relates to the level of probability for a specified F value. A nondirectional test of, say, A and B treatments yields an observed F that can be due to A being better than B or B being better than A. If expectations are that A is better than B, and the test shows A to be better than B, what is the probability that the observed F is due to sampling error? If the observed F in the two-tailed test (which most tables report) is due to sampling error only five times in 100 ($p = .05$), then the probability for A being better than B in a directional test is .025 for the same F value. In a directional test, the probability of chance variation is one-half the two-tailed table value.

Given a directional expectation but using a nondirectional test and a nondirectional F table can result in erroneously failing to reject the Statistical Hypothesis and not accepting the correct directional Research Hypothesis because the reported probability is twice as large as it should be. If the F is observed eight times in 100 with the usual tables, then the correct probability for a supporting finding (assuming the weighting coefficient is in the hypothesized direction) is half that tabled value—four times in 100. Good results may be rejected because they apparently do not quite reach the predetermined alpha level (as reported in tables and computer output) when indeed the chance probability of the test results is erroneously indicated to be twice as large as it should be. Such attention to what may appear to be an academic point is really quite important when sample sizes are limited. On the other hand, with large sample sizes, the size of R^2, as well as the statistical significance should be considered.

For the specific Research Hypothesis in the isotonic-isometric example, the slope question was directional ($b_2 > b_1$), thus if the results are in the hypothesized direction, the reported probability from typical tables or typical computer output should be divided by two. One must state a directional Research Hypothesis and check to make sure the hypothesized weights are in the hypothesized direction before reporting the "directional probability."

Recommendations for directional hypothesis testing. We strongly believe that past findings and theoretical expectations should dictate the Research Hypothesis. If one is groping in the dark about the field of investigation, then the nondirectional test might yield a set of variables that may be relevant for understanding a particular criterion. Following a hunch is one of the inductive phases of inquiry, and directionality may not be called for. Indeed, if one is searching for relevant variables, inferential statistics in itself may not be needed. Bivariate correlations and sample means are fine descriptors for expressing hunches.

On the other hand, if prior knowledge is available in a researcher's domain, we firmly believe that the directional Research Hypothesis is not only to be preferred but is the appropriate hypothesis to test. The continuing emphasis on directional Research Hypotheses will appear throughout the remainder of this text. It should be noted that the regression approach can be applied to either directional or nondirectional Research Hypotheses. The choice of either directional or nondirectional Research Hypotheses is not a statistical choice; it depends on what the researcher wants to conclude from the analysis.

Present statistical developments limit the directional Research Hypothesis to a single restriction on a Full Model. Thus, there will only be one degree of freedom for the numerator of the F ratio. If the degrees of freedom in the numerator of the F ratio is greater than one, then a directional Research Hypothesis has not been tested.

Directional interaction. The notions of directional hypothesis testing are as applicable to interaction hypotheses as to any other kind of hypothesis. In the example illustrated by Figure 7.7, the researcher expected the difference at Q4 to be different from the average of Q1, Q2, and Q3. In addition, the researcher may have expected the superiority of Drug 1 over Drug 2 at Q4 to be *greater* than at the average of Q1, Q2, and Q3. This specific expectation calls for a directional test of significance. For both the directional and nondirectional hypotheses, the Full Model, restrictions, and Restricted Model are the same. The only difference is in the expectation, or the "want."

The Full Model for testing the interaction just discussed is:

$$\text{(Model 7.12) } Y1 = a_1C1 + a_2C2 + a_3C3 + a_4C4 + a_5C5 + a_6C6 + a_7C7 + a_8C8 + E_{12}$$

where:

    $Y1$ = the criterion variable;
    $C1$ = 1 if Drug 1 and IQ Level 1, 0 otherwise;
    $C2$ = 1 if Drug 2 and IQ Level 1, 0 otherwise;

C3 = 1 if Drug 1 and IQ Level 2, 0 otherwise;
C4 = 1 if Drug 2 and IQ Level 2, 0 otherwise;
C5 = 1 if Drug 1 and IQ Level 3, 0 otherwise;
C6 = 1 if Drug 2 and IQ Level 3, 0 otherwise;
C7 = 1 if Drug 1 and IQ Level 4, 0 otherwise; and
C8 = 1 if Drug 2 and IQ Level 4, 0 otherwise.

What the researcher wants is:

$$(a_7 - a_8) > [(a_1 - a_2) + (a_3 - a_4) + (a_5 - a_6)] / 3$$

The antithesis of the above want is:

$$(a_7 - a_8) = [(a_1 - a_2) + (a_3 - a_4) + (a_5 - a_6)] / 3$$

Imposing the above restriction on the Full Model results in an extremely complicated Restricted Model, which will not be presented. The directional Research Hypothesis, of course, demands that $(a_7 - a_8)$ be greater than the quantity on the right-hand side of the restriction.

If a researcher is going to treat interaction as an interesting phenomenon in its own right, then the interaction should be expected to occur in a certain specified way. A quick search in one's field of research will indicate that many researchers have expectations about how the interaction will occur. Because the probabilities recorded in most statistics books and printed by most computer programs for the *F* test are two-tailed probabilities, adjustments on the probability are called for if a directional Research Hypothesis is tested.

Directional interaction computer problem. Applied Research Hypothesis 7.2 is designed to fit the directional interaction notions just presented. (As with all computer problems in this text, the vectors are in Appendix A.) Neither the Full Model nor the Restricted Model calls for the unit vector when testing for directional interaction. As discussed earlier, most computerized regression programs provide the unit vector automatically; therefore, the researcher must consider it to be in the model. For the directional interaction hypothesis in Applied Research Hypothesis 7.2, the unit vector is linearly dependent upon the two treatment vectors. Therefore, because the computer program automatically includes the unit vector, one of the method vectors (X11) is omitted from both the Full Model and the Restricted Model. The R_f^2 and R_r^2 obtained by the computer program will be the same R_f^2 and R_r^2 as with the two method vectors instead of the unit vector and only one method vector.

Many computer programs use a matrix inversion solution and will not run if linear dependencies are included in the model. If such a program is being used, one of the two treatment vectors should not be included in the model. Computer programs using an iterative solution will simply assign a zero weight to one of the treatment vectors. Our preference is to conceive the models with all the "necessary" vectors in them (because this seems to make the models more communicative), which often means initially not including the unit vector. Once the models are constructed and deemed to reflect the Research Hypothesis, then the unit vector can be included and linearly dependent vectors omitted. For instance, in the two-group directional interaction hypothesis, the two Y-intercepts can be found by adding the unit vector weight to the weight for the respective treatment group.

Applied Research Hypothesis 7.2

Directional Research Hypothesis: For a given population, as X8 increases, the relative effectiveness of Method A (X10) as compared to Method B (X11) on the criterion of interest (X9) will linearly increase.

Statistical Hypothesis: For a given population, as X8 increases, the difference between Method A (X10) and Method B (X11) on the criterion of interest (X9) will remain the same.

Full Model: $X9 = a_0U + a_{10}X10 + a_{15}X15 + a_{16}X16 + E_1$

where:
$\qquad X15 = X10 * X8;$ and
$\qquad X16 = X11 * X8.$

Want: $a_{15} > a_{16}$; restriction: $a_{15} = a_{16}$

Restricted Model: $X9 = a_0U + a_{10}X10 + a_8X8 + E_2$

alpha = .001
$R_f^2 = .98; R_r^2 = .87$
$F_{(1,56)} = 319$
Computer probability = .0001; directional probability = .00005

Interpretation: Since the weighting coefficient a_{15} is numerically larger than a_{16}, the directional probability is appropriate and the following conclusion can be made: As X8 increases, the relative effectiveness of Method A (X10) as compared to Method B (X11) on the criterion of interest (X9) will linearly increase.

INTERACTION TERMS AS COVARIATES

Any variable may be used as a covariate, as long as that variable is not influenced by the treatments. If one considers interaction to be a phenomenon in its own right, then interaction terms could be used as covariates. Some rationale should be used for including this covariate, otherwise many interactions (second degree, third degree, etc.) might be included that would drain the degrees of freedom and spuriously generate nonsignificance. The regression models for the covariance analogue are presented in chapter 8. Basically, the covariates appear in both the Full Model and the Restricted Model, with the treatment vectors appearing in only the Full Model.

INTERACTION BETWEEN TWO CONTINUOUS PREDICTORS

Normally when one uses a continuous variable as a predictor, one uses the originally scaled values of that predictor in a model, such as in Model 7.13. This model is actually expressing the linear correlation between Y and P.

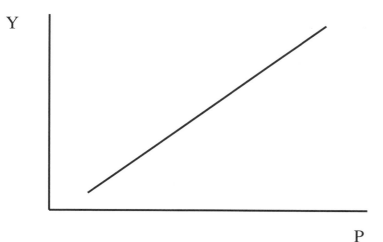

Figure 7.8.
A single straight line, fit by Model 7.13.

$$(\text{Model } 7.13)\ Y\ =\ a_0U + a_1P + E_{13}$$

This relationship is shown in Figure 7.8 as a straight line. When multiple predictors are used to predict the criterion, most researchers still choose to use the originally scaled variables. With more than one predictor, the graphic representation of the relationship between the criterion and predictors may require more than two dimensions. Several such situations are discussed in this section. It must be remembered that the research question dictates the model and graphic representation of the model.

Model 7.13 was represented in Figure 7.8 as a two-dimensional straight line. It also could be represented in three dimensions. Suppose the researcher has knowledge of another variable, Q, for all subjects. If the relationship between Y and P is expected to be the same for all values of Q, then the model to express this relationship is still Model 7.13; but to show the constancy of the relationship over values of Q, Figure 7.9 would be used. The criterion, Y, is on the vertical axis, and the values for P and Q are along the sides of the cube. The predicted scores on Y form a flat plane defined by the values of Y, P, and Q. Figure 7.9 illustrates that Q is not needed as the plane is at the same height on Y for all values of Q (for any value of P). That is, at the front right side of the cube, the plane is parallel to the Q axis. Also, at the back left side of the cube, the plane is parallel to the Q axis.

If one were to model the data with two dimensions—both P and Q—the weight for Q would be zero. This value would indicate that the slope of the plane along Q is zero and that, therefore, Q would not add to the predictability of Y. That is, the Statistical Hypothesis (that the weight for Q is equal to zero) could not be rejected.

Suppose, however, that the relationship between Y and P is not expected to be constant across values of Q. Then Q also would have to be included in the model as a predictor. Model 7.14 shows this state of affairs.

$$(\text{Model } 7.14)\ Y\ =\ a_0U + a_1P + a_2Q + E_{14}$$

Figure 7.10 shows that this model allows the predicted Y scores to form a flat plane that is tilted such that the criterion Y value for a given P value is different for each Q

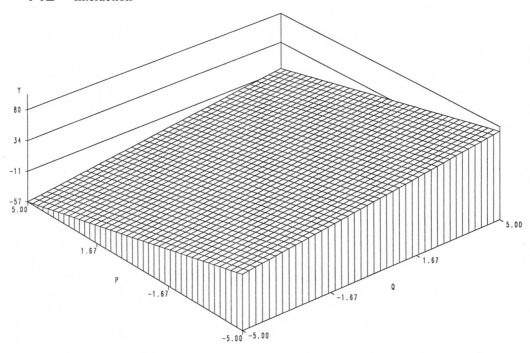

Figure 7.9.
A plane not varying over Q, fit by Model 7.13.

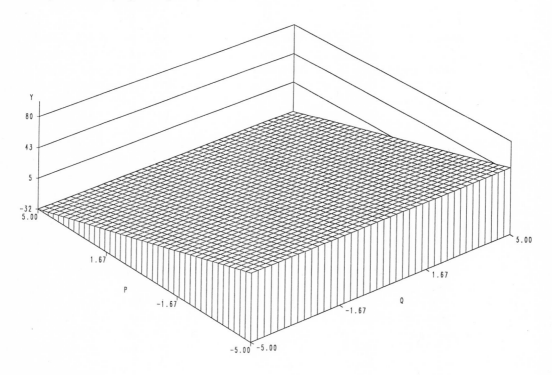

Figure 7.10.
A plane varying over P and Q, fit by Model 7.14.

value. (This can also be seen by noting that at each corner of the cube the plane has a different value on Y.)

As was discussed in a previous section, one may have reason not to want to use the original values of the predictors. One may, for example, expect a particular relationship (say, a flat plane) to exist for values below a particular point (R) on the Q variable and expect another relationship (say, a flat plane) to exist for values at and above R on Q. Model 7.15 would allow this expectation.

(Model 7.15) $Y = a_0U + a_1U1 + a_2(P * U1) + a_3(Q * U1) + a_4U2 + a_5(P * U2) + a_6(Q * U2) + E_{15}$

where:

$U1 = 1$ if the score on Q is below the value of R; and
$U2 = 1$ if the score on Q is at or above R.

Model 7.15 allows for two planes, one for Q values below R and another flat plane for Q values at and above R. These two planes would not have to intersect at R as depicted in Figure 7.11, although if the data were systematic, the intersection would likely be at R. Notice that the interactions in the models in Figures 7.11 and 7.12 are again represented by multiplication. In Figure 7.12, the multiplication is of two continuous variables.

If the researcher expects that an interaction exists between P and Q as they relate to Y, a model such as Model 7.16 could be used.

(Model 7.16) $Y = a_0U + a_1P + a_2Q + a_3(P * Q) + E_{16}$

This model is represented not by a flat plane or by a combination of flat planes, but instead by a twisted plane of the type pictured in Figure 7.12. Each edge of the plane is straight, and each line drawn across the values of Q is straight, yet the plane is twisted to allow it to intersect the four corners of the cube at values different from those that would be possible if the plane were flat.

COMPARISON OF CATEGORICAL AND CONTINUOUS INTERACTION

One major weakness of ANOVA analyses is the artificial categorization of continuous variables. Researchers can usually obtain continuous data and usually want to infer along some continuum, but they often categorize their data before analyzing it. Indeed, phenomena in the real world usually follow systematic functions rather than discrete leaps and bounds. The GLM approach readily allows one to investigate continuous variables and specifically to investigate the interaction between categorical variables and continuous variables.

Suppose that one wanted to treat IQ as a continuous variable rather than artificially categorizing it into four levels. Suppose also that Figure 7.13a is the pictorial representation of the suspected interaction, and Figure 7.13b is the state of affairs allowing no interaction between drugs and IQ as they affect the criterion. The regression formulation allowing for interaction would be:

(Model 7.17) $Y1 = a_1D1 + a_2D2 + b_1S1 + b_2S2 + E_{17}$

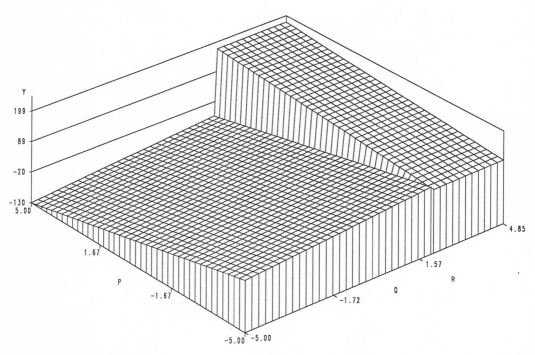

Figure 7.11.
One plane below R on Q and another plane above R, fit by Model 7.15.

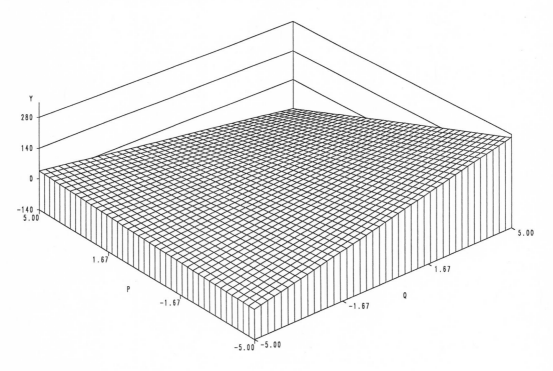

Figure 7.12.
A twisted plane depicting interaction between two continuous predictors, fit by Model 7.16.

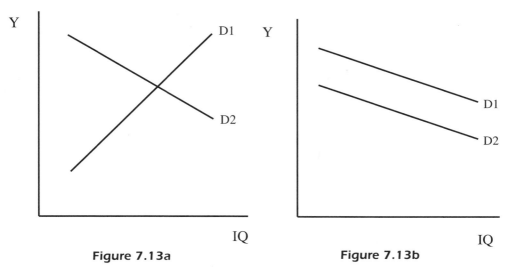

Figure 7.13a   **Figure 7.13b**

Figure 7.13.
States of affairs allowing linear interaction (7.13a) and not allowing linear interaction (7.13b) with two drugs.

where:

D1 = 1 if have Drug 1, 0 otherwise;
D2 = 1 if have Drug 2, 0 otherwise;
Q5 = IQ for each subject;
S1 = D1 * Q5 = IQ if have Drug 1, 0 otherwise; and
S2 = D2 * Q5 = IQ if have Drug 2, 0 otherwise.

The slope of the Drug 1 line would be b_1, and the Y-intercept would be a_1. The slope of the Drug 2 line would be b_2 and the Y-intercept would be a_1. If these two lines are not interacting, they will be parallel, which means that the two slopes must be equal, resulting in the restriction: $b_1 = b_2$. Setting the two slopes equal to a common slope, b_5, results in the following Restricted Model:

$$(\text{Model } 7.18)\ Y1 = a_1 D1 + a_2 D2 + b_5 Q5 + E_{18}$$

The F test of significance between Models 7.17 and 7.18 would have $df_n = (4 - 3)$ and $df_d = (N - 4)$. Note that the vectors that allow for interaction can be found by multiplying the drug variable by the continuous IQ variable.

Suppose now that a third drug is under consideration, as in Figure 7.14. The regression formulation allowing for interaction with the three drugs would be:

$$(\text{Model } 7.19)\ Y1 = a_1 D1 + a_2 D2 + a_3 D3 + b_1 S1 + b_2 S2 + b_3 S3 + E_{19}$$

where:

D1 = 1 if have Drug 1, 0 otherwise;
D2 = 1 if have Drug 2, 0 otherwise;
D3 = 1 if have Drug 3, 0 otherwise;
Q5 = IQ for each subject;
S1 = D1 * Q5 = IQ if have Drug 1, 0 otherwise;

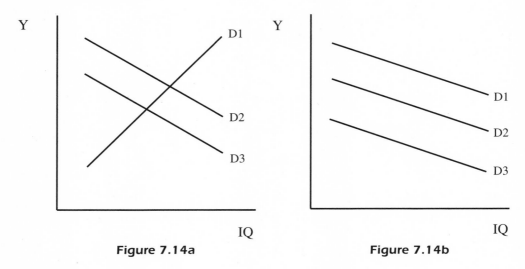

Figure 7.14a   **Figure 7.14b**

Figure 7.14.
States of affairs allowing linear interaction (7.14a) and not allowing linear interaction (7.14b) with three drugs.

$S2 = D2 * Q5 = IQ$ if have Drug 2, 0 otherwise; and
$S3 = D3 * Q5 = IQ$ if have Drug 3, 0 otherwise.

The slope of the Drug 3 line would be b_3, and the Y-intercept would be a_3. If these three lines are not interacting, they will be parallel, which means that the three slopes must all be equal, resulting in the restriction: $b_1 = b_2 = b_3$. Setting all three slopes equal to a common slope, say b_5, results in the following Restricted Model:

$$(\text{Model } 7.20)\ Y1 = a_1D1 + a_2D2 + a_3D3 + b_5Q5 + E_{20}$$

For the F test of significance between Models 7.19 and 7.20 there would be an $F_{(6-4,N-6)}$. Note that the vectors that allow for interaction can be found by multiplying the drug variables by the continuous IQ variable.

Now suppose that the three drugs were actually three different dosage levels of the same drug, and we now wish to treat the drug variable as a continuous variable rather than artificially categorizing it into levels. The interaction Research Hypothesis now would be:

There is an interaction between IQ and drug dosage level, over and above the separate linear effects of IQ and drug dosage level.

A more precise and insightful way of stating the Research Hypothesis would be:

The combined effect of IQ and drug dosage level is needed to predict the criterion, over and above the separate linear effects of IQ and drug dosage level.

Testing the interaction between the continuous variable of IQ and that of drug dosage level would be accomplished by constructing Models 7.21 and 7.22.

$$(\text{Model } 7.21) \; Y1 \; = \; a_0 U + a_4 D4 + b_5 Q5 + c_4 (D4 * Q5) + E_{21}$$

where:

Y1 = criterion;
U = unit vector;
D4 = drug dosage level for each subject;
Q5 = IQ for each subject; and
D4 * Q5 = product of drug dosage level and IQ for each subject.

The product term (D4 * Q5) allows for the linear interaction between drug dosage level and IQ. The restriction on Model 7.21 that would not allow interaction to occur would be to set c_4 equal to zero, having the effect of dropping (D4 * Q5) out of the model:

$$(\text{Model } 7.22) \; Y1 \; = \; a_0 U + a_4 D4 + b_5 Q5 + E_{22}$$

Note that the unit vector appears in both Models 7.21 and 7.22. When all variables in the regression model are continuous, the unit vector is not generally a redundant piece of information and therefore must be counted as one of the linearly independent vectors. Furthermore, the interaction term in Model 7.21 (D4 * Q5) is not *linearly* dependent upon the other vectors in the model. The interaction term is a function of these variables, but it is not a *linear* function; therefore, it is *linearly* independent. (See Appendix B-3 for proof.) There are four linearly independent vectors in Model 7.21 and three in Model 7.22, hence the test of significance for this comparison is: $F_{(4-3, N-4)}$.

The product of two continuous variables has been labeled the *moderator variable* in the literature (Saunders, 1956). The multiplication of two continuous variables is simply an extension of the multiplication of two categorical variables (already referred to in the literature as *interaction*), so it is unfortunate that new terminology (moderator variable) was introduced. For heuristic purposes, the single concept of interaction should suffice, although any variable that increases predictability has a legitimate place in a model, whether it has a special name or not.

Some authors (Goldberg, 1972) indicate that few interactions between continuous variables will be found. Goldberg's work (he called them *configural variables*) is with highly correlated test items, which one would not necessarily expect to interact with one another. Others (DuCette & Wolk, 1972; Wood & Langevin, 1972) have found limited, selected instances in which a moderator variable increases prediction. Whether any variable, whether interaction or not, is necessary is an empirical question.

Table 7.2 indicates the advantages of treating variables as continuous rather than as artificially categorized. When variables are treated as continuous, more degrees of freedom are obtained in the denominator, and at the same time, fewer degrees of freedom exist in the numerator. The extreme situation appears when both variables are treated as continuous, resulting in one degree of freedom in the numerator, corresponding to the one restriction made on the Full Model (Model 7.21). So, if there is statistical significance in comparing Model 7.21 to Model 7.22, one knows which restriction is generating the significance. Inspection of the weight for that source will then indicate if the results are in the hypothesized direction. But when more than one restriction is made, as in comparing Model 7.19 to Model 7.20, and significance is found, the source of the significance cannot be pinpointed. These comments are

Table 7.2

Comparison of Interaction Possibilities

	Models from text	Specific df_n	Specific df_d	General df_n	General df_d
Both predictor variables categorical	7.4a vs. 7.5	6	$N - 12$	$(r - 1)\ (c - 1)$	$N - (c * r)$
One predictor variable continuous and one categorical	7.19 vs. 7.20	2	$N - 6$	$(k - 1)$	$N - (k * 2)$
Both predictor variables continuous	7.21 vs. 7.22	1	$N - 4$	1	$N - 4$

Note: N = number of subjects; r = number of categories of one predictor variable; c = number of categories of the other predictor variable; and k = number of categories of the categorical variable.

applicable to any test in which more than one restriction is made—not solely to a problem with interaction hypotheses.

One limitation regarding the interaction of continuous variables is that many more subjects are required to get stable estimates. Essentially, the number of "inside" cells is equal to the product of the range of scores on one variable and the range of scores on the other variable. Thus, if Variable A has 20 different score values and Variable B has 14 different score values, the product would be 280. The traditional rule of thumb of 30 subjects per cell would require 8,400 subjects. We argue that, because the variables are probably somewhat continuously related, fewer than 8,400 subjects are necessary. But the point is that more subjects are necessary when considering continuous data than when considering categorical data. We discuss the number of subjects needed in a particular analysis more fully at the end of chapter 8.

8

STATISTICAL CONTROL OF POSSIBLE CONFOUNDING VARIABLES

One goal of research is the isolation of the unique effects of variables. To show that a particular variable is influencing the criterion in a particular way, other competing variables, confounding variables, or potential explainers of the data must be eliminated. Competing explainers can be eliminated, to various degrees, by theoretical (logical) argument, by research design, or by statistical control. Theoretical argument is usually the weakest defense, whereas research design is usually the strongest. Unfortunately, it is not always possible in the behavioral sciences either to "argue away" or to "design away" the competing explainers. Statistical control, through the analysis of covariance, can be a very useful tool for isolating the effects of variables.

CONTROL OF CONFOUNDING VARIABLES THROUGH THEORY

One might argue based on theory, or common sense, or logic, that a certain variable cannot be a confounding variable. For example, a researcher could argue that hair color, number of letters in one's name, and Social Security number are unrelated to math achievement. That researcher could then ignore these variables when researching math achievement. Usually, though, researchers do not go through such a thought process—they just choose to investigate certain variables and by default ignore others. Whether those ignored variables are relevant or not is ultimately an empirical question, but for the time being they are ignored.

CONTROL OF CONFOUNDING VARIABLES THROUGH RESEARCH DESIGN

Randomization

Research conducted in the laboratory is directed toward determining the effects of a particular independent variable upon some outcome state. Some examples of laboratory research questions are:

1. Is titanium a better luminescent filament than carbonized cotton?
2. Do pigs fed 6 grams of complete protein daily weigh more than pigs fed 2.5 grams of complete protein daily after 60 days of feeding?
3. Do dogs raised in isolation for 80 days after birth respond to noxious stimuli less effectively than dogs raised in a natural kennel environment?

To answer these questions, the researcher attempts to control for all known contaminants in the research design. In the dog question, breed differences surely may contaminate the results. If the isolated dogs were Doberman pinschers and the control dogs were mongrels picked up at the pound, the observed differences, if any, in sensitivity to noxious stimuli might be contaminated by breed differences in sensory responsivity.

The researcher may control for the effects of a contaminating variable (here, breed) by choosing only one "level" of that variable (one breed) or by choosing several levels (breeds) and then, in either case, randomly assigning half the members of each breed to each treatment. If only one level (breed) is selected, that is the only level to which the results can be safely generalized. In the dog question, the researcher may select a specific breed and randomly split litters of puppies into experimental and control groups. The ANOVA on the responsivity criterion would then yield an F value that gives the researcher an estimate of how likely it is that the observed differences are due to sample variation within that breed. By using the split litter of a single breed, the population is that specific breed, and thus large response deviations between the two treatment samples would not be contaminated by the possibility of their having originally come from two different breed populations.

In laboratory research, the researcher may have most of the relevant (potentially contaminating) variables under physical control. Such control also may be achieved in the laboratory for complex manipulations of more than one variable. Applied research in the behavioral sciences, though, is frequently conducted in natural settings where physical control of contaminants is either costly or impossible.

Matching

Consider the case of a curriculum specialist who wants to test the effectiveness of a new curriculum designed to improve reading ability. She expects students exposed to her material to perform better than students exposed to the old method. Suppose she has a sample of 200 students available for random assignment. Past research may suggest that, in relation to her criterion score of reading, several variables are known to be related to performance, such as (a) girls score better than boys, (b) children from middle-class homes tend to perform better, (c) high-IQ children score higher, and (d) past reading ability is positively related to the criterion of interest. These four variables are possible sources of contamination if the subjects are not assigned randomly to treatments or if the random assignments, by chance, place more of one kind of student in one of the two treatment groups. To avoid such contaminating effects (i.e., nonequivalent random samples), one may attempt a matched-pairs assignment procedure—such as getting (a) two girls, (b) from middle-class homes, (c) with high IQ, and (d) high initial reading ability—and then randomly assigning one to the new treatment and the other to the old treatment. If such a procedure is followed rigorously for all ranges of the possible contaminants, the researcher will have two groups of matched pairs. With the limited original sample of 200 children, however, one would typically find that, after obtaining, say, 30 pairs, there are no "real" pairs left. Some of the rest may pair up on one or two of the variables, but they would not pair on the other variables. The mind boggles at the effort required to form matched pairs for more than one or two variables, and successful matching on all 200 subjects would surely be unlikely.

Given that matching is accomplished for 30 pairs, one may conduct the study, but then the results would be generalizable only to the peculiar population that the 30 sample pairs represent. Experimental control such as this parallels the rigor of the laboratory, but does it really solve this curriculum specialist's question? Of her 200 subjects, only 60 subjects were used, and they are not necessarily the same on the other relevant variables as those nonmatched children. The population to which the results can be generalized is not readily apparent. If the 200 students represent the population to which she wants to generalize, the selected matched sample does not really represent the population to which she wants to generalize, and therefore she cannot answer her question.

Once researchers obtain their matched samples, they usually proceed as if those samples were randomly assigned. Thus a *t* test (or Generalized Research Hypothesis 4.1) would have been used by the researcher in the above example. But more is known about the subjects than that they were given the different methods. That there *are* matched pairs can be included in the analysis. The resulting design is analogous to the dependent-groups design discussed in chapter 10. Here, the "person vectors" are the "pair vectors."

Designing out the Competing Explainer

Another way to control for a possible competing explainer is to design the variable into the study. If IQ is known to influence the criterion, the effects of IQ can be controlled for. Because the variable is known to have an effect on the criterion, it usually is not tested. For the curriculum study, the blocking variable of IQ thus changes the design from a one-way design (two curriculum methods) to a two-way design (curriculum methods and IQ). The analysis procedure discussed in the two-way ANOVA in chapter 4 is applicable. Note, though, that IQ "groups," "levels," or "blocks" have been created, thus lumping together many different IQ scores and considering their effects to be similar. The covariance design, to be discussed next, does not lump all IQ scores together but treats the possible contaminating variables as continuous variables.

CONFOUNDING VARIABLES UNDER STATISTICAL CONTROL

Rationale

An alternate approach to the control difficulty in the proposed curriculum study would be to match the two groups the best one can and then statistically control for the contamination that was not under the researcher's control. Given a rough match between the experimental and control groups, one may find that one group has a few more high-IQ students, fewer girls, fewer middle-class children, and a slight difference in initial reading ability when compared with the other group. These differences are contaminants whose magnitude of effect on the criterion would be unknown if not accounted for in the research question and subsequent analysis. The researcher really wants to know, "Over and above the influences of IQ, gender, entering reading ability, and social class, is the innovative procedure superior to the traditional procedure as measured by posttest reading-ability scores?"

The question can be cast into the following Research Hypothesis:

Over and above the influences of IQ, gender, entering reading ability, and social class, the innovative procedure is superior to the traditional procedure as measured by the posttest reading-ability scores.

The competing hypothesis would be the following Statistical Hypothesis:

Over and above the influences of IQ, gender, entering reading ability, and social class, the innovative procedure is equal to the traditional procedure as measured by the posttest reading-ability scores.

Full Model, Restricted Model, and F Test

The Full Model that reflects the Research Hypothesis has posttest reading ability as the criterion, and the predictor set includes four covariates (IQ, gender, entering reading ability, and social class) and two treatment vectors. The Full Model would be:

$$(\text{Model 8.1}) \ Y1 = a_1 U1 + a_2 U2 + c_1 IQ + c_2 G + c_3 RA + c_4 SC + E_1$$

where:

$Y1$ = posttest reading ability;
$U1$ = 1 if $Y1$ is from a member of the innovative treatment, 0 otherwise;
$U2$ = 1 if $Y1$ is from a member of the traditional treatment, 0 otherwise;
IQ = IQ score of each individual represented in $Y1$;
G = 1 if the subject is female, 0 otherwise;
RA = entering reading-ability score;
SC = numerical value on social-class index;
E_1 = the error vector, the difference between the observed criterion and the predicted criterion $(Y1 - \hat{Y}1)$; and
a_1, a_2, c_1, c_2, c_3, and c_4 are regression weights calculated to minimize the sum of the squared elements in vector E_1.

Note that the weights for the covariates are labeled c_1, c_2, c_3, and c_4. One could use any letter to represent these weights, yet some researchers may find that for mnemonic purposes the c helps to suggest that these weights are associated with the covariates.

The following shows how the Full Model might look in vector form.

Subject	Y	=	aU1	+	bU2	+	cIQ	+	dG	+	eRA	+	fSC	+	E
1	9		1		0		120		1		7		2		?
2	7		1		0		115		0		6		1		?
3	5		1		0		100		1		3		3		?
4	8		0		1		120		0		7		1		?
5	7		0		1		120		1		6		3		?
6	4	= a	0	+ b	1	+ c	105	+ d	1	+ e	3	+ f	2	+	?
.	.		.		.		.		.		.		.		?
.	.		.		.		.		.		.		.		?
.	.		.		.		.		.		.		.		?
200	4		0		1		105		1		3		2		?

Look at the row of scores for subject number one: She had a posttest score of 9, was a member of the innovative treatment ($U1 = 1$), was not a member of the traditional treatment ($U2 = 0$), had an IQ of 120, was a girl ($G = 1$), had a reading-ability pretest score of 7, and was in the second level of social class ($SC = 2$).

Subject number four had a posttest score of 8, was not in the innovative treatment ($U1 = 0$), was in the traditional treatment ($U2 = 1$), had an IQ of 120, was a boy ($G = 0$), pretested at 7 ($RA = 7$), and was from a home of the highest social class ($SC = 1$).

The restriction implied by the Statistical (or competing) Hypothesis of interest is $a_1 = a_2$ (the Research Hypothesis implies that a_1 is greater than a_2). Setting $a_1 = a_2$, both equal to a common weight a_0, one obtains the Restricted Model:

$$\text{(Model 8.2)} \quad Y1 \; = \; a_0 U1 + a_0 U2 + c_1 IQ + c_2 G + c_3 RA + c_4 SC + E_2$$

where:

$a_0 \; = \;$ the common weight for both groups that reflects the equality of a_1 and a_2; and

$E_2 \; = \;$ the new error vector that is the difference between the observed criterion score and the predicted criterion, predicted from the variables in the Restricted Model ($Y1 - \hat{Y}1$).

Because U1 and U2 are mutually exclusive vectors whose elements are ones and zeros, and because they share a common weight, one can simplify the Restricted Model to:

$$\text{(Model 8.3)} \quad Y1 \; = \; a_0 U + c_1 IQ + c_2 G + c_3 RA + c_4 SC + E_2$$

where:

$U \; = \;$ $U1 + U2$ and yields the unit vector (U) with ones for everyone in the study.

The Restricted Model forces both groups to have the same constant, and thus any predictability is due solely to the covariates and the regression constant, a_0. Almost all scales in the social sciences are arbitrarily scaled, so the regression constant is almost always employed to adjust all the predicted scores up or down to have the same mean as the criterion. Strictly speaking, the unit vector in the Restricted Model is a covariate and should be specified in the Research Hypothesis. Customary usage has led researchers always to place the unit vector in the Full and Restricted Models. As with most customs, this one can be deviated from—as discussed in later chapters. Unless otherwise indicated, the unit vector is assumed to be in both the Full and Restricted Models.

The R_f^2 is the proportion of the observed sample criterion variance accounted for by group membership and the covariates. The R_r^2 is the proportion of the observed sample criterion variance accounted for by the covariates alone. Any loss in R^2 between the Full and Restricted Models will be the proportion of *unique* sample criterion variance accounted for by knowledge of which treatment the subject received (over and above the effects of the covariates).

The F-test equation is, as always, the general F formula:

$$F_{(m1 - m2, N - m1)} \; = \; \frac{(R_f^2 - R_r^2) / (m1 - m2)}{(1 - R_f^2) / (N - m1)}$$

The number of linearly independent vectors in the Full Model (m1) is six. Because there are five linearly independent vectors in the Restricted Model (the regression constant and the four covariates), m2 = 5. With 200 subjects, the degrees of freedom are 1 and 194. Assuming that $R_f^2 = .53$ and $R_r^2 = .45$, then substituting, yields:

$$F_{(1, N-6)} = \frac{(.53 - .45) / (1)}{(1 - .53) / (N - 6)} = \frac{.08/1}{.47/194}$$

The value of .08 is the proportion of the *sample* criterion variance uniquely due to knowledge of group membership, over and above the covariate knowledge. When this is divided by the numerator degrees of freedom, the result is a proportional *estimate* of the *population* criterion variance using the unique knowledge of group membership. One may want to label this $\hat{v}_u$, where the u indicates that the proportionally estimated population variance is due to *unique* knowledge of group membership.

The value .47 is the proportion of the *sample* criterion variance unaccounted for by the variables in the Full Model and is called *error variance*. When this value is divided by the denominator degrees of freedom, one obtains a proportional *estimate* of the *population* criterion variance that is the most stable estimate. (See chapter 2 for a review.)

$$F_{(1,194)} = \frac{.08/1}{.47/194} = \frac{.08}{.0024} = 33.33$$

One can determine how often an *F* of 33.33 or larger is observed due to sampling variation. To answer the Research Hypothesis, the researcher should set the alpha level before she starts her analysis. The value one selects depends upon cost, inconvenience, and other matters. For the innovative curriculum, the curriculum specialist may decide that conversion to the new procedure is a bother but that, if the observed adjusted criterion mean is due to sampling variation fewer than 25 times in a 1,000 (alpha = .025), she would be willing to convert to the new program. The tabled *F* value for an alpha of .025 is 3.88. That is, an *F* value of 3.88 or larger (with 1 and 194 degrees of freedom) will be observed 2.5% of the time when the Statistical Hypothesis is really true. The observed *F* of 33.33 meets her decision-point criterion; and if upon inspection of the weights a_1 and a_2 her Research Hypothesis is supported, that is, if $a_1 > a_2$, then she accepts the Research Hypothesis and rejects the competing Statistical Hypothesis. Given these findings, she is on *fairly* safe ground in converting to the innovative reading program. Furthermore, the curriculum specialist may report her findings in an appropriate journal so her research community can benefit from her study. However, with a demonstrated R^2 gain of 8%, cost-conscious educators may be less enthusiastic. One might do well to set a minimum R^2 criterion, as well as an alpha level.

If one chooses to use R^2 as the criterion, then it would be more appropriate to use the adjusted R^2, the adjustment being made because the R^2 is a sample-specific index and is thus a function of the number of variables and the number of subjects. If the number of subjects is small, or if the number of variables is large, the adjusted R^2 can be substantially lower than the sample R^2. When the adjusted R^2 is negative, one definitely should not rely on that model. Several adjustments are available and were summarized by Newman, McNeil, Seymour, and Garver (1978). SAS uses the following adjustment:

$$\text{adjusted } R^2 = 1 - [(1 - R^2)(N - 1) / (N - m1)]$$

where:

 N = the number of subjects; and
 m1 = the number of predictor variables in the model.

Utility of Analysis of Covariance

Note should be made here that in this text we take a position somewhat different from others' regarding the legitimacy of the covariance analysis. Some statisticians would take the position that lack of random assignment disallows a meaningful conclusion. Our position is that research and decisions must be made in the real world. Random assignment of groups is ideal, but insight can be gained when this is not possible. Our emphasis on replication (discussed in full in chapter 11) is a check on any bias that might occur from not having equivalent groups. *Analysis of covariance cannot completely fix a badly designed study; but controlling for confounding variables is better than ignoring them.* The reader should realize that if $a_2 > a_1$ in the previous curriculum example (i.e., the traditional procedure yielded the higher posttest mean score), she should not report the results as "significant in the opposite direction" because that was not the question under investigation. Indeed, given the apparent debilitating influence of the innovative program, the researcher should be surprised because her careful planning, based upon past knowledge, was <u>not</u> supported. Several questions may be worth pursuing, such as:

1. Is the criterion measure appropriate?
2. Did the teachers sabotage the program?
3. Is the method interacting with one or more of the covariates?
4. Is the innovative treatment really not that good after all?
5. Were the data collected and analyzed properly?

If, upon tracking down possible contaminants, she finds no explanation, the researcher may then replicate the study (with the new Research Hypothesis opposite to the original). Given that the new Research Hypothesis can be accepted, she should publish her results to alert the reading research community to a possible flaw in the theoretical knowledge in her field, assuming that anyone cares that the "innovative treatment" is significantly worse than the traditional procedure.

What Does the Over and Above Analysis Do?

The problem just presented in detail is rather complex, but it is just such complexity of applied research that makes the analysis of covariance useful. For a conceptual understanding of covariance, consider the following simple problem.

Suppose one wanted to test the influence of the innovative reading program discussed, and 100 boys and 100 girls were randomly selected for each method. (Gender cannot be a competing explainer in this method of selecting subjects because the same proportion of boys to girls is in each method.) The researcher also knows initial reading ability will be related to the criterion. If the two groups have moderately different means on the reading pretest, any observed posttest difference between groups is likely to be influenced by those pretest mean differences.

An inspection of Figure 8.1 shows that, on the pretest, Group 1 had a mean of 4.9 and Group 2 had a pretest mean of 4.2. Group 1 is also superior to Group 2 on the posttest ($\bar{X}_1 = 5.7$; $\bar{X}_2 = 4.9$), but both groups improved. If pretest differences were not statistically taken into account one may conclude that Group 1 is .8 of a point better than Group 2; however Group 1 was initially .7 of a point better on the pretest. On inspection of the two lines of best fit one should note that, across the range of scores where the two groups overlap, there is only about a .1 difference in favor of Group 1. These pretest group differences should be statistically controlled for to determine if Method 1 is superior to Method 2, over and above the criterion variance that the pretest accounts for. The Research Hypothesis is:

> Method 1 is superior to Method 2 on posttest reading scores, over and above pretest scores.

This hypothesis can be stated in several ways; the following are two alternate wordings that say the same thing:

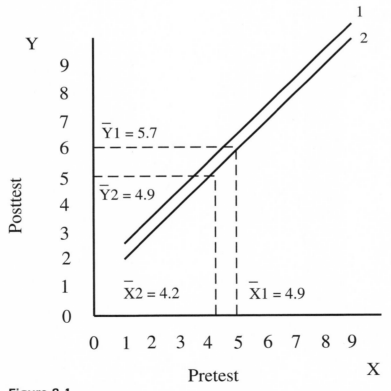

Figure 8.1.
The lines of best fit representing Group 1 and Group 2 across the range of pretest scores on reading as they relate to posttest scores.

Method 1 is superior to Method 2 on posttest reading scores when the pretest differences are statistically controlled;

or

Method 1 is superior to Method 2 on posttest reading scores when the pretest differences are covaried out.

The competing Statistical Hypothesis is:

Method 1 is *not* superior to (is as effective as) Method 2 on posttest reading scores, over and above the pretest scores.

The Full Model is:

$$(\text{Model 8.4}) \; Y1 \;=\; a_1 U1 + a_2 U2 + c_1 X1 + E_4$$

where:

$U1$ = 1 if the criterion score is from an individual in Method 1, 0 otherwise;
$U2$ = 1 if the criterion score is from an individual in Method 2, 0 otherwise;
$Y1$ = posttest (criterion) score;
$X1$ = pretest score; and
a_1, a_2, and c_1 are regression weights calculated so as to minimize the sum of the squared elements in vector E_4.

The Research Hypothesis implies $a_1 > a_2$, and the Statistical Hypothesis demands $a_1 = a_2$. Thus the Restricted Model is:

$$(\text{Model 8.5}) \; Y1 \;=\; a_0 U + c_1 X1 + E_5$$

where:

a_0 = the common (no difference) weight; and
U = the unit vector ($U1 + U2$).

There are three linearly independent vectors in the Full Model ($U1$, $U2$, and $X1$) and two linearly independent vectors in the Restricted Model (U and $X1$). Suppose that $R_f^2 = .60$ and $R_r^2 = .595$, the F ratio would then be:

$$F_{(m1 - m2, N - m1)} = \frac{(.60 - .595) / (3 - 2)}{(1 - .60) / (197)} = 2.5$$

With 1 and 197 degrees of freedom, a directional F of 3.8 or greater is observed 25 times in 1,000. The observed F of 2.5 is smaller than 3.8. Thus with an alpha of .025 one would fail to reject the Statistical Hypothesis ("Over and above pretest differences, Method 1 and Method 2 are equal on posttest reading"). The researcher might still maintain a belief (appropriately) that the Research Hypothesis is true but that the relatively small treatment effect examined with a sample of only modest size led to a relatively small F. Indeed, the Research Hypothesis may yet be true, but based upon the study, the Research Hypothesis cannot be accepted.

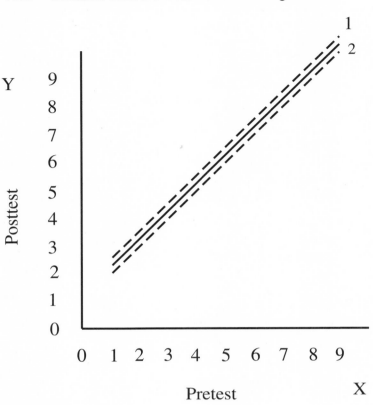

Figure 8.2.
The lines of best fit between pretest and posttest reading scores without knowledge of group membership. The dashed lines are the superimposed lines of best fit shown in Figure 8.1.

The solid line in Figure 8.2 shows the relationship of the pretest scores with the posttest scores without knowledge of group membership. The dashed lines are taken from Figure 8.1. It should be obvious that the moderate pretest differences are accounting for the observed posttest group differences; therefore, knowledge of group membership is almost totally redundant with knowledge of pretest scores. When this is observed, one can conclude that, over and above pretest scores, observed group differences are chance differences due to sampling variation.

The analysis of the difference between the two treatments, statistically controlling for the effects of one confounding variable, is presented as Generalized Research Hypothesis 8.1. An actual data analysis (using data given in Appendix A) for this kind of question is presented in Applied Research Hypothesis 8.1.

Assumptions of the Analysis of Covariance

Analysis of covariance (the over and above question) makes one assumption in addition to those made for the ANOVA. This added assumption is that the slope of the line for each of the k groups is the same across the range of the covariate. The assumption is imposed by the linear model and can easily be seen in the following linear equation:

$$Y1 = a_1 U1 + a_2 U2 + a_3 U3 + \ldots + a_k Uk + c_1 X1 + E_1$$

Generalized Research Hypothesis 8.1

Directional Research Hypothesis: For a given population, Method A is better than Method B on the criterion, over and above the covariable.

Nondirectional Research Hypothesis: For a given population, Method A and Method B are differentially effective on the criterion, over and above the covariable.

Statistical Hypothesis: For a given population, Methods A and B are not differentially effective, over and above the covariable.

Full Model: $Y1 = a_1G1 + a_2G2 + c_1C1 + E_1$ (m1 = 3)

Want (for directional Research Hypothesis): $a_1 > a_2$; restriction: $a_1 = a_2$
Want (for nondirectional Research Hypothesis): $a_1 \neq a_2$; restriction: $a_1 = a_2$
Restricted Model: $Y1 = a_0U + c_1C1 + E_2$ (m2 = 2)

      Automatic unit vector analogue
      Full Model: $Y1 = a_0U + a_2G2 + c_1C1 + E_1$ (m1 = 3)
      Want (for directional Research Hypothesis): $a_2 < 0$; restriction: $a_2 = 0$
      Want (for nondirectional Research Hypothesis): $a_2 \neq 0$; restriction: $a_2 = 0$
      Restricted Model: $Y1 = a_0U + c_1C1 + E_2$ (m2 = 2)

where:
    $Y1$ = criterion;
    U = 1 for each subject;
    $G1$ = 1 if subject received Method A, 0 otherwise;
    $G2$ = 1 if subject received Method B, 0 otherwise;
    $C1$ = covariable score; and
    a_0, a_1, a_2, and c_1 are least squares weighting coefficients calculated so as to minimize the sum of the squared values in the error vectors.

Degrees of freedom numerator = (m1 - m2) = (3 - 2) = 1
Degrees of freedom denominator = $(N - m1) = (N - 3)$

where:
    N = number of subjects;
    m1 = pieces full; and
    m2 = pieces restricted.

There are k groups and k + 1 weights. Each group has its own Y-intercept, yet there is only one weight (c_1) associated with the covariate (X1). Since a one-unit increase on X1 will yield a c_1 increase on Y1, regardless of the group with which the score is associated, all lines by necessity are parallel (have the same slope), as in Figure 8.3.

The assumption just presented is referred to in statistical texts as *homogeneity of regression lines*. Violations of ANOVA assumptions have been shown to be inconsequential when N is large, and therefore F in ANOVA is said to be robust. However,

Applied Research Hypothesis 8.1

Directional Research Hypothesis: For a given population, X12 is better than X13 on the criterion X2, over and above the covariable X14.

Statistical Hypothesis: For a given population, X12 and X13 are equally effective on the criterion X2, over and above the covariable X14.

Full Model: $X2 = a_0U + a_{12}X12 + a_{14}X14 + E_1$ ($m1 = 3$)

Want: $a_{12} > 0$; restriction: $a_{12} = 0$

Restricted Model: $X2 = a_0U + a_{14}X14 + E_2$ ($m2 = 2$)

alpha = .05
$R_f^2 = .99$; $R_r^2 = .95$
$F_{(1,57)} = 222$
Computer probability = .0001; directional probability = .00005

Interpretation: Since the weight for X12 is larger than that for X13 in the Full Model, the directional probability can be referred to. Since the calculated probability is less than alpha, the Statistical Hypothesis can be rejected and the Research Hypothesis accepted. (See Appendix D-8.1.)

if one violates the assumption of homoscedasticity in analysis of covariance (ANCOVA), then one should not pursue the ANCOVA question. While some statistical texts would have researchers throw in the towel at this point, we encourage researchers to look at the resulting interaction and to treat interaction as a valuable phenomenon in its own right. If the lines are not parallel, then interaction exists and can be tested with the procedures outlined in chapter 7.

Consider Figures 8.4a and 8.4b. Suppose Group 1 received one treatment, Group 2 received another treatment, and X1 was a potential covariate of interest. If one obtained the lines of best fit for each group independently ($Y1 = a_1U1 + b_1X1 + E_1$ [for Group 1] and $Y1 = a_2U2 + b_2X1 + E_1$ [for Group 2]), and the plots looked like Figure 8.4a, it would be apparent that the two slopes were not equal, that b_1 was not equal to b_2. The two groups do not have a common slope across the potential covariate.

If the assumption of the common slope is made, however, the Research Hypothesis might be:

Treatment 1 is superior to Treatment 2 on the criterion Y1, across the range of variable X1.

The Full Model would be:

$$(\text{Model 8.6}) \quad Y1 = a_1U1 + a_2U2 + b_1X1 + E_6$$

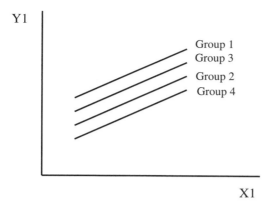

Figure 8.3.
Four parallel lines that represent homogeneous slopes across X1 for the four groups.

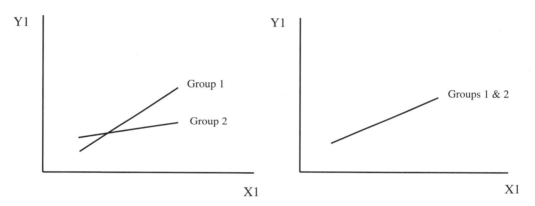

Figure 8.4a.
A possible data configuration.

Figure 8.4b.
The two lines of best fit for data in Figure 8.4a are superimposed when forcing the assumption that the slopes are common.

The Full Model would yield weights (a_1, a_2, and b_1) that (for these data) would yield lines of best fit as depicted by the line in Figure 8.4b. (The lines from Figure 8.4a would be superimposed in Figure 8.4b.)

Thus, when the competing hypothesis for the above Research Hypothesis is tested, the results would suggest that the Research Hypothesis is untenable, and the conclusion would be the failure to reject the Statistical Hypothesis, which is:

> Treatment 1 is equal to Treatment 2 on the criterion, across the range of variable X1.

The restriction implied by the Statistical Hypothesis on Model 8.6 is: $a_1 = a_2$, both equal to a common weight a_0. Therefore, the Restricted Model is:

$$(\text{Model 8.7}) \; Y1 \; = \; a_0U + b_1X1 + E_7$$

Holding tenable the Statistical Hypothesis is reasonable if one is willing to consider the treatment effects averaged over the whole range of the covariate. But averaging the effects results in ignoring a very systematic state of affairs, as pictured in Figure 8.4a. The assumption of homogeneity of regression slopes is not tenable, and therefore the data should be looked at differently. Some statisticians shun the ANCOVA for the very reason just presented. The covariate, however, is not necessarily the problem. One also may fail to reject a Statistical Hypothesis if an important covariate is overlooked or, in fact, any time the Full Model does not describe the data well (i.e., does not achieve a high R^2). Indeed, no analysis—including an ANCOVA—should be made without research forethought. The reader should note that Figure 8.4a represents a state of affairs that was discussed extensively in chapter 7 as interaction, and the test for interaction is just the test one needs to check the assumption of homogeneity of regression slopes.

Testing for Nonparallel Slopes

Suppose the researcher has a new method of treatment and he wants to test the superiority of the new method over the old. Furthermore, he "knows" initial ability influences posttest performance, and he wants to ask the question: "Over and above the influence of pretest ability, is Treatment 1 superior to Treatment 2 in relation to posttest performance?"

First he must consider, "Is this really the best question?" Is there anything about Treatments 1 or 2 that could make one of the treatments superior for a certain range of the pretest scores but not for another range? If upon examination of the two treatment procedures he has a suspicion that "the new method will 'really' work better for the individuals who score high on the pretest but not so well for the individuals who score low on the pretest"; and, furthermore, if his suspicion is incorrect (when tested), he may then want to know, "Is the new treatment (Treatment 1) superior to the old treatment (Treatment 2), over and above pretest ability measures?" In this case, the researcher would have a two-stage analysis.

1. Stage 1: Test for directional interaction. If interaction is found to exist, the researcher would then plot the lines of best fit to determine the tenability of the *directional* hypothesis.
2. Stage 2: If *no* directional interaction is found, then the researcher would test the directional hypothesis that the new treatment is superior to the old treatment when pretest ability is covaried.

The sequence of hypothesis testing for the two-group case follows. The Stage 1 Research Hypothesis would be:

Treatment 1 will be increasingly more effective in producing positive posttest performance across the levels of pretest ability than will Treatment 2.

The Stage 1 Statistical Hypothesis would be:

Differential treatment effects on posttest performance will be constant across the range of pretest ability.

To test the Research Hypothesis for Stage 1, the Full Model must have two weights associated with pretest ability (one for each treatment):

$$(Model\ 8.8)\ Y1\ =\ a_1 T1 + b_1(T1 * X1) + a_2 T2 + b_2(T2 * X1) + E_8$$

where:

$Y1$ = the criterion of posttest performance;

$T1$ = 1 if the score on the criterion is from a subject given Treatment 1, 0 otherwise;

$X1$ = the pretest ability score;

$T2$ = 1 if the score on the criterion is from a subject given Treatment 2, 0 otherwise;

$(T1 * X1)$ = the pretest score if the individual is from Treatment 1, 0 otherwise;

$(T2 * X1)$ = the pretest score if the individual is from Treatment 2, 0 otherwise;

E_8 = the error vector; and

a_1, a_2, b_1 and b_2 are regression weights calculated to minimize the sum of the squared values in E_8.

To reflect the Statistical Hypothesis, the two lines of best fit must be forced to be parallel. This can be done by setting the two slopes, $b_1 = b_2$, both equal to a common weight, b_3, resulting in the Restricted Model for Stage 1:

$$(Model\ 8.9)\ Y1\ =\ a_1 T1 + a_2 T2 + b_3 X1 + E_9$$

where:

all variables are specified as for Model 8.8; and a_1, a_2, and b_3 are new regression weights calculated to minimize the sum of the squared elements in vector E_9.

If difficulty is encountered in understanding this brief section, the reader should review chapter 7 regarding interaction. The F test will compare R_f^2 and R_r^2. The degrees of freedom numerator is the number of linearly independent vectors in the Full Model (m1) minus the number of linearly independent vectors in the Restricted Model (m2). The denominator degrees of freedom is N - m1. Model 8.8 has four linearly independent vectors, and Model 8.9 has three linearly independent vectors (thus [m1 - m2] = [4 - 3] = 1).

Graphically, if the Research Hypothesis associated with Stage 1 is accepted, the plots may look like either Figure 8.5a or Figure 8.5b.

In Figure 8.5a, the lines of best fit do *not* cross in the range of interest, but an interaction exists because the higher the score on X1, the greater the difference (d) between Treatment 1 and Treatment 2 on the criterion, favoring Treatment 1 ($d_b > d_a$). Figure 8.5b shows the case where the lines of best fit cross within the range of interest. If the subject had an X1 score of better than 4, Treatment 1 yields superior criterion performance; but if an individual's X1 score is below 4, Treatment 2 yields superior criterion performance.

Stage 1 tested for directional interaction. If the researcher had wished to test for homogeneity of regression slopes, the nondirectional interaction would have been tested. Statisticians recommending such a test (of homogeneity of slopes) are doing

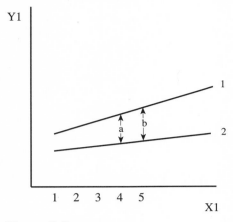

Figure 8.5a.
Data supportive of directional interaction with Treatment 1 always being superior to Treatment 2.

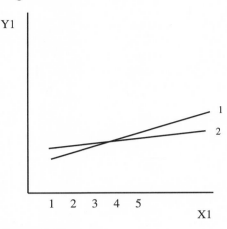

Figure 8.5b.
Data supportive of directional interaction with Treatment 1 not always being superior to Treatment 2.

so with the intent of *wanting not to find significance* and therefore hope the data support the Restricted Model over the Full Model. If this is the researcher's intent, then we suggest making sure that the Restricted Model is a preferred model by setting an alpha level such as .60 or .70. When the Research Hypothesis describes the expected state of affairs, the researcher generally sets the alpha level at somewhere around .01 or .05. The situation is conceptually inverted when the Statistical Hypothesis describes the expected state of affairs, so the appropriate alpha level would be closer to .60 or .70 (Cohen, 1970).

In Stage 1 as presented (a directional hypothesis for interaction), the researcher's expectation was that Treatment 1 would have a greater slope than Treatment 2. If the data indicated, on the other hand, that Treatment 2 has a greater slope than Treatment 1, the Research Hypothesis of Stage 1 cannot be accepted, but one would not want to act as though the slopes were homogeneous. Whenever the results turn out to be in the direction opposite to that hypothesized, the researcher must go back to the theory "drawing board" to attempt to discover the anomaly. Therefore, if the slopes were quite discrepant in the direction opposite to that hypothesized in Stage 1, the following Stage 2 analysis would not be appropriate.

If in Stage 1 it was found that there was no interaction, then Stage 2 would proceed with the following Stage 2 directional Research Hypothesis:

Treatment 1 is superior to Treatment 2 in producing positive posttest performance when the influence of pretest ability is covaried.

The Statistical Hypothesis for Stage 2 would be:

The effects of Treatments 1 and 2 on posttest performance are equal when the influence of pretest ability is covaried.

To test the Stage 2 Research Hypothesis, the Full Model must have two weights associated with treatment (a_1 and a_2) plus a common weight (b_3) for the covariate X1. Thus, the Full Model for Stage 2 is:

$$(\text{Model 8.10}) \ Y1 \ = \ a_1 T1 + a_2 T2 + b_3 X1 + E_{10}$$

The reader should note that this model is the same as the model associated with the Statistical Hypothesis in Stage 1. To reflect the Statistical Hypothesis for Stage 2, the weights for treatments must be set equal ($a_1 = a_2$), both equal to a common weight, a_3. Since T1 and T2 are mutually exclusive vectors, which when added yield the unit vector, the Restricted Model for Stage 2 can be expressed as:

$$(\text{Model 8.11}) \ Y1 \ = \ a_3 U + b_3 X1 + E_{11}$$

where:

a_3 and b_3 are regression weights calculated to minimize the sum of the squared elements in the new error vector, E_{11}.

The Full Model allows for two parallel lines, and the Restricted Model forces the two lines into one common line for both treatments. If statistical significance is obtained when Model 8.10 is compared with Model 8.11, and if a_1 is greater than a_2, then the Stage 2 Research Hypothesis is supported. The two lines of best fit will be parallel, and Treatment 1 will be more effective than Treatment 2 (a_1 - a_2 units more effective). If $a_1 = a_2$ or $a_1 < a_2$, the Research Hypothesis is untenable and Treatment 1 is not more effective than Treatment 2 when pretest scores are covaried.

The F test used to test the Stage 2 Research Hypothesis is the same generalized F test:

$$F_{(m1-m2, N-m1)} \ = \ \frac{(R_f^2 - R_r^2) / (m1 - m2)}{(1 - R_f^2) / (N - m1)}$$

R_f^2 is given by Model 8.10 and R_r^2 is provided by Model 8.11. There are three pieces of information in Model 8.10, and the sole restriction on that model results in the two pieces of information in Model 8.11; therefore, $m1 = 3$ and $m2 = 2$. That one restriction results in the one degree of freedom in the numerator.

Homogeneity of Regression Slopes with More than One Covariate

When a researcher deals with such complexities as the first covariate example in this chapter, several covariates may be used, and tests for homogeneity of regression slopes might not be made. If a researcher suspects interaction with any of the covariates, then that interaction should be tested because ignoring that interaction inflates the unknown variance. On the other hand, if a researcher has good reason not to expect interaction, parallel slopes can be assumed (or one could test for homogeneity) across the several covariates. Some group bias on the covariates of interest will be adjusted by using the over and above test. If treatment is interacting as shown in Figure 8.5a, the assumed parallel lines in the analysis of covariance will give a larger error term, which *may* or *may not* mask the superiority of the treatment. The major risk is that one may discard a new treatment that is very good beyond a certain point on the covariate but not much better (and possibly inferior) at other levels. In other situations, one would accept one treatment as better when the Statistical Hypothesis is rejected because of a difference in covariate means. The approach presented here is the best procedure at this time for dealing with groups that are not initially equivalent. The procedure is not entirely satisfactory, but it is better than assuming that the groups are equivalent.

Given the interaction illustrated in Figure 8.5b, forcing parallel lines will almost completely mask the effect that Treatment 1 is better for people who pretest beyond 3 and inferior to Treatment 2 for individuals below 3. Whether one tests for homogeneity in an ANCOVA is a judgment matter and depends upon expectations of how the treatments will function for individuals at different points across the covariate.

Application of ANCOVA to Discrimination

Williams, Williams, and Roman (1988) looked at salary differences between the sexes, adjusting for certain relevant variables (those that should be related to salary). If there is no sexual discrimination, then the gender variable will not add to the R^2, over and above those relevant variables. (Such an analysis also could test for racial or age discrimination.) But the one problem with this analysis is the choice of the covariates. The researcher wants to include all of the relevant variables but not those that are confounded with gender (such as rank if there has been previous sexual discrimination).

Kinion (1990) applied the GLM to investigate the differences between two leadership styles. She tested hypotheses about (a) the differences between the two groups; (b) the differences between the two groups, over and above the covariables of corporate office title, age, and gender; and (c) the effects of education (seven levels of education, such as elementary, secondary, two years of postsecondary) on the presence of a health-promotion program.

Differential Treatment Effects on More than One Dependent Variable

We restrict our discussion of the GLM in this text to only one dependent or criterion variable, and we do so for three reasons. First, this is an introductory text on the GLM. Second, much of the conceptual clarity provided by the GLM is lost when expanding the mathematics to more than one dependent variable (Cohen, 1990). Third, as Newman (1988) has argued, all of the analyses of multiple dependent variables are basically univariate analyses that can be accomplished with the GLM. The careful reader will note that, when multiple variables are used to predict a dichotomous criterion, one could mathematically reverse the predictors and criterion and conceptualize the model as one dichotomous predictor (treatment) predicting multiple dependent variables. Colliver, Verhulst, and Kolm (1987) illustrated how the GLM can be used to test Research Hypotheses concerning two (or more) treatments having differing effects on two (or more) dependent variables. You may recognize this as an interaction question, with treatment interacting with the "kind of outcome." They conceptualized the dependent variables as repeated measures, thus incorporating person variables (see Generalized Research Hypothesis 10.2). Because the dependent variables are different, the researchers standardized the dependent variables. Then they tested for interaction to see if the treatments resulted in differing effects on the various dependent variables.

Geometric Interpretation

The ANCOVA models can be thought of in terms of lines, as depicted in Figures 8.1 through 8.5. If there is an interaction between the groups and the covariate, as in Figure 8.4a, then a line is needed for each group (two lines in Figure 8.4a). Each line needs a Y-intercept and a slope; thus, if two lines are needed, four parameters must be determined (requiring four pieces of information in the Full Model). If there is no interaction (the lines are homogeneous) then they all have the same slope—but possibly different intercepts. This situation is depicted in Figure 8.3. The parallel-line model would be called for here. With four groups, one would need four intercepts and only one slope because the four lines are parallel.

The traditional ANCOVA Research Hypothesis compares the k-parallel-line model with the single-line model. Figure 8.1 contains two parallel lines, and Figure 8.2 contains only one solid line. The data can be modeled both ways; the question is, "Which way is best, in terms of high R^2 and few pieces of information?" As indicated in Exhibit 8.1, part 2(b), there are $k + 1$ pieces of information in the Full Model and two pieces of information in the Restricted Model, part 2(c). If there are two groups, as in Figure 8.1 and Figure 8.2, then $k = 2$, and pieces full = 3, and pieces restricted = 2.

Note in Exhibit 8.1 how the Restricted Model for the "test for homogeneity" is the Full Model for the ANCOVA test. Likewise, the Restricted Model for the ANCOVA test is the Full Model for the correlation test. This demonstrates the connection between these tests. In addition, one should begin to realize that any model can be either a Restricted Model or a Full Model, it all depends upon the Research Hypothesis. For any Full Model, one must realize that there are "fuller" models, but one assumes that the fuller models are not better models. What is usually the case is that some of these fuller models are better. If the R^2 is less than 1.00 for the Full Model, then there are better fuller models. With this Full Model, the researcher just has not obtained the necessary predictor information or has not expressed the data with the right functional relationship (the topic of the entire next chapter).

Exhibit 8.1

1. Interaction between covariate and groups: test for homogeneity
    Full Model: k-lines model
        (a) Line for each of k groups (k intercepts, k slopes)
    Restricted Model: parallel-lines model
        (b) k parallel lines (k intercepts, 1 slope)
2. Difference between groups, over and above covariate: traditional ANCOVA test
    Full Model: parallel-lines model
        (b) k parallel lines (k intercepts, 1 slope)
    Restricted Model: single-line model
        (c) one line (1 intercept, 1 slope)
3. Covariate related to criterion: test for correlation
    Full Model: single-line model
        (c) one line (1 intercept, 1 slope)
    Restricted Model: unit vector model
        (d) one horizontal line (1 intercept, no slope)

Relationships between the ANCOVA models

INTRODUCTION TO POWER

Power is the ability of the statistical analysis to find significance if in fact significance is there. There are various ways to increase power:

1. Increase the number of subjects.
2. Change alpha from, say, .05 to .10.
3. Increase the difference between the Statistical Hypothesis and reality (e.g., if $\mu =$ 50, then a Research Hypothesis of $\mu > 45$ is less powerful than a Research Hypothesis of $\mu > 40$).
4. Make the error term smaller:
    a. by using measures that are more reliable and more valid;
    b. by including individual differences if they are in the design;
    c. by blocking on known, relevant variables;
    d. by including nonlinear and interaction predictor variables if thought relevant; and
    e. by covarying initial differences between groups.

Some of these concepts have already been discussed; others will be discussed in the next chapter. The material that follows has been adapted from a paper by Newman and Benz (1980), which goes into more detail, particularly on the calculation of the required number of subjects and the combination of study results.

What follows is a discussion of how to use Cohen's (1977) power analysis tables (see Appendix G). There are four parameters that one must be aware of when conducting a power analysis:

alpha	α
sample size	N
effect size	f^2
power	power

If one knows any three of the four, one can solve for the fourth.

Alpha

Alpha (α) is the probability of making a Type I error. It is under the control of the researcher and is generally set at .05, .01, or .001.

Effect Size

Effect size can be thought of as how far apart the means of two groups are in terms of standard deviation units (e.g., 1.2 standard deviations, .50 standard deviations). Another way of looking at it is in terms of the proportion of variance accounted for. In correlational analyses, the r^2 (if using Pearson correlation)—or R^2 (if using the GLM)—would provide this information; and in ANOVA designs, the ω^2 would provide the information (ω^2 being the symbol used in ANOVA to represent the proportion of variance accounted for).

Cohen (1977) uses f^2 to represent effect size and arbitrarily defines three effect sizes: large (>.35), medium (.15 to .35), and small (<.15). Effect size in reality is set depending on how well the researcher knows her field of research and what she is looking for. Effect sizes that are considered large in one instance may be considered small in another.

Power

Power is the probability of detecting a population fact when in fact that population fact exists. Because the probability of not detecting a population fact is defined as the probability of making a Type II error, power is then one minus the probability of making a Type II error. For example, if the power of a test is .76, this means that 76 times out of 100 the statistical procedures (given the researcher's choice of N, alpha, and effect size) will be capable of detecting the population fact if it exists.

Sample Size

N is the total number of subjects used in the study.

Calculating Power

To use Appendixes G-1 through G-6, two variables need to be determined. First, the df_n needs to be computed. Remember, df_n is equal to (m1 - m2), the pieces of information in the Full Model minus the pieces of information in the Restricted Model. Cohen labels this value as U. Second, the variable L needs to be calculated. The variable L has no direct meaning but it is used as one of the entries for Appendixes G-1 through G-6.

$$L = f^2 * V$$

where:

$\quad\quad$ f^2 = effect size; and

$\quad\quad$ V = $df_d = (N - m1) = (N$ minus pieces of information in Full Model).

There is a separate table for each of the three different alpha values .01, .05, and .10. Assume we have 100 subjects ($N = 100$) and we want to be able to detect a medium-size effect ($f^2 = .15$). Assume also that we have 10 linearly independent variables (including the unit vector) in the Full Model. We are interested in asking the following question: "Do these 10 variables account for variance in the criterion, over and above no information?" Let us assume our alpha level is set at .01. We now can determine power.

$f^2 = .15$

$V = (N - m1) = (100 - 10) = 90$

$L = f^2 * V$
$L = (.15) * (90)$
$L = 13.5$

$U = (m1 - m2) = (10 - 1) = 9$

alpha = .01

Since alpha is .01, we would use Appendix G-1. We enter the table at a U of 9. We look for an L value of 13.5, which falls between L values of 12.00 and 14.00, requiring an interpolation to obtain an estimated power of 49.5.

If we are interested in doing this problem at an alpha of .05, we would use Appendix G-2. Looking at U = 9, L is between 12.00 and 14.00, and we have an estimated power of 72. We can see that as alpha becomes less stringent (.01 to .05 to .10), the power increases from 49% to 72% to 81%.

Solving for N

Given the same research question, we can determine the N size for a given effect size. Solving the previous power equation for N yields: $N = (L / f^2) + m1$. For this problem we have (a) alpha = .05; (b) m1 = 10; (c) f^2 = .02; (d) U = 9; and (e) power = .80. (Cohen recommends a power of .80 if no other information is given, which is a comparable rationale to setting alpha equal to .05.) To determine the L size for an alpha of .05, we use Appendix G-5. We enter the table for a given power and a particular U. Since power is set at .80 and U is 9, our L is 15.65. Using the above formula, we solve for N:

$$N = (15.56 / .02) + 10$$
$$N = 788$$

So, 788 subjects would be required to detect a small effect size (.02) with an alpha of .05 with this question (using 10 predictor variables). Whenever one solves for N and the value has a decimal, one always rounds upwards, so if N had been, say, 792.5, N would be equal to 793. If an alpha of .01 were desired, the number of subjects required would be 1,082.

When deciding on the desired effect size, one may adopt the large, medium, or small effect sizes arbitrarily identified by Cohen, or one may look at the effect size reported in the literature for that topic. One could randomly sample 10 articles, calculate the effect sizes, take the average, and use that as a yardstick, a starting point. However, most research contains a test of significance (such as a t test, z test, F test, or chi-square test) instead of effect size. How does one change these different tests of significance to effect sizes? The astute GLM reader is aware that these analyses could have initially been accomplished with the GLM; fortunately, formulae exist to change these reported test statistics to effect sizes.

A study reporting a t test also may report the amount of variance accounted for with a point biserial correlation coefficient; a study employing an F test may report an eta coefficient; and one employing a chi-square test may report a phi or contingency coefficient (see Table 8.1). These all represent the proportion of variance accounted for and can be represented as r_m.

All of these tests of significance can be transferred into the effect size, f^2. This is done through the intermediary components of S and Q^2. Once S and Q^2 have been computed, the value r_m can be computed as:

$$r_m = Q^2 / (Q^2 + S)$$

The square of r_m, r_m^2, is the proportion of variance accounted for. The effect size in the literature can then be calculated by the following formula:

Table 8.1
Information Needed to Calculate r_m

Statistic	Q	Q^2	S	Correlational Analysis
t test	t	NA	df	point biserial
z test	z	NA	N	point biserial
ANOVA	NA	$F * (df_n)$	df_d	eta
Chi-square	NA	χ^2	N	phi or contingency
Regression	NA	$F * (df_n)$	df_d	Multiple R

$$f^2 = \frac{r_m^2}{1 - r_m^2}$$

What is interesting is that when r_m is small, the effect size and r are almost identical. But when r_m is large, effect size approaches infinity as r_m approaches 1.

Suppose that a search of the literature produced three studies with these results:

Study #1: $t = 4$, $df = 84$;

Study #2: $F = 3$, $df_n = 2$, $df_d = 94$;

Study #3: $\chi^2 = 4$, $N = 96$.

The following shows how to convert the results from these three studies to r_m.

Study #1 (t test): t to r_m (where: $t = 4$ and $df = 84$)

$$r_m = \frac{Q^2}{Q^2 + S} = \frac{4^2}{4^2 + 84} = \frac{16}{100} = .16$$

Study #2 (ANOVA): F to r_m (where: $F = 3$, $df_n = 2$, and $df_d = 94$)

Please note that since the F in Table 8.1 is under Q^2, we do not have to square the F as we did for the t in Study #1.

$$r_m = \frac{Q^2}{Q^2 + S} = \frac{(3 * 2)}{(3 * 2) + 94} = \frac{6}{100} = .06$$

Study #3 (chi-square): χ^2 to r_m (where: $\chi^2 = 4$, and $N = 96$)

$$r_m = \frac{Q^2}{Q^2 + S} = \frac{4}{4 + 96} = \frac{4}{100} = .04$$

Each of these r_m can then be transformed to an effect size through the following formula:

$$f^2 = \frac{r_m^2}{1 - r_m^2}$$

	r_m	r_m^2	f^2
Study #1 (*t* test):	.16	.0256	.03
Study #2 (ANOVA):	.06	.0036	.004
Study #3 (chi-square):	.04	.0016	.0016

The average effect size for the above three studies could be calculated to provide guidance to the researcher on the choice of effect size from three apparently disparate types of statistical results. It should be clear, though, that the effect size for these three studies all fall within Cohen's "small" definition. Here we have evidence of what has been alluded to in previous chapters. One can obtain statistical significance, but there may not be any practical significance. Finally, one also could use the effect-size equation to calculate the r_m^2 from a published effect size:

$$f^2 = \frac{r_m^2}{1 - r_m^2}$$

$$f^2 * (1 - r_m^2) = r_m^2$$

$$f^2 - (f^2 * r_m^2) = r_m^2$$

$$f^2 = r_m^2 + (f^2 * r_m^2)$$

$$f^2 = r_m^2 * (1 + f^2)$$

$$\frac{f^2}{1 + f^2} = r_m^2$$

ANCOVA to Obtain Increased Power

Researchers are urged to use ANCOVA even when random assignment is made and even when the group means of the covariate are identical. Mueller (1990) discusses this use of ANCOVA and states concerns for using the technique when groups have not been randomly assigned. We, in this text, take the position that the statistical technique is unaware of whether the subjects have been randomly assigned. Making a covariance adjustment is better than not making one. Ultimately, research must be conducted with manipulated *and* intact groups. The ANCOVA can provide insight until that time.

In cases where pretest ability is known, the inclusion of these data in an over and above analysis will usually provide a better estimate of within-group variance ($\hat{v}_w$). Usually people who score high on the pretest will score relatively high on the posttest, and those who score low on the pretest will usually score fairly low on the posttest. The correlation between pretest and posttest is often greater than zero. Therefore, the R^2 of a model containing knowledge of both treatment and pretest scores will be larger than an R^2 of a model using only knowledge of treatment. The over and above test still tests the unique contribution of the independent variable (e.g., treatment), but the proportional estimate of the population variance within $[(1 - R_f^2)/(N - m1)]$ will be smaller at the expense of having only one degree of freedom (due to including the covariate) and at the cost of collecting the covariate score. Essentially, the reasoning is, "Why throw away knowledge regarding the sample when one has it?" If the task of the researcher is to attempt to account for as much of the variance as possible (R^2 as close to 1.0 as possible), then one might go further and recommend that the researcher include many covariates that account for nonrandom criterion variance. Chapter 1 was written with this viewpoint.

A word of caution is in order here. The ideal covariate is one that is *not correlated* with the overlap between the criterion and the predictor(s) but *is correlated* with the error in those predictions. If the covariate is correlated with the predictor part of the overlap between predictors and criterion, then the predictors' effects are confounded with the covariate. In this case, the predictor effects will be attributed to the covariate, and what might have originally been an effective predictor may now be wiped out. Such are the trials and temptations of research in the behavioral sciences.

9

NONLINEAR RELATIONSHIPS

Some researchers assume the GLM deals only with *straight* lines. In this chapter, we illustrate many nonrectilinear (also called nonlinear or curvilinear) forms that one may encounter in research and analyze with the GLM. Observed nonlinear sample data may be due to at least two conditions. First, the theoreticical expectation is reasonable regarding a nonlinear function existing among the predictor(s) and the criterion (e.g., the curvilinear relationship in Newton's law, $d = \frac{1}{2}gt^2$). Second, the scaling of the X and Y variables is arbitrary, so departures from rectilinearity may be a scaling artifact. The two major sections of this chapter deal with these two conditions.

Before entering these two sections, a discussion regarding homoscedasticity and heteroscedasticity is provided. Those readers who are tempted to skip this section should instead at least skim the material because an understanding of these ideas may be needed to communicate with statisticians. In this discussion, we treat the violations of assumptions as the basis for good information for use in prediction rather than as conditions that preclude statistical analysis.

HOMOSCEDASTICITY AND HETEROSCEDASTICITY

In chapter 2 the assumption of homogeneity of variance was discussed in relation to ANOVA. Given two samples that receive different treatments, least squares procedures assume that each sample is from a common population, and therefore these samples come from "populations" with equal variance. Violation of this assumption usually does *not* upset the inferences made when using the F distribution. When deriving a line of best fit, homogeneity of variance is reflected by equal variance about the line of best fit for each scale point on the X axis. This equal variance about the line of best fit is what is called *homoscedasticity*. The scattergram in Figure 9.1 illustrates a case in which homogeneity of variance is observed in the sample and is thus a reasonable expectation for the population. Note that at scale point 10 on the X axis the observed Y scores are distributed in about the same way as they are for observations at scale point 20 on the X axis. Given such a data plot, one can assume that homoscedasticity exists.

When scattergrams depart from the ideal and distributions are not symmetric about the hypothesized line of best fit, most statisticians will admonish the researcher

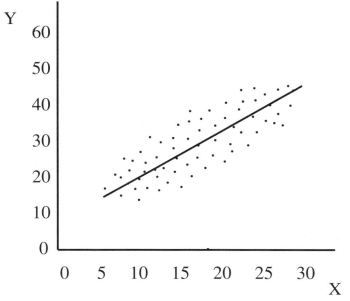

Figure 9.1.
A representation of a line in which the Y observations are homogeneously distributed about the line for all values of X.

to be cautious in interpreting that line because the observed heteroscedasticity suggests the line has different errors across scale values on the X axis. (Appendix D-9.1 shows how PROC PLOT can be used in SAS to illustrate any distribution.) Figure 9.2 illustrates three heteroscedastic cases.

If one views the scattergrams in Figure 9.2 from a straight-line point of view, one may be upset because an assumption is violated, and the R^2 is small due to heteroscedasticity. From the position of a researcher, systematic departures around the line of best fit as illustrated ought to be seen as a starting point for inquiry—it is very likely that in all three cases the straight-line model constructed to fit the data is not appropriate. Given the systematic departures from homoscedasticity in Figure 9.2, one might suspect that a theoretically unexpected interaction (see chapter 7) between the X variable and another unknown variable(s) yields the odd distributions.

For the sake of simplicity, suppose that all three cases illustrated are due to what is called a *treatment-by-aptitude interaction* and that the researcher was unaware of the "treatment." Consider Figure 9.2a. If one were interested in the relationship of ability (X) to criterion performance (Y), one might ask, "Why is there so much variability at the extremes of the ability continuum?" In the hypothetical case under consideration, the sample may be drawn from two different classes of students with two teachers. Can the results be due to teacher differences? For example, perhaps Teacher A is great for students who are below average, and Teacher B is great for students who are above average. If one ignored teacher effects, the line in Figure 9.2a would be the outcome. If one expanded the simple linear equation ($Y = aU + bX + E_1$) to include two lines, one for students of Teacher A and one for students of Teacher B, Figure 9.3 might be the outcome state. As depicted in Figure 9.3, the heteroscedasticity is removed because the systematic teacher effects are accounted for, and the R^2 will be dramatically increased from zero to an R^2 substantially greater than zero.

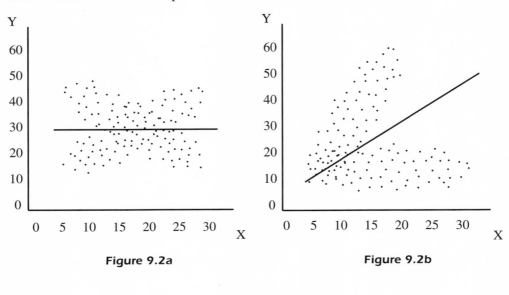

Figure 9.2a   **Figure 9.2b**

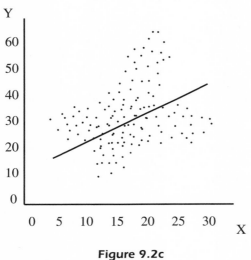

Figure 9.2c

Figure 9.2.
Three hypothetical cases of heteroscedasticity: Figure 9.2a, in which $R^2 = 0$; and Figures 9.2b and 9.2c, in which $R^2 > 0$ but with different forms.

The apparent heteroscedasticity in Figure 9.2b might be due to an ordinal interaction (two or more lines that are *not* parallel, that do *not* cross in the range of interest). Figure 9.2c may be due to one treatment that yields strong predictability of Y from X and another treatment that yields little relationship between X and Y. Figure 9.4 illustrates these two conditions that might account for the observed heteroscedasticity in Figures 9.2b and 9.2c.

So far this discussion of heteroscedasticity has dealt with straight lines of best fit. Consider the conditions observed in Figures 9.5 and 9.6. Figures 9.5a and 9.6a suggest that unequal variability across X might be due to a nonrectilinear relation-

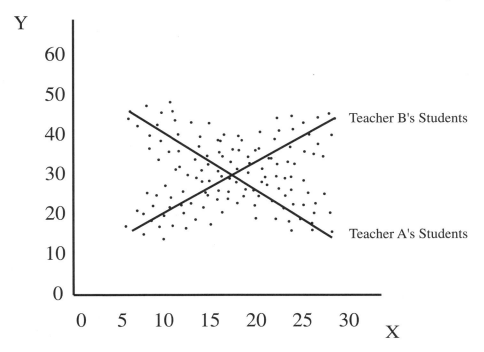

Figure 9.3.
Data from Figure 9.2a, in which the observed variability in extremes is explained by teacher effects. The ellipses about the two lines reflect homoscedastic data points.

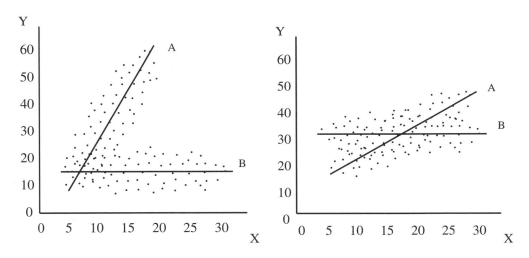

Figure 9.4a.
Representation of Figure 9.2b where the variability at the higher end of the X axis is revealed to be due to a systematic effect, in which Treatment A yields high scores at the upper end of X, whereas Treatment B yields very little effect on criterion performance across the total range of X.

Figure 9.4b.
An interaction in which Treatment B has no differential effect across X, but Treatment A yields a very strong relationship between X and Y.

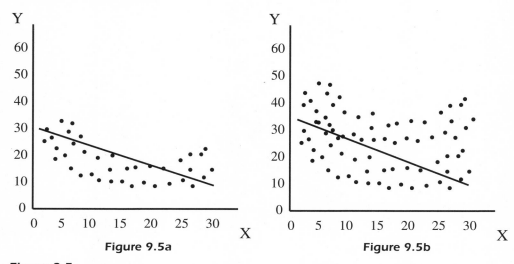

Figure 9.5a

Figure 9.5b

Figure 9.5.
Scattergrams 9.5a and 9.5b illustrate heteroscedasticity that suggests that a straight line is not the best functional relationship between X and Y

ship between X and Y. Figures 9.5b and 9.6b illustrate a situation in which Treatments A and B have two different nonrectilinear relationships between X and Y. In both instances, the apparent heteroscedasticity is due to the nonlinear relationship between X and Y. Whether the curves represented in Figure 9.6 are due to a theoretical expectation or are due to poor scaling is undetermined in this discussion. The rest of this chapter deals with these two notions.

In summary, homoscedasticity is statistically desirable because the R^2 is enhanced. Furthermore, the homoscedasticity assumption implicitly assumes that the proper form of regression equation has been fitted. When the equal variability about the line of best fit is lacking, the best research strategy may be to search for unknown variables that may account for the variability and then include those variables in the analysis. Ideally, if one has all the relevant predictor variables, there will be no variability about the line of best fit, an R^2 of 1.0 will be observed, and this whole section would be superfluous.

FITTING EXPECTED NONLINEAR FUNCTIONAL RELATIONSHIPS

The Second-Degree Relationship

In many psychomotor skill-learning conditions, the theory postulates a positively accelerating curvilinear relationship. For example, for computer keyboard data entry of records, hours of practice and data-entry production (measured in punchcards completed) might take a form such that little gain in data-entry production is observed for the first few hours of practice, and then a spurt is observed. Figure 9.7 represents this expectation; the solid line represents the curvilinear fit. As shown, the straight line may do a fairly decent job of representing the curved data; but if the curved line is the best fit, then at the extremes (1-2 and 9-10 hours of practice) errors

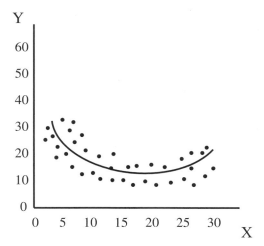

Figure 9.6a.
The removal of unequal variability about the line of best fit given in Figure 9.5a by fitting a curved line to the data.

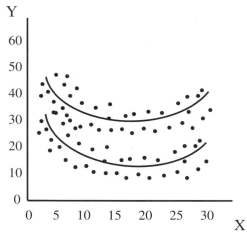

Figure 9.6b.
The removal of unequal variability by fitting two curved lines.

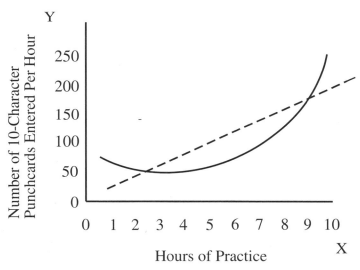

Figure 9.7.
The line of best fit that represents a curved (solid line) and a rectilinear (dashed line) fit to the data.

of underprediction with the straight-line model will be observed (the solid line is above the dashed), and the middle hours of practice will yield errors of overprediction (the dashed line is above the line of best fit).

A linear regression equation that includes a second-order polynomial of the hours of practice vector X will fit the solid (curved) line in Figure 9.7 fairly well. The linear model is:

$$\text{(Model 9.1)} \quad Y = aU + bX + sX^2 + E_l$$

where:

    Y = number of 10-character punchcards entered in an hour;

    X = number of hours of practice; and

    X^2 = the square of the elements in vector X.

    [Note: In this and succeeding models, the unit vector (U), the weighting coefficients (here, a, b, and s) and the error vector (here, E_1) will not generally be defined.]

And the vectors, representing the second-degree curvilinear model, would be:

Person	Y	=	aU	+	bX	+	sX^2	+	E
1	50		1		1		1		?
2	51		1		2		4		?
3	75	= a	1	+ b	5	+ s	25	+	?
4	170		1		9		81		?
.	.		1		.		.		?
.	.		1		.		.		?
.	.		1		.		.		?
60	250		1		10		100		?

Note that the first score represented in Y is 50 data cards entered in an hour with one hour's practice ($X = 1$; $X^2 = 1$). The 60th score in Y is 250 and $X = 10$; thus, 250 punchcards were entered in an hour for the person with 10 hours' practice ($X^2 = 10 * 10 = 100$).

Suppose a researcher had available 60 naive students in a data-entry class and wanted to determine if the surge in performance represented a likely nonchance event. One could monitor each student's performance after each hour of practice and then use a repeated-measures test for curvilinearity. To remain conceptually simple, the present example will not include the repeated-measures design (chapter 10 discusses the repeated-measures design).

To test the expected second-degree curvilinear relationship, the researcher may randomly assign six students to each of the 10 practice conditions (6 for one hour, 6 for two hours, etc.). The researcher would then give the practice and record the number of records entered for each student. The Research Hypothesis may be:

For the specified population, there is a positive second-degree curvilinear relationship between hours of practice and number of 10-character data cards entered, over and above a positive rectilinear relationship.

The example Statistical Hypothesis is:

For the specified population, there is *no* positive second-degree curvilinear relationship between hours of practice and number of 10-character data cards entered, over and above a positive rectilinear relationship.

Model 9.1 (above) is the appropriate Full Model, and the vectors are as described previously. The Research Hypothesis states $s > 0$ and the Statistical Hypothesis sets $s = 0$; therefore, the Restricted Model is:

$$(\text{Model } 9.2) \ Y \ = \ aU + bX + E_2$$

There are three linearly independent vectors in the Full Model and two in the Restricted Model. Suppose $R_f^2 = .75$ and $R_r^2 = .60$; the same general F ratio would be used to test the Research Hypothesis:

$$F_{(m1 - m2, N - m1)} \ = \ \frac{(R_f^2 - R_r^2) \ / \ (m1 - m2)}{(1 - R_f^2) \ / \ (N - m1)}$$

$$F_{(1,57)} \ = \ \frac{(.75 - .60) \ / \ (3 - 2)}{(1 - .75) \ / \ (60 - 3)} \ = \ \frac{.15/1}{.25/57} \ = \ \frac{.15}{.0044} \ = 34.1$$

An F of 7.12 with 1 and 57 degrees of freedom, when the weights are in the hypothesized direction, is observed fewer than 5 times in 1,000 ($p < .005$). If the researcher had set alpha at .01 and the weights b and s are positive, then the Statistical Hypothesis must be rejected and the Research Hypothesis accepted. The weights need to be positive because the researcher expects a *positive* second-degree curvilinear relationship, over and above a *positive* linear relationship. Indeed, if s were negative, the curve would go up and then descend, an inverted U that would be upsetting to this researcher. Continued practice would yield less production beyond a certain point. Such a finding would go contrary to all expectations regarding skill learning and if observed should call for an immediate replication so that a strong empirical base can be referred to.

The reader may note that 10 groups were formed (one for each of the whole hours of practice), yet these 10 groups were cast into one continuous vector with values 1 through 10. One would expect that 5.5 hours of practice would yield production somewhere between the production observed by the 5- and 6-hour practice groups and so on for all half-hour periods; therefore, it is reasonable to consider practice to be represented on a continuum.

The sign of the linear component does not have to be hypothesized as Applied Research Hypothesis 9.1 illustrates. The second-degree component in Applied Research Hypothesis 9.1 only accounts for an additional 6% of the criterion variance, but that 6% is highly significant because only 8% was left unaccounted for by the rectilinear fit. (See Appendix D-9.1 for SAS setup.)

Inverted U-Shaped Curve: The Expected Relationship

In many areas of research, the relationship between X and Y is expected to be a specific curvilinear relationship. For example, arousal (central neural excitement) has been found to yield an inverted U-shaped relationship to cognitive performance on moderately difficult tasks. That is, those who have either low or high arousal levels score poorly, and those who are moderately aroused produce the highest responses. Figure 9.8 represents the inverted U relationship in which individuals who score 1 or 10 on arousal (X) score 5 on a given cognitive performance task. Subjects with moderate arousal levels (5 and 6) score above 20 on the cognitive performance

Applied Research Hypothesis 9.1

Directional Research Hypothesis: For a given population, there is a positive second-degree effect of X3 on X2, over and above the linear effect of X3.

Statistical Hypothesis: For a given population, there is <u>not</u> a positive second-degree effect of X3 on X2, over and above the linear effect of X3.

Full Model: $X2 = a_0U + a_3X3 + a_{16}X16 + E_1$ (pieces full = m1 = 3)

where:
$X16 = X3 * X3.$

Want: $a_{16} > 0$; restriction: $a_{16} = 0$

Restricted Model: $X2 = a_0U + a_3X3 + E_2$ (pieces restricted = m2 = 2)

alpha = .01
$R_f^2 = .98$; $R_r^2 = .92$
$F_{(1,57)} = 159$
Computer probability = .0001; directional probability = .00005

Interpretation: Since the weighting coefficient for the second-degree component (X16) is positive and $p <$ alpha, the Research Hypothesis is tenable: "For a given population, there is a positive second-degree effect of X3 on X2, over and above the linear effect of X3."

task. The theory claims that low-aroused and high-aroused subjects do poorly on the cognitive performance task for different reasons. The person who is barely aroused has insufficient excitation for focusing on the task. On the other hand, the highly aroused individual has such a high level of excitation that trivial stimuli distract the person from the primary cognitive task. Moderate levels provide sufficient central neural excitation to focus on the task but not so much that it yields distraction.

Given such theoretical expectations, in an experimental situation the researcher might hypothesize:

There is an inverted U relationship between measured arousal level and performance on a cognitive task of moderate difficulty for a specific population of humans.

The researcher might then select a sample of subjects who represent a specific population (e.g., 16-year-old boys in high school) and provide the moderately difficult task while monitoring the individuals' arousal levels. The research expectation might provide the following Research Hypothesis:

Among 16-year-old high school boys, there is an inverted U relationship between measured arousal level and cognitive performance on a task of moderate difficulty.

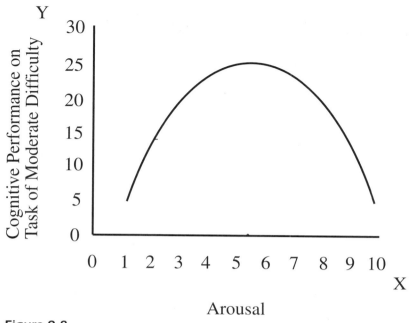

Figure 9.8.

The expected inverted U relationship between arousal and cognitive performance.

The inverted U relationship is a very specific expectation. To obtain the inverted U in the upper right-hand quadrant as in Figure 9.9a, the weight associated with the arousal vector must be positive, and the weight associated with the vector containing squared arousal scores must be negative. The Y-intercept, a, may be positive, negative, or zero. A plot of the data must be made to figure out if the change in direction is within the range of observed X scores.

The following vector display provides an example to show the weights for an inverted U relationship with a perfect fit of the data:

Subject	Y	=	aU	+	bX	+	sX²	+	E
1	0		1		0		0		0
2	4		1		1		1		0
3	6	= 0	1	+ 5	2	+ (-1)	4	+	0
4	6		1		3		9		0
5	4		1		4		16		0
6	0		1		5		25		0

If the three elements for subject 1 in vectors U, X, and X^2 are multiplied by the appropriate weights, the predicted score is zero and the observed score is zero. If the same procedure is used for all six subjects, the predicted score goes up and then down as X increases. Inspection of the values for Y in the vector display confirms that the change in direction is within the range of observed X scores.

The GLM allowing for a second-degree relationship is:

$$(\text{Model } 9.3)\ Y = aU + bX + sX^2 + E_3$$

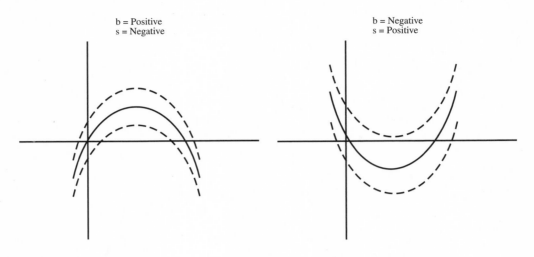

b = Positive
s = Negative

b = Negative
s = Positive

Figure 9.9a

Figure 9.9b

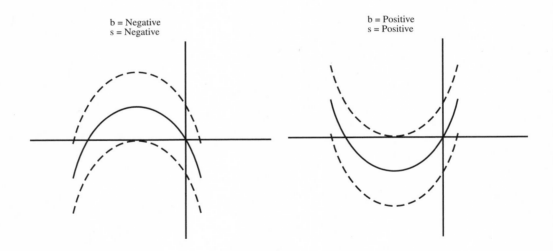

b = Negative
s = Negative

b = Positive
s = Positive

Figure 9.9c

Figure 9.9d

Figure 9.9.
Changes in the second-degree curve as weights for a, b, and s change sign (the solid line represents a = 0, while the upper dashed line represents a › 0, and the lower dashed line represents a ‹ 0).

Figures 9.9a, 9.9b, 9.9c, and 9.9d show several second-degree relationships that may be fit with the second-degree model. A few general observations can be made from these curves:

1. The numerical value of a can be found where the curve passes through the Y axis at X = 0.
2. The numerical value of b is the slope of the curve at X = 0. Notice that the curves in Figures 9.9a and 9.9d have a positive slope when crossing the Y axis.
3. When s is negative, the open end of the curve is down; when s is positive, the open end is up.
4. When s and b are of the same sign (both positive or both negative) the maximum (or minimum) is at a negative value on the X axis. When s and b are of opposite signs (one negative and one positive) the maximum (or minimum) is at a positive value on the Y axis.

Figure 9.9a represents an inverted U outcome state. If the values for X range from zero to the right-hand side of the figure, then the inverted U relationship is obtained. However, if the X values range from zero to a small value of X, what can one say? Figure 9.10 is a representation of Figure 9.9a over the range of scores from zero to one such small value, 4, where a positive b and a negative s does not yield an inverted U. If one observed such a plot in the arousal study, one might conclude that the population the sample represents does *not* have individuals with high levels of arousal. Indeed, if a larger population is defined, one that included arousal levels up to the right-hand scale value as in Figure 9.9a, one might expect to find the inverted U relationship holding.

All curves in Figure 9.9 show all four quadrants. In most research in the behav-

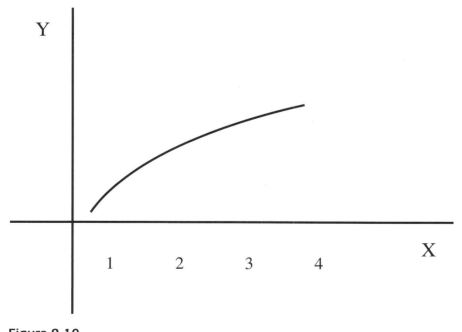

Figure 9.10.

Data from Figure 9.9a with the range of observed scores limited from 0 to 4; the second-degree curve with weights a and b both equal to a positive value, and s is a negative value.

ioral sciences, X and Y values are positive; therefore, most researchers would be concerned solely with the upper right-hand quadrant.

Testing the Inverted U

To test the Research Hypothesis, "Among 16-year-old high school boys, there is an inverted U relationship between measured arousal level and cognitive performance on a task of moderate difficulty," three conditions must be met to yield an affirmative answer.

The linear equation that allows for an inverted U relationship is:

$$(\text{Model } 9.4)\ Y = aU + bX + sX^2 + E_4$$

where:

   Y = cognitive performance on the task of moderate difficulty;
   X = the continuous predictor variable (arousal level); and
   X^2 = a vector containing the square of the elements in vector X.

Condition 1: The weight (b) associated with the linear component (X) must be positive.
Condition 2: The weight (s) associated with the second-degree polynomial (X^2) must be negative.
Condition 3: If Conditions 1 and 2 are met, then a plot of the line of best fit must show that those individuals with the lowest and highest scores on the X variable have somewhat the same (minimum) score on Y, and the maximum score on Y is approximately midway between the lowest and highest scores on X. To meet Conditions 1 and 2, two Research Hypotheses and two Statistical Hypotheses must be cast. The directional Research Hypothesis for testing Condition 1 is:

> Among 16-year-old high school boys, there is a positive linear relationship between measured arousal level and performance on a cognitive task of moderate difficulty, over and above the second-degree polynomial of arousal and the regression constant.

Given the Full Model (Model 9.4), the Research Hypothesis implies that $b > 0$. The associated Statistical Hypothesis is:

> Among 16-year-old high school boys, there is *no* linear relationship between measured arousal level and cognitive performance on a task of moderate difficulty, over and above the second-degree polynomial of arousal and the regression constant.

The Statistical Hypothesis implies that $b = 0$, resulting in the following Restricted Model:

$$(\text{Model } 9.5)\ Y = aU + sX^2 + E_5$$

If $b > 0$ and the directional probability is less than the stated alpha, the first condition is met. If not, then the inverted U expectation is untenable.

Given an affirmative answer to Condition 1, then Condition 2 must be verified. The directional Research Hypothesis for testing Condition 2 is:

> Among 16-year-old high school boys, there is a negative relationship between the second-degree polynomial of arousal and cognitive performance, over and above the positive linear component of arousal and the regression constant. (This Research Hypothesis states s < 0.)

The Statistical Hypothesis is:

> There is *no* relationship between the second-degree polynomial of arousal and cognitive performance, over and above the positive linear component of arousal and the regression constant.

The Full Model is the same as Model 9.4, and the restriction implied by the Statistical Hypothesis is s = 0, resulting in the following Restricted Model:

$$\text{(Model 9.6) } Y = aU + bX + E_6$$

If the weight s is negative and the directional probability is less than the stated alpha, then Condition 2 is met. A plot of the line of best fit could yield many relationships, but all will be some form of the three lines depicted in Figure 9.11.

Figure 9.11a shows a case in which the inverted U relationship exists, but the upper end of the arousal continuum is not present in the population. The dashed line as extrapolated, does yield the inverted U. Given such a finding, the researcher may wish to seek a population that includes the higher end of the arousal scale; or the researcher may conclude that for the specified population an inverted J relationship exists (no highly aroused subjects are observed in the population of 16-year-old high school boys).

Figure 9.11b is the same as Figure 9.11a, except that no lowly aroused 16-year-old high school boys are observed. Figure 9.11c represents the idealized inverted U relationship. Given Figure 9.11c, all three conditions are met, and one may conclude, "Among 16-year-old high school boys, an inverted U relationship exists between measured arousal level and cognitive performance on a moderately difficult task."

Note of caution! Testing for the signs of the regression weights must be done within the context of making one restriction at a time upon the same Full Model (Model 9.4 in this example) because the sign of each weighting coefficient was posited within the context of the other variables in the Full Model. An inverted U requires both a linear component and a second-degree component. (Whether a nonzero Y-intercept is necessary depends on the Research Hypothesis and was of no interest in this application.)

In the case just presented, if the sign for b is tested using a Full Model that does not contain X^2 ($Y = aU + bX + E_2$), then setting b = 0 would yield a nonsignificant F value because there is no linear trend independent of the second-degree polynomial. (See the dashed line in Figure 9.11c.)

Similarly, the second-degree polynomial must be tested as a restriction on the Full Model containing the linear component X, because a restriction of s = 0 on a model not containing X ($Y = aU + sX^2 + E_3$) might not be significant if a perfect inverted U relationship exists. Figure 9.12 shows two possible lines using $Y = aU + sX^2 + E_3$. When b is *not* included in the Full Model, the curve is symmetrical about

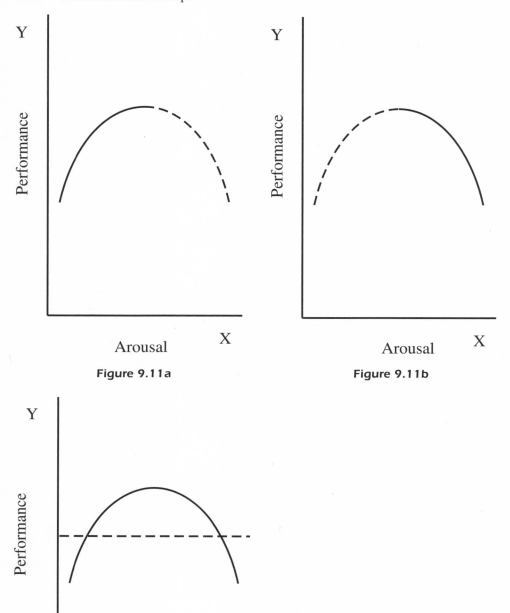

Figure 9.11a

Figure 9.11b

Figure 9.11c

Figure 9.11.
Three possible outcome states from the second-degree model: $Y = aU + bX + sX^2 + E_1$, when b is positive and s is negative.

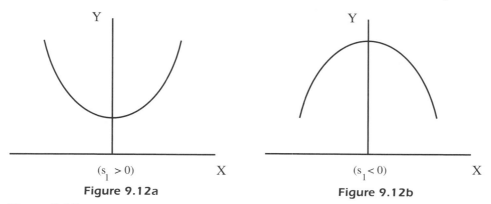

Figure 9.12a   Figure 9.12b

Figure 9.12.

Curves must be symmetric about the Y axis if the linear component is not in the model.

the Y axis and thus will not make a good fit when the maximum point departs mark-edly from the Y axis (e.g., see Figure 9.11c). Applied Research Hypothesis 9.2 pro-vides an application of these notions on the data in Appendix A.

If more than one continuous predictor variable is used in the Full Model, one should *not* interpret the magnitude of the regression weight because small variations from sample to sample may change these numerical values greatly, though the signs should be stable. This is because continuous vectors tend to be correlated, and small sample variations may yield large changes in the weighting coefficients. (The weights could be used in a predictive situation as the predicted criterion would be approxi-mately the same from one set of weights to another.) The magnitude of the R^2 will re-main somewhat constant, however. McNeil and Spaner (1971) present a more generally applicable argument for including correlated predictor variables in regression models.

Third-Degree Polynomial Line Fitting

Curved lines of best fit may take many forms, and the second-degree polyno-mial just presented is only one. In many psychomotor studies, an S-shaped curve is expected. For example, consider the problem in this chapter regarding the relation-ship between data-entry production and hours of practice. After several hours of practice, a surge in production might be observed. Given further practice, improve-ment in production will eventually level off due to the limits of the method, the person entering the data, and the data-entry machine. Figure 9.14 illustrates a pos-sible outcome state.

To fit the data for the first 10 hours of practice, a simple second-degree poly-nomial is adequate ($Y = aU + bP + sP^2 + E_1$). Beyond 10 hours of practice, however, the curve is expected to change direction. To investigate such a situation, the re-searcher who conducted the previous study could use those data and could select 50 more subjects, randomly assign these to five practice groups (hours of practice 11, 12, 13, 14, and 15), and test the Research Hypothesis, "For naive data-entry stu-dents, the number of 10-character cards entered per hour slowly increases, spurts, and then levels off with increased hours of practice."

A linear regression model that includes the regression constant (U), hours of practice (P), hours of practice squared (P^2), and hours of practice cubed (P^3) will allow the line of best fit to reflect the hypothesized S-shaped curve:

$$\text{(Model 9.7) } Y = aU + bP + sP^2 + tP^3 + E_7$$

Applied Research Hypothesis 9.2

General Hypothesis: For a given population, there is an inverted U relationship between X5 and the criterion X2.

Condition 1 Research Hypothesis: For a given population, there is a positive linear relationship between X5 and X2, over and above the predictability of the second-degree polynomial of X5 and the regression constant.

Condition 1 Statistical Hypothesis: For a given population, there is no linear relationship between X5 and X2, over and above the predictability of the second-degree polynomial of X5 and the regression constant.

Full Model for Condition 1: $X2 = a_0U + a_5X5 + a_{15}X15 + E_1$ (m1 = 3)

where:

$X15 = X5 * X5.$

Want: $a_5 > 0$; restriction: $a_5 = 0$

Restricted Model for Condition 1: $X2 = a_0U + a_{15}X15 + E_2$ (m2 = 2)

alpha = .01
$R_f^2 = .95$; $R_r^2 = .38$
$F_{(1,57)} = 755$
Computer probability = .0001; directional probability = .00005

Interpretation: Reject Condition 1 Statistical Hypothesis and accept Condition 1 Research Hypothesis.

Condition 2 Research Hypothesis: For a given population, there is a negative relationship between the second-degree polynomial of X5 and X2, over and above the predictability of X5 and the regression constant.

Condition 2 Statistical Hypothesis: For a given population, there is no relationship between the second-degree polynomial of X5 and X2, over and above the predictability of X5 and the regression constant.

Full Model for Condition 2: $X2 = a_0U + a_5X5 + a_{15}X15 + E_1$ (m1 = 3)

Want: $a_{15} < 0$; restriction: $a_{15} = 0$

Restricted Model for Condition 2: $X2 = a_0U + a_5X5 + E_3$ (m2 = 2)

alpha = .01
$R_f^2 = .95$; $R_r^2 = .55$
$F_{(1,57)} = 535$
Computer probability = .0001; directional probability = .00005

Interpretation: Reject Condition 2 Statistical Hypothesis and accept Condition 2 Research Hypothesis. Based upon the plot of X5 against X2, pictured in Figure 9.13, the data reflect a J-shaped curve rather than a U-shaped curve. There is clearly a maximum, but the line of best fit did not continue downward enough to justify re-ferring to this curve as U shaped. Therefore the General Hypothesis cannot be accepted.

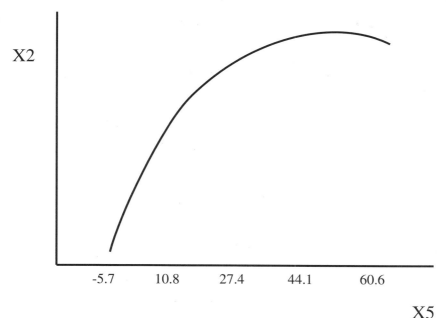

Figure 9.13.
The obtained (J-shaped) relationship between X5 and X2.

where:

Y = number of 10-character cards entered per hour;
P = hours practiced;
P2 = squared elements of P; and
P3 = cubed elements of P.

For the particular S-shaped curve in Figure 9.14 to be obtained, t must be negative (t < 0) and s positive (s > 0) because the criterion scores are lower for large values of P. Hence the highest-powered term, P^3, must have a negative weight. To get the inflection point, the second-degree term must be opposite to that of the third-degree term. If the curve is horizontal at the Y axis, then the linear weight (b) will be zero. If the curve is slanted downward from left to right, b will be negative, and if slanted upward, b will be positive. The sign and magnitude of the linear term and the Y-intercept are not relevant to the determination of the S-shaped curve (analogous to the sign and magnitude of the intercept being irrelevant to whether there was a second-degree term, as in Figure 9.9). One may test each of the weights by eliminating (restricting out) one weight at a time from the Full Model. If each of the two variables adds to the predictability and each of the weights has the appropriate sign, then the Full Model is the preferred model to fit the data. If t = 0, then the expected leveling has not yet occurred, and the researcher might wish to extend the study to figure out when practice should stop.

When dealing with higher-order relationships, the signs of the weights become difficult to specify without resort to analytic geometry. If one has a mastery of analytic geometry, one may be able to specify the expected signs of weights and go about testing expectations. Without such training, the researcher may use empirical procedures to develop models and test expectations. Indeed, it can be difficult to

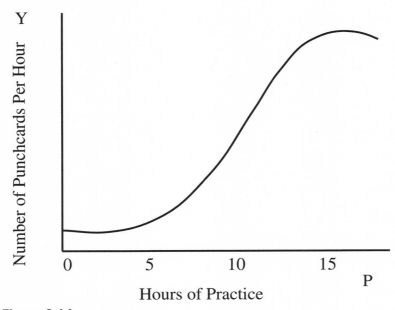

Figure 9.14.
An S-shaped curve resulting from the functional relationship between hours of practice and data-entry production per hour.

ascertain the curve of best fit. Finding the *best* curve usually requires a data-snooping exercise. Curve fitting is complicated even more by the prospect that powers other than whole numbers might be operating. Suppose, for instance, that the exponent is unknown in Equation 9.1:

$$Y = aX^m \qquad (9.1)$$

Taking the logarithm of both sides of Equation 9.1 results in:

$$\log_e Y = \log_e a + m(\log_e X) \qquad (9.2)$$

If $\log_e X$ and the unit vector are used to predict $\log_e Y$, then the following model results:

$$(\text{Model } 9.8) \quad \log_e Y = aU + b\log_e X + E_8$$

The unknowns in Equation 9.1 thus can be found empirically by Model 9.8; b is the numerical value of the exponent m, and a is $\log_e a$.

Many researchers have been mistakenly given some misconceptions regarding curve fitting. First, not all the lower powers need to be included. That is to say, the data may be well fit by the second-degree term only.

Second, the lower-powered terms are not the "simplest" variables. Complexity should be viewed as the number of predictor variables needed to account satisfactorily for criterion variance (McNeil & McShane, 1974). Researchers have for too long first investigated linear relationships because they thought that those were the simplest.

Third, the functional fit over the range of sample values is applicable to that range. Generalizations beyond that range are good guesses, but they are based on the assumption that the same functional relationship occurs as in the range originally studied (an empirical question, which additional data gathering can answer). The third-degree function in Figure 9.14 is essentially linear from 7 to 12 hours of practice. And it is a second-degree function from 0 to 10 hours of practice.

Hypothesis Seeking Through the Descriptive Use of Multiple Regression

As stated in chapter 1, the task of the researcher is to build theoretical models and cast linear equations that will encompass the variables and functional relationships that account for a large proportion of observed criterion variance. Ultimately one hopes to approximate an R^2 of 1.00. In the quest to increase predictability, one may use linear equations descriptively to find the most parsimonious predictive model. Using the data presented in Figure 9.14, one could develop a regression model with 15 linearly independent vectors:

$$(\text{Model } 9.9)\, Y \;=\; a_0U + a_1X1 + a_2X2 + \ldots + a_{15}X15 + E_9$$

where:

    Y = the criterion;
    $X1$ = 1 if the criterion is from a person with one hour of practice, 0 otherwise;
    $X2$ = 1 if the criterion is from a person with two hours of practice, 0 otherwise;
    .
    .
    .
    $X15$ = 1 if the criterion is from a person with 15 hours of practice, 0 otherwise.

Model 9.9 will be the best fitting model (in terms of R^2) for any set of data points because each X-axis value is allowed to have its own mean. The model yields maximum curvilinearity and is referred to as the *eta coefficient model*. Though the eta coefficient model yields maximum curvilinearity, it has several drawbacks. First, there are many weighting coefficients to be calculated, and therefore the likelihood of being able to replicate those weights on successive samples is low. Second, the eta coefficient model does not allow for inferences about Y values for values between the specified X-axis values, whereas using continuous predictor variables does allow for such generalizations (McNeil, 1970b).

If the S-shaped curve is the relationship the data take, the third-degree model (Model 9.7) will account for almost as much criterion variance as the eta coefficient model, but Model 9.7 will do so in a more parsimonious fashion because only four pieces of information are needed, whereas 15 are needed in the eta coefficient model. Furthermore, the elimination of a or b may yield as predictive a model as Model 9.7 and will be a more parsimonious model, containing only three pieces of information:

$$(\text{Model } 9.10)\, Y \;=\; bP + sP^2 + tP^3 + E_{10}$$

where the restriction a = 0 is placed on Model 9.7.

$$(\text{Model } 9.11)\, Y \;=\; aU + sP^2 + tP^3 + E_{11}$$

where the restriction b = 0 is placed on Model 9.7.

Through this process, the model one finds to be the most predictive yet parsimonious model (as determined by the researcher) should be used as the basis for a cross-validation study (see chapter 11). While it is the best model for the sample data, it may not turn out to be generalizable to the population. It has a lowered likelihood of being generalizable because it was not hypothesized beforehand, and it was discovered by searching through the sample data.

In a case where a researcher has knowledge of gender, treatment, and some continuous covariate—but little expectation about how these variables interrelate in the prediction of criterion performance—the following procedure may prove valuable in formulating a hypothesis that the researcher can then test in a future study. Suppose there are three treatment conditions, two sexes, and an ability measure, plus a criterion measure. One may cast bivariate plots between the ability measure and the criterion for each of the possible groups (males in Treatment 1; males in Treatment 2; males in Treatment 3; females in Treatment 1; females in Treatment 2; females in Treatment 3). The plots might look like Figures 9.15a through 9.15f.

Each of the lines of best fit is the most parsimonious when considering only that combination of sex and treatment; however, an inspection of the figures might lead to a more parsimonious model for the entire sample. There are 13 linearly independent vectors that describe the six lines (note that for Figure 9.15e a second-degree polynomial is included, but the Y-intercept is zero, thus $a_5 = 0$; and Figure 9.15f has three linearly independent vectors since $a_6 \neq 0$). The males and females who had Treatment 1 look like they have the same slope, but they have different Y-intercepts. Thus $b_1 = b_2$ seems like a reasonable restriction. Furthermore, the curves for males and for females who had Treatment 3 seem the same, but $a_5 = 0$ for males and $a_6 > 0$ for females; thus $b_5 = b_6$ and $s_5 = s_6$ seem like reasonable restrictions.

Model 9.12 is one possible model reflecting all six lines in one linear equation. It was derived from an inspection of the data rather than from a prior Research Hypothesis. The expected equalities can be verified by imposing the appropriate restrictions on Model 9.12 and then comparing the resulting R^2 values for each model.

(Model 9.12) $Y = a_1 U1 + b_1 X1 + a_2 U2 + b_2 X2 + a_3 U3 + b_3 X3 + a_4 U4 + b_4 X4 + b_5 X5 + s_5 X5^2 + a_6 U6 + b_6 X6 + s_6 X6^2 + E_{12}$

where:

   Y   = the criterion;
   U1 = 1 if Y comes from a male in Treatment 1, 0 otherwise;
   X1 = ability if Y comes from a male in Treatment 1, 0 otherwise;
   U2 = 1 if Y comes from a female in Treatment 1, 0 otherwise;
   X2 = ability if Y comes from a female in Treatment 1, 0 otherwise;
   U3 = 1 if Y comes from a male in Treatment 2, 0 otherwise;
   X3 = ability if Y comes from a male in Treatment 2, 0 otherwise;
   U4 = 1 if Y comes from a female in Treatment 2, 0 otherwise;
   X4 = ability if Y comes from a female in Treatment 2, 0 otherwise;
   X5 = ability if Y comes from a male in Treatment 3, 0 otherwise;
   X5² = ability squared if Y comes from a male in Treatment 3, 0 otherwise;
   U6 = 1 if Y comes from a female in Treatment 3, 0 otherwise;

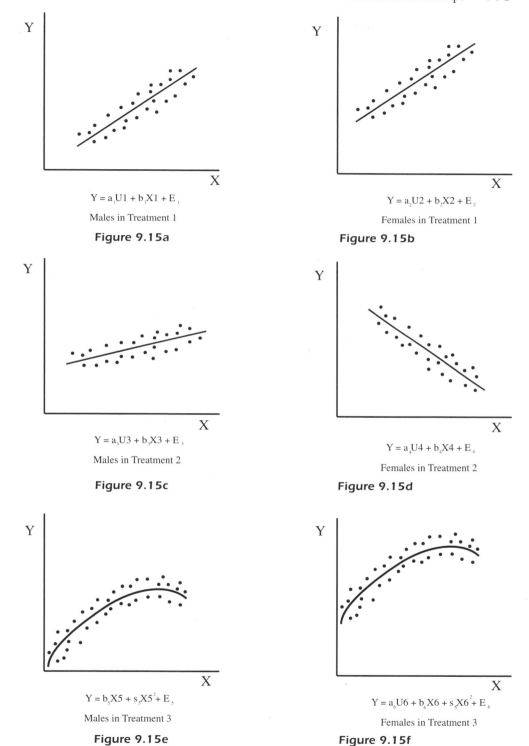

$Y = a_1U1 + b_1X1 + E_1$

Males in Treatment 1

Figure 9.15a

$Y = a_2U2 + b_2X2 + E_2$

Females in Treatment 1

Figure 9.15b

$Y = a_3U3 + b_3X3 + E_3$

Males in Treatment 2

Figure 9.15c

$Y = a_4U4 + b_4X4 + E_4$

Females in Treatment 2

Figure 9.15d

$Y = b_5X5 + s_5X5^2 + E_5$

Males in Treatment 3

Figure 9.15e

$Y = a_6U6 + b_6X6 + s_6X6^2 + E_6$

Females in Treatment 3

Figure 9.15f

Figure 9.15.
Six groups and six lines that might reflect the lines of best fit (see Model 9.12 for definition of vectors).

X6 = ability if Y comes from a female in Treatment 3, 0 otherwise; and

X6^2 = ability squared if Y comes from a female in Treatment 3, 0 otherwise.

Because empirical line fitting as seen in this context is an attempt to describe parsimoniously the data, no Research Hypothesis is necessary. One may successively cast models that reflect $b_1 = b_2$, $b_5 = b_6$, and $s_5 = s_6$ and compare the R^2 of those models with Model 9.12.

In the context of Model 9.12, one may investigate the possibility that $b_1 = b_2$, both equal to b_7, a common slope. If vectors X1 and X2 are added to form X7, then b_7 is a common slope for males and females under Treatment 1. The resulting model is:

(Model 9.13) $Y = a_1U1 + a_2U2 + b_7X7 + a_3U3 + b_3X3 + a_4U4 + b_4X4 + b_5X5 + s_5X5^2 + a_6U6 + b_6X6 + s_6X6^2 + E_{13}$

There are 13 linearly independent vectors in Model 9.12 and 12 in Model 9.13; thus, Model 9.13 is more parsimonious. If Model 9.13 is judged by the researcher to be as predictive (that is, if it has as high an R^2) as Model 9.12, then the researcher would prefer Model 9.13 to Model 9.12, and Model 9.13 becomes the preferred model to be compared against further models.

The researcher may then wish to investigate the possibility that $b_5 = b_6$, both equal to b_8, a common weight; and $s_5 = s_6$, both equal to s_8, a common weight. These two restrictions can be made one at a time, sequentially, or simultaneously. If done together, a large loss in R^2 would not tell the researcher whether one, the other, or both of the restrictions caused the loss in R^2. On the other hand, if the R^2 loss is small from Model 9.13 (or Model 9.12) to Model 9.14, then one could conclude that $b_5 = b_6$ and $s_5 = s_6$.

To gamble a bit, because the researcher has a hunch that both restrictions are reasonable, she may apply both restrictions at once. By adding vectors X5 and X6 to get vector X8 and adding X5^2 and X6^2 to get X8^2, the weights b_8 and s_8 will give the common linear and second-degree slopes for males and females given Treatment 3. The resulting model for comparison with Model 9.13 is:

(Model 9.14) $Y = a_1U1 + a_2U2 + b_7X7 + a_3U3 + b_3X3 + a_4U4 + b_4X4 + b_8X8 + s_8X8^2 + a_6U6 + E_{14}$

There are 12 linearly independent vectors in Model 9.13 and 10 in Model 9.14. If the R^2 value for Model 9.14 is the same as or close to (as judged by the researcher) the R^2 value for Model 9.13, then the researcher would conclude that Model 9.14 is a better description of the data than Model 9.13 because Model 9.14 is as predictive but more parsimonious than Model 9.13. The researcher could then formulate a hypothesis based on Model 9.14 and then test this hypothesis on new, "unsnooped" data.

When using the empirical line-fitting procedure just described, one should be aware that one is data snooping and hypothesis seeking. Under such circumstances, the obtained R^2 must be seen as a *descriptive* statistic, which can be used for future hypothesis testing. *One should not use the snooped data inferentially.* A replication with new subjects is necessary before generalizing to the population. (A more extensive discussion of data snooping appears in chapter 11.)

RESCALING OBSERVED NONLINEAR RELATIONSHIPS

In the preceding section of this chapter, we discussed transforming (squaring, cubing, etc.) variables to reflect the hypothesized nonlinear functional fit between the predictor set and the criterion. In those instances, it was expected that the actual constructs (those underlying phenomena that the predictors and the criteria were measuring, however imperfectly) were related to one another in a nonlinear manner. When a nonlinear relationship between variables is observed, however, it may not be due to the constructs' being related nonlinearly; it could be due instead to either variable being a poor measure (or map) of the construct it represents.

Our concern in this final section of the chapter is with transforming a variable such that the numbers more accurately reflect the construct under consideration. Because constructs, by definition, cannot be observed, researchers must realize that the measurements they use to indicate those constructs are arbitrary numbering systems that someone has decided are "good."

The "goodness" of the arbitrary scoring could be ascertained by observing the overlap between the numbers and the construct. Figure 9.16 represents what researchers usually assume to be the state of affairs, whereas Figure 9.17 represents what may be the more likely state of affairs. The process of transforming variables is an attempt to reach as closely as possible the state of affairs in Figure 9.16.

There is no conventional or obvious way to transform a variable such that the numbers will better match the construct. A thorough knowledge of one's area of research will certainly help. A well-thought-out nomological net also would be of some assistance. But because the transformations must ultimately be tested by successive empirical verifications, perhaps the most efficient method would be through data snooping. Data snooping was briefly discussed earlier and will be more fully developed in chapter 11. We first give several examples to illustrate the need for transformation of variables—a process from now on referred to as *rescaling*.

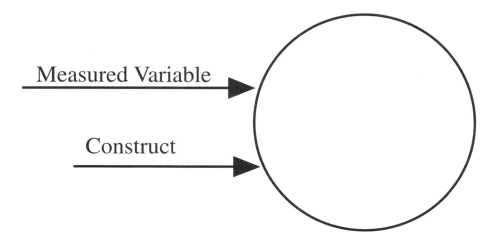

Figure 9.16.
The assumed overlap between the measured variable and the construct that variable measures—100% overlap.

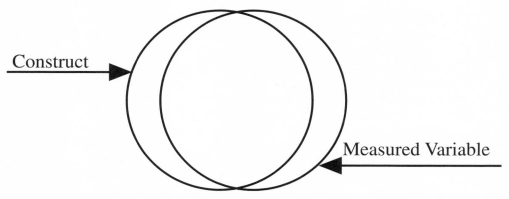

Figure 9.17.
A more likely correspondence than in Figure 9.16 between the measured variable and the construct.

Rescaling Example One: Difficulty of Items

The numbers assigned to measure a construct should reflect that the scale points are in order (e.g., a score of 5 reflects more of the construct than does a score of 4 and less than does a score of 6) and should reflect the distance between scale points (e.g., the construct distance between scores of 4 and 6 is the same as the construct distance between scores of 8 and 10).

The arbitrarily scored scale in Figure 9.18 has seven items, but it covers eight units on the construct. (Note: It must be emphasized here that attainment of actual

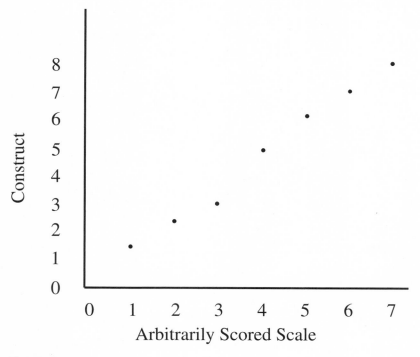

Figure 9.18.
An arbitrarily scored scale that does not accurately reflect construct distances between scale points.

construct scores—the criterion in this part of chapter 9—would most likely not be possible.) If a unit weight is assigned to each item on the scale, it will not accurately map the construct. Persons who get scores of 4, 5, 6, or 7 on the arbitrarily scored scale really have 5, 6, 7, or 8 units of the hypothetical construct. In order for the arbitrarily scored scale to map the construct, one unit would have to be added to the arbitrarily scored scale if the arbitrarily scored scale score is greater than three. Another possibility would be to add another item to the seven-item test.

Rescaling Example Two: Difficulty of Items

Suppose that the scale developer had done a relatively good job in writing items and assigning numbers to them, with the exception of the very difficult items. Only a few difficult items were included, and they were really difficult. Figure 9.19 might be the result of this scale. Items 1 through 5 each represent one unit on the construct, but items 6 through 10 cover a distance from more than 5 to 25 on the construct. Assigning a unit weight to each item will not map the construct accurately; the arbitrarily scored scale must be modified. Either additional items have to be added to the instrument, or the scale has to be transformed. Since the "right" items would have to be added, transforming the existing scale is easier. If the arbitrarily scored scale is five or below, it does not have to be transformed. If the arbitrarily scored scale is greater than five, then the following SAS data transformation would map the arbitrarily scored scale onto the construct:

IF X > 5 THEN X = (4.0 * X) + -15;

The values 4 and -15 were obtained by ascertaining the line of best fit over the range from 5 to 10. This line has a slope of 4 and, if extended to the Y axis, would have a Y-intercept of -15.

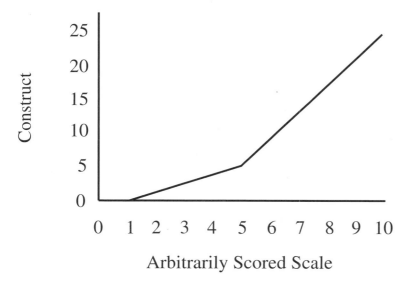

Figure 9.19.
Relationship between an arbitrarily scored scale and the construct when only a relatively few difficult items are included.

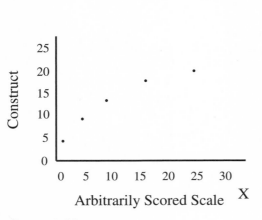

Figure 9.20.
Data depicting a ceiling effect.

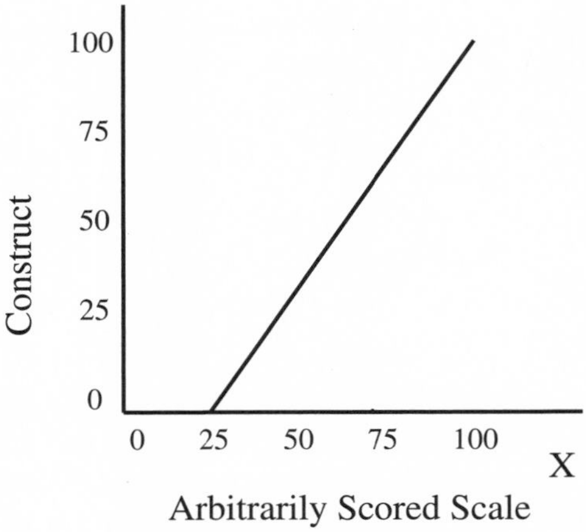

Figure 9.21.
The possible relationship between a 100-item
multiple-choice test and the construct.

The reader should note in the previous two examples that the rescaling suggested would simply result in all data points falling on a straight line. The rescaled arbitrarily measured scale rectilinearly maps the criterion. The resultant scoring system is also arbitrary but has an added feature—a rectilinear fit with the construct. That perfect mapping with the construct is the goal of this discussion.

Rescaling Example Three: Ceiling Effect

The term *ceiling effect* implies that there is not enough discrimination at the high end of the arbitrarily scored scale, as in Figure 9.20. One could overcome this problem by including additional difficult items or by rescaling the existing arbitrarily scored scale.

The construct can be better mapped by taking the square root of X and then multiplying by 4 as indicated by the following SAS statement:

X = X**.5 * 4;

The multiplication by 4 is not a necessary step because the first step (that of taking the square root) results in the scores falling on a straight line. The multiplication by 4 simply results in the rescaled scores numerically matching the (arbitrarily scored) construct.

Rescaling Example Four: Guessing Effect

If a 100-item, 4-choice, multiple-choice test is developed, some 25% of the items can be correctly answered solely by guessing. One could therefore argue that any score of 25 or below reflects a construct score of zero. The remaining scores also could be adjusted for the guessing effect. The required adjustments to make the data in Figure 9.21 reflect the construct are in the following SAS statements:

IF X < 25.1 THEN X = 0;
IF X > 25.0 THEN X = (4 / 3) * (X - 25);

These data transformations set the rescaled score equal to zero if the original score was less than or equal to 25. If the original score was greater than 25, the rescaled score is adjusted for guessing.

Rescaling Example Five: Conceptual Rescaling

Haupt (1993) investigated drug use of 4th, 5th, and 6th graders. One question the students were asked to answer was, "How often do you take part in any school clubs, sports groups, or other activities?" The answers were originally scored as never = 1, sometimes = 2, I used to = 3, and a lot = 4. Because club participation was used to infer the amount of time spent in structured activities, the scoring for responses "sometimes" and "I used to" were reversed to reflect the continuum of time in structured activities depicted in Figure 9.22.

Another example of conceptual rescaling was provided by Presley and Huberty (1988). They make the point that, when scores are to be weighted for the purposes of summing together, the standard deviations must be considered. The best way to do that is to transform all scores so that they have a common mean and common standard deviation. So if you wanted to construct a course grade by giving a 40% weight to the midterm exam and a 60% weight to the final, then the rescaling would be (in a SAS statement):

GRADE = (.40 * ZMIDTERM) + (.60 * ZFINAL);

The means and standard deviations can be found using PROC MEANS in SAS. Once those values are determined, a second run can be made using that information.

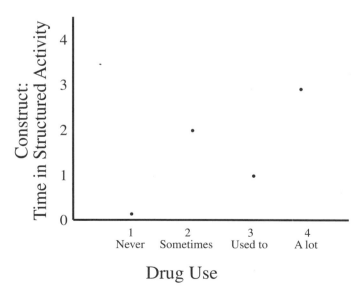

Figure 9.22.
An arbitrarily scored scale that needs rescaling based on conceptual notions about the construct.

Assume that the first run produces a midterm mean of 83.6, with a standard deviation of 10.12; and a final mean of 85.4, with a standard deviation of 8.63. Then the necessary SAS statement for the second run would be:

ZMIDTERM = (MIDTERM - 83.6) / 10.12;
ZFINAL = (FINAL - 85.4) / 8.63;
CRGRADE = (.40 * ZMIDTERM) + (.60 * ZFINAL);

The Spearman Rank Correlation

Researchers are often encouraged to rank their data when they have no faith in the interval property of their data and then to use the Spearman rank correlation if an index of relationship is desired. The lowest score is assigned a value of one, the next lowest a value of two, and so on. If one wanted to find the rank correlation between, say, X1 and X2, one could determine the rank of each score on each variable and then obtain the R^2 for the following model:

$$(Model\ 9.15)\ R1\ =\ aU + bR2 + E_{15}$$

where:
   R1 = the ranked X1 values; and
   R2 = the ranked X2 values.

The square root of the R^2 of Model 9.15 would be the desired numerical value for the Spearman rank correlation. The important point to realize is that assigning the ranks of 1, 2, and so on is one way of rescaling.

Square Root Transformation

Researchers are sometimes admonished, when they discover unequal variances in several groups, to make a square root transformation. Depending upon the nature of the data, this procedure may serve the purpose of equalizing the (sample) variances. Two points should be made. First, the statistical assumption is about the population variances not about the sample variances. Second, blindly applying a square root transformation will not always produce the desired results. The kind of transformation required on the criterion to obtain homogeneous variances depends on the data and will not always be a square root function. Indeed, one should realize from previous chapters that the unequal variances might be due not to scaling considerations but to inherent interactions.

Nonlinear Transformations of the Criterion

The Spearman rank correlation and square root transformations actually make transformations on the criterion variable. They are both routine approaches that are applied somewhat mechanically to all variables—predictors and criterion. One could, though, make a specific transformation of the criterion, either for the purpose of rescaling or to reflect the functional fit between the predictor(s) and the criterion.

The Pythagorean theorem is a good example of the utility of transforming the criterion. The reader may remember that the Pythagorean theorem is $C^2 = A^2 + B^2$. In terms of the GLM:

$$(\text{Model } 9.16)\ C^2\ =\ aA^2 + bB^2 + E_{16}$$

where:

C^2 = square of the hypotenuse of a right triangle;

A^2 = square of the length of one side of the triangle;

B^2 = square of the length of the other side of the triangle; and a and b are both equal to 1.

McNeil, Evans, and McNeil (1979) pointed out several interesting aspects about this regression solution:

1. The R^2 is 1.00, and thus the error components are 0 for each subject (each triangle).
2. The linear component of A, the linear component of B, and the unit vector are all absent from this best model.
3. An R^2 of 1.00 could not have been achieved if the criterion had not been squared.
4. The transformations on the criterion and on the predictors are not for rescaling purposes (as the lengths were measured by ruler, a very good interval way to measure distance). The transformations were made to match the functional fit.

Functional Fit or Rescaling?

When transformations on the arbitrarily scored variables are made, the researcher is either rescaling variables or allowing for a given functional fit. Which of the two is being done is not important from a predictive point of view—the crucial concern is whether the R^2 has been increased by the transformation. A simple example may be of value here. Suppose that Model 9.17 yields an R^2 of .60 but that Model 9.18 yields an R^2 of 1.00.

$$(\text{Model } 9.17)\ Y\ =\ aU + bX + E_{17}$$

$$(\text{Model } 9.18)\ Y\ =\ aU + cX^2 + E_{18}$$

These two models cannot be compared by the general F test because one is not a restriction of the other; indeed, both have two predictor pieces of information. These two models can be compared, however, on the basis of degree of predictability. Based upon the goal of predictability, Model 9.18 would be the preferred model because it yields a higher R^2 value. If the researcher felt that X was a good rectilinear mapping of the construct, then Model 9.18 would reflect a functional fit, as discussed in the early part of this chapter. If the squaring of X was a result of acknowledging the arbitrariness of the scaling of the original measure, then Model 9.18 would reflect the rescaling notions discussed in the latter part of this chapter.

As discussed earlier, the criterion can be transformed, although Model 9.19 would yield a different R^2 than would Model 9.18.

$$(\text{Model } 9.19)\ \sqrt{Y} = aU + bX + E_{19}$$

Whether the transformation represents a functional fit or rescaling is important from an "understanding" point of view. Suppose that a given criterion is predicted quite well by a given predictor, as in Figure 9.23. The researcher's theory says

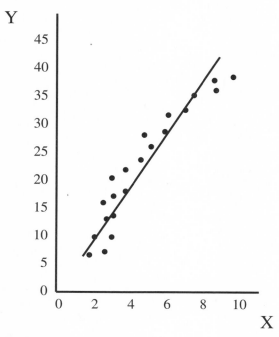

Figure 9.23.
An observed rectilinear fit over the range of X values 0–10.

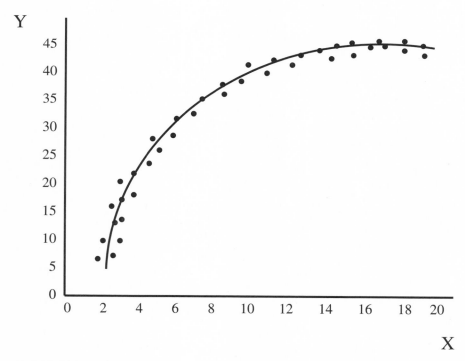

Figure 9.24.
An observed nonlinear fit over the range of X values 0–10.

that the same functional fit ought to be the case for higher values of the predictor. As shown in Figure 9.24, though, the linear fit observed over the lower values of the predictor does not generalize to higher values. Indeed, the relationship over the entire range of values in Figure 9.24 is better depicted by a second-degree curve. If higher values of the predictor variable were investigated, perhaps the functional fit would be other than a second-degree one. The functional fit that is claimed for predictor values outside the range originally investigated must be empirically verified with those new values of the predictor variable.

The important concern of this chapter is to emphasize the notion that numbers put on measurements are arbitrary. Initial attempts at numbering most likely do not do a good job of mapping the construct. Given that the researcher has rescaled the variable and feels secure that the rescaled variable maps the construct, the concern then is to reflect (with regression models) the hypothesized functional fit. Using the X variable allows for only a linear fit, whereas other fits may be inherent in the data. Given that the scaling is arbitrary, there is no more reason to expect a linear fit than a nonlinear fit. X is actually X^1, and realizing that should make it clear that there are an infinite number of other exponents that could be investigated (e.g., X^2, X^3, $X^{1.5}$, $X^{.3}$, $X^{.313}$). Researchers or students often ask us to justify our choice when we use, say, $X^{1.3}$. We respond, "Why do you use $X^{1.000}$?" The linear component is a default value that in the age of computers should be tested for its value not automatically assumed to be of import. If the linear component is used, then some justification should be provided—just as justification should be provided for any other exponent.

Reliability and Validity

Reliability is the extent to which a variable yields the same value on repeated measures. *Validity* is the extent to which a variable measures what it is supposed to measure. The regression approach can be used to determine the magnitude of both.

Test-retest reliability is simply the correlation between two administrations of the same test. Applied Research Hypothesis 9.3 provides practice with this kind of question. Note that the Restricted Model is forced to have an R^2 of .81 because that is the R^2 specified by the Statistical Hypothesis. The *F* would most likely have to be computed by hand using the same general *F* formula (most computer programs do not allow for the insertion of a specified R^2 value).

Reliability is of minor concern, though, as validity is the ultimate test of any variable. The purpose of rescaling is to get the measure to be a valid indicator of the construct. The aim of functional curve fitting is to get the predictor set to be a valid indicator of the criterion. Most measurement texts correctly say that a variable can be highly reliable yet have no validity for a particular construct. Shoe size can be reliably measured, yet it is not considered a valid measure of IQ.

Furthermore, many constructs meaningfully fluctuate over a short period. Breathing rate changes from one hour to the next; anxiety fluctuates from one task to another; and reading interest fluctuates from one comicstrip to another. The concern in each case should be, "Does the measure accurately reflect the criterion?"

It is our position that too much attention has been placed on the reliability of measures, and not enough attention has been placed on the validity. Because some form of reliability is a necessary but not sufficient condition for validity, and because validity is the ultimate goal, more attention should be placed on validity.

Applied Research Hypothesis 9.3

Directional Research Hypothesis: For a given population, the test-retest reliability of X2 (the correlation between X2 and X3) is greater than .90.

Statistical Hypothesis: For a given population, the test-retest reliability of X2 is equal to .90.

Full Model: $X2 = a_0U + a_3X3 + E_1$

Want: $R^2 > .81$; restriction: $R^2 = .81$

Restricted Model: $R_r^2 = (.90)^2 = .81$

alpha = .02
$R_f^2 = .92$; $R_r^2 = .81$
$F_{(1,58)} = 80$; required $F_{(1,40)} = 9.31$

Interpretation: Since the F must be calculated by hand, the approximate F of 80 must be compared to the tabled F value of 9.31. Since the calculated F is larger than the tabled F, and since the correlation between X2 and X3 is greater than .90 (indicated also by the weight $a_3 > 0$), the Research Hypothesis is tenable: "For a given population, the test-retest reliability of X2 is greater than .90."

Note: Most F tables have values for selected degrees of freedom. If the desired denominator degrees of freedom does not exist, then one should move in the conservative direction and refer to an F with fewer denominator degrees of freedom.

The Criterion as an Approximation of the Construct

In the past, a cause-and-effect relationship has been interpreted for data that, although highly significant, yielded small R^2 values. One of the positions we take in this text is that researchers can more fruitfully spend their time finding variables that will increase the R^2 rather than trying to understand how that small proportion of criterion variance has been predicted. Related to this problem, though, is the adequacy of the overlap of the criterion measure being used with the construct one has chosen to consider. Too often the criterion is equated with the construct, when in reality it must be realized that *the criterion is only an approximation of the construct*. Unfortunately, there is no way at the present to know what overlap exists between the construct and the criterion variable. Figure 9.25b is the state of affairs assumed by most researchers, whereas Figure 9.25a is more likely the state of affairs; that is, there is only a partial overlap between the criterion and the construct. One may do well to continue trying to account for the criterion variance beyond that accounted for by predictor variables 1, 2, and 3 in Figure 9.25a because of the remaining overlap between construct and criterion that has yet to be accounted for.

Figures 9.25c and 9.25d also may be the existing state of affairs. In Figure 9.25c, the common overlap between the criterion and the construct has yet to be

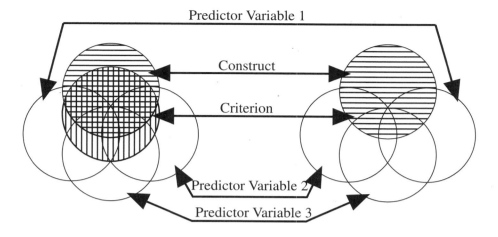

Criterion
Approximates Construct

Figure 9.25a

Criterion and
Construct Isomorphic

Figure 9.25b

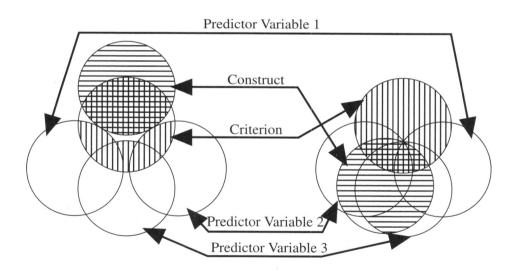

Predicted Variance Not
Overlapping with
Construct Variance

Figure 9.25c

All of the Common Variance
Between Criterion and
Construct Is Accounted For

Figure 9.25d

Figure 9.25.
Some possible consternating relationships between criterion, construct, and predictor variables.

accounted for. That portion of the criterion that is not predictive of the construct has (unfortunately) been well accounted for. Any interpretation regarding the explainers (let alone causers) of the construct in this situation will be grossly mistaken. If the predictor variables should really map the construct, according to one's theory, then the predictor variables probably should be rescaled.

Figure 9.25d presents an entirely different picture. All of the common variance between the criterion and the construct has been accounted for. To attempt to build up the predictability of the criterion would in this case be wasted effort because the criterion is mapping no more of the construct. Interpretations of causality in this case may be of some value, but one must note that complete accounting of the construct (as indexed by R^2) will not be possible. The situation in Figure 9.25d may call for the rescaling of the criterion measure. If various rescaling measures do not map the construct better, then the researcher should either turn to another criterion measure or realize that the construct is multidimensional. In this latter case, multiple criterion measures would be necessary, an extension of multiple regression notions. Many multivariate procedures do exist but are extensions beyond the scope of this text. The interested reader may refer to Bray and Maxwell (1985), Hand and Taylor (1987), Harris (1985), Stevens (1986), or Tatsuoka (1988).

Nonlinear Interaction

The models and figures to be discussed now deal with those instances in which the researcher expects a second-degree relationship between one of the predictors and the criterion. Model 9.20 allows for a first-degree and a second-degree relationship between Q and Y, while the relationship between P and Y is linear.

$$\text{(Model 9.20) } Y = aU + bP + cQ + dQ^2 + E_{20}$$

This relationship is pictured in Figure 9.26; the predicted Y scores form a curved plane, with curved edges along the Q sides and straight edges along the P sides.

Model 9.21 adds to the above model an interaction between P and Q:

$$\text{(Model 9.21) } Y = aU + bP + cQ + dQ^2 + e(P * Q) + E_{21}$$

The graphic representation of Model 9.21 is shown in Figure 9.27. The curved plane formed by the predicted criterion scores has curved edges along the Q sides and straight lines with differing slopes along the P sides.

Curve fitting, when using more than one predictor variable, is referred to as *fitting a response surface*. Interest may focus on maximums, minimums, similarity of one group's surface to another group's, or simply on the nature of the surface.

Suppose that two continuous predictors (X and Y) each have a U-shaped relationship with the criterion Z. The model would require second-degree terms in both X and Y:

$$\text{(Model 9.22) } Z = aU + bX + sX^2 + cY + dY^2 + E_{22}$$

If the data perfectly matched the above model, the plot would look like the SAS-produced plot in Figure 9.28. Notice that the X face of the cube is U shaped, as is the Y face of the cube.

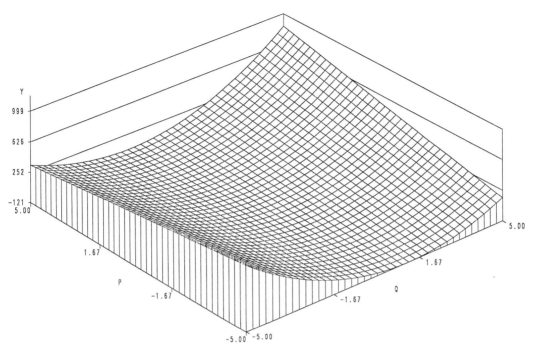

Figure 9.26.
A plane curved on one axis, fit by Model 9.20.

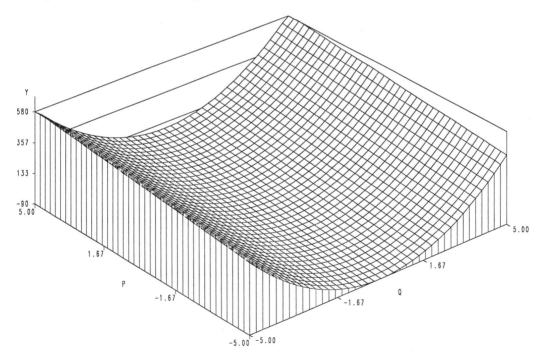

Figure 9.27.
A twisted plane curved on one axis, fit by Model 9.21.

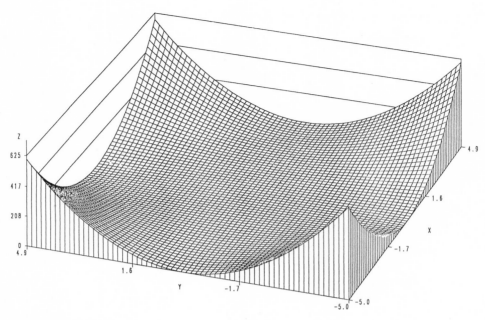

Figure 9.28.
A plane curved on both axes, fit by Model 9.22.

These models and figures are only examples of the infinite types of relationships one may expect to find and are presented as a means of helping the researcher to visualize those relationships. We must again emphasize, though, that being able to draw a figure of a model is not a necessary step. If a model yields a high degree of predictability, that is sufficient. Indeed, figures in more than three dimensions are a little difficult to communicate—let alone comprehend.

Interpretation of Complex Interactions Versus Predictability

Whenever one finds a significant interaction, one should try to graph that interaction and try to interpret it. If one has hypothesized a directional interaction before the fact, then the interpretation of the significant interaction should not be difficult. If one did not suspect interaction, then a replication study should substantiate the results before one starts to have faith in that interaction. (Actually, this applies not just to interaction; any nonhypothesized result needs replication before being accepted.) One may continually find an interaction term that accounts for a large proportion of the criterion variance; that is, the goal of predictability has been satisfied in achieving a large R^2. But why this particular interaction exists may not be explainable at the time it is found. To throw away this high degree of predictability because of a failure to comprehend it seems a bit egotistical. On the other hand, obtaining premature closure on interpretations of data may be one of the most drastic mistakes researchers make. One could argue that insightful interpretations are made as a function of the data telling us something new rather than as a function of our preconceived notions twisting the data so that they make sense to us.

Interacting Variables: One Way to View all Variables

A strong case was made in chapter 7 and again in this chapter for the consideration of interaction variables. All variables can be considered interacting variables if one remembers that any number to the zero power is equal to one. Any variable that is a multiplication of two or more other variables, say, P * Q, is an interaction variable. By convention, the powered superscript is dropped when it is equal to one; therefore P * Q also could be represented as $P^1 * Q^1$. Also, $P^2 * Q^1$ is an interaction variable, and $P^0 * Q^1$ can be considered an interaction variable. But, again, any number to the zero power is equal to one; therefore, $P^0 * Q^1 = Q$. Thus Q can be conceptualized as the interaction of P to the zero power and Q to the first power. Similarly, U can be conceptualized as Q^0, $P^0 Q^0$, or $P^0 Q^0 R^0$, and so on.

Mendenhall (1968, p. 94) presents the regression model allowing for interaction in a two-by-three ANOVA design as:

$$(\text{Model } 9.23)\ Y = aU + bP + cQ + dQ^2 + e(P * Q) + f(P * Q^2) + E_{23}$$

where:

P = two distinct scores representing the two levels of the first independent variable; and

Q = three distinct scores representing the three levels of the second independent variable.

Model 9.23 can be further conceptualized as interaction variables as in Model 9.24:

$$(\text{Model } 9.24)\ Y = a(P^0 * Q^0) + b(P^1 * Q^0) + c(P^0 * Q^1) + d(P^0 * Q^2) + e(P^1 * Q^1) + f(P^1 * Q^2) + E_{24}$$

Each term in this model is one form of interaction between one independent variable and the other independent variable. Models 9.23 and 9.24 are conceptually and mathematically equivalent.

One could extend this notion further by conceptualizing polynomial terms as interactions. For instance, P^2 is equivalent to P * P. That is to say, the second-degree term is an interaction between the two linear terms. (Jack Byrne, in a personal communication, first brought this to our attention.) We do not view this way of conceptualizing variables as necessary for all variables, but there might be some value in it for some readers.

These notions at least explain to us why the weight for the unit vector has conventionally been depicted as a_0. If one looks at, say, a third-degree model, one sees a pattern developing:

$$(\text{Model } 9.25a)\ Y = a_0 U + a_1 X + a_2 X^2 + a_3 X^3 + E_{25}$$

If we realize that X can be represented as X^1 and U as X^0, then Model 9.25a could be rewritten as:

$$(\text{Model } 9.25b)\ Y = a_0 X^0 + a_1 X^1 + a_2 X^2 + a_3 X^3 + E_{25}$$

The subscript for each weighting coefficient for each vector is the same as the power of that vector. The subscript associated with the weight for the unit vector is related to the power that the predictor variable has been taken to—namely zero.

Trend and Time Series

Economists for many years have been concerned with time-series analyses, in which data (usually representing some one entity) are repeatedly measured over several months or years. The amount of time separating the repeated measures is, of course, up to the researcher. The time span could be days, hours, or even smaller units, such as minutes or seconds.

With data of this sort, the concern is not so much with finding causal factors as it is with ascertaining trends that exist in the data. The only variables usually measured are the criterion variable and the time at which the criterion was observed. One usually does *not* consider time by itself to be a causative factor; but given the pattern or trend of the criterion scores over an extended period, one *may* be able to predict the criterion value at a given later time period.

Few time series can be reflected solely by a straight line, but the reader should not be concerned about that now. Higher-order polynomials can be tested to see if those kinds of trends exist. Often the nature of the criterion being investigated can yield a hint about what kinds of trends to investigate.

Suppose that subjects are given 10 trials on each of five successive days. One might anticipate a similar trend within each day, superimposed on some overall trend, as in Figure 9.29. One would then want to allow for an overall linear trend in the data, as well as a common daily second-degree trend. The model allowing these trends would be:

$$(\text{Model } 9.26)\ Y\ =\ aU + bX + cX^2 + dT + E_{26}$$

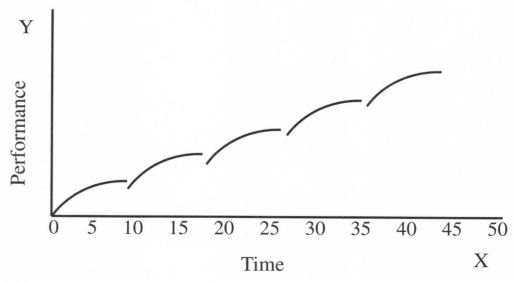

Figure 9.29.
Similar trend for 10 trials within each day, superimposed on overall linear trend.

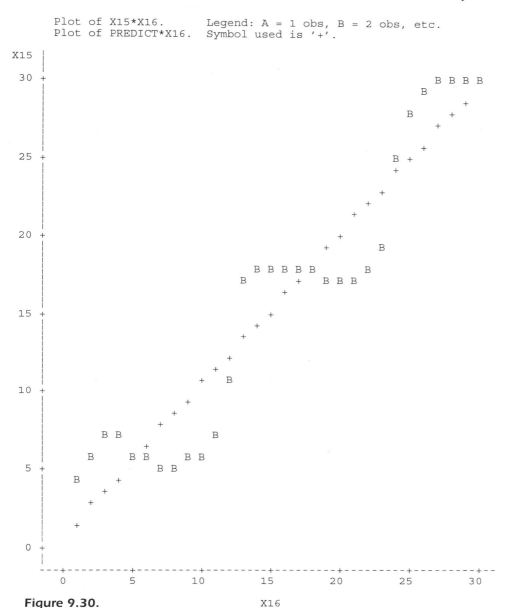

```
        Plot of X15*X16.      Legend: A = 1 obs, B = 2 obs, etc.
        Plot of PREDICT*X16.  Symbol used is '+'.

X15 |
    |
 30 +                                              B B B B
    |                                            B
    |                                                  +
    |                                        B       +
    |                                              +
 25 +                                    B +
    |                                    +
    |                                 +
    |                               +
    |                             +
 20 +                         +
    |                       +          B
    |                 B B B B  B      B
    |               B       +  B B B
    |                       +
 15 +                 +
    |                +
    |              +
    |            +
    |          +
 10 +         +
    |        +
    |      +
    |    +     B
    |  B B           B
    |     +
    | B     B B     B B
  5 +        B B
    | B       +
    |    +
    |  +
    |
  0 +
    |
    - - +--------+--------+--------+--------+--------+--------+--------+--
        0        5       10       15       20       25       30
```

Figure 9.30.   X16

Sinusoidal relationship between month and the criterion, superimposed upon the linear relationship.

where:

    Y   = performance;
    T   = trial number;
    D   = day number; and
    X   = T - ([D - 1] * 10) (representing trial within day).

Figure 9.30 illustrates some hypothetical data to clarify these ideas. Fitting a straight line to the data yields an R^2 of, say, .89 that is significantly different from zero (using an alpha of .01). Therefore, we can conclude that there is a linear trend

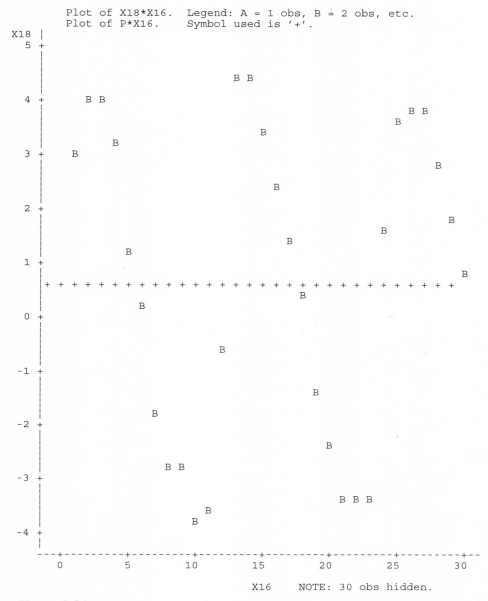

```
           Plot of X18*X16.   Legend: A = 1 obs, B = 2 obs, etc.
           Plot of P*X16.     Symbol used is '+'.
   X18 |
     5 +
       |
       |
       |                                B  B
     4 +      B  B
       |                                                       B  B
       |                                                    B
       |            B                        B
     3 +      B                                                   B
       |
       |                                  B
     2 +                                                             B
       |                                                  B
       |                              B
     1 +         B
       |
       | + + + + + + + + + + + + + + + + + + + + + + + + + + + + + +
       |                                      B
     0 +                          B
       |
       |
    -1 +
       |                                    B
       |            B
    -2 +
       |                                 B
       |
    -3 +         B  B
       |                                 B  B  B
       |                  B
       |               B
    -4 +
       |
       ---+---------+---------+---------+---------+---------+---------+--
          0         5         10        15        20        25        30

                              X16    NOTE: 30 obs hidden.
```

Figure 9.31.
Sinusoidal relationship between month and the criterion after the linear trend has been taken out of the criterion.

over months. Note, though, that the data points are somewhat discrepant from the line of best fit and also that the deviations of the data points from the line of best fit are systematic. For some values of X16 (months), the observed criterion values are much higher than the linearly predicted values. For other months, the observed criterion values are much less than the linearly predicted values. That the R^2 is somewhat less than 1.00 suggests that a linear fit is not the best fit for the data. The systematic values of the errors in prediction suggest that there are some other fits to be found—that the criterion is not fluctuating solely in a linear fashion over time.

One way of attempting to discover the other trend(s) in the data is to rotate the figure such that the straight line going through the data points is horizontal. The problem now is to discover the remaining trend, after the linear trend has been extracted (or to discover the trend in the data, over and above the linear trend). The data in Figure 9.31 appear to follow a trigonometric function, over and above the linear function. That supposition is tested in Applied Research Hypothesis 9.4.

The careful reader will note that the data were snooped upon when we extracted the linear term and looked at the deviations from the linear trend. If an extremely high R^2 has been obtained with the Full Model in Applied Research Hypothesis 9.4, the researcher is obligated to the replication process. Once the functional relationship has been replicated, the researcher is in a position to predict to future months. Such a prediction assumes that the relevant influences remain the same as in the empirically analyzed months. Unforeseen events (such as wars, changes in policy, and changes in consumer habits) will have their (unpredictable) effect on the magnitude of the criterion. These latter confounding variables are threats to the *external* validity of the study not the *internal* validity. If the study has been carefully conceived and executed, then we can feel confident about the results being what they are. The application of those results to future years (the only reason for doing the study in the first place) will be accurate as long as all the variables stay the same.

Applied Research Hypothesis 9.4

Directional Research Hypothesis: For a given population, there is a sinusoidal trend in the prediction of X15, over and above the linear trend.

Statistical Hypothesis: For a given population, there is not a sinusoidal trend in the prediction of X15, over and above the linear trend.

Full Model: $X15 = a_0 U + a_{16} X16 + a_{17} X17 + E_1$ (m1 = 3)

where:

$\quad$ X17 $\quad$ = $\quad$ SIN (X16 * 6.283/12);
$\quad$ X16 $\quad$ = $\quad$ the consecutive months numbered 1–30;
$\quad$ 6.283 = $\quad$ twice the value of pi; and
$\quad$ 12 $\quad$ = $\quad$ the period of the expected trend (12 months in the period).

Want: $a_{17} \neq 0$; restriction: $a_{17} = 0$

Restricted Model: $X15 = a_0 U + a_{16} X16 + E_2$ (m2 = 2)

alpha = .001
$R_f^2 = .98$; $R_r^2 = .80$
$F_{(1,57)} = 492$
Computer probability = .001; directional probability = .0005

Interpretation: Since the directional probability is less than alpha, one can reject the Statistical Hypothesis and hold as tenable the Research Hypothesis.

10

DETECTION OF CHANGE

Researchers usually want not just to describe a situation at a particular time but to detect when changes occur and determine what caused those changes, so that in the future those causal factors (or causers) can be deliberately manipulated to bring about desirable changes. An educational researcher may find that anxiety is a predictor of mathematics performance, that low-anxious students usually do better on mathematics tests than do high-anxious students. But in addition, that same researcher wants to "upset the prediction" for the high-anxious students, so that they will do well on mathematics tests. To do this, causes of the differential performance must be identified—is it only anxiety; is it some aspect of the test or the testing situation; or is there another explanation?

ASCERTAINING CAUSALITY

One of the major goals of scientific inquiry is to identify causal relationships because one aim of the human endeavor is to "make things better" (by one's own definition) by upsetting the prediction that exists and causing a more desirable outcome. Researchers may be able to predict that some children will learn more slowly than others, or that certain people will contract cancer, or that certain fields will yield fewer crops. But these researchers want to upset those predictions: Those children who are expected to learn less would learn as well; those people who are expected to get cancer would not contract the disease; and those fields that are expected to yield few crops would yield more.

To upset prediction, the causal factors must be identified and then changed through manipulation. What follows might be the process of identifying a causal factor (or causer). Initially, a situation is described, and the researcher then gets some ideas about what causers are operating. The researcher then formulates a hypothesis (directional, if there is enough information) regarding what factors predict the outcome of interest. The worker relies on intuition, theory, and logical tenets to design a study to test the hypothesis, wanting this study to eliminate as many competing explanations of the results as possible, and then conducts the study and tests the hypothesis. If the directional hypothesis is supported, there is evidence of the factors that are predicting the outcome of interest. A directional hypothesis regarding factors *that might cause* the outcome is formulated, and the researcher must now design a study that will eliminate further competing explanations and in which the hypothesized causers can be manipulated to see if a change in the causers brings about the expected change in the outcome. The ultimate test for having discovered a causer rests upon successive replications of the manipulated causer, all resulting in the hypothesized change in the criterion. Causality cannot be found out in a single study. There will always be too many competing explainers of the results. Campbell and Stanley (1966) have a well-organized categorization of possible competing explainers. These possible competing explainers must be eliminated, either by the design of the study or by logical analysis. The design of the study cannot rule out all of the possible explainers, so logical analysis must always be included to ascertain causality. Some researchers suggest they are *not* trying to find out causal relationships, but a little questioning about their purposes reveals a causative purpose.

DIRECTIONAL RESEARCH HYPOTHESES

The utility of a directional Research Hypothesis becomes even more apparent when trying to detect where change occurs due to what causers. A new teaching method is developed not simply to change the rate of learning but to increase the rate. A new diet is developed not just to change the weight of overweight people but to reduce their weight. A new traffic regulation is put into effect not merely to change the accelerating rate of accidents but to reduce the trend. If researchers are ultimately going to upset prediction, then they must know the direction in which the causative factors elicit their effect.

RANDOM ASSIGNMENT

The major tool for letting the design of the study eliminate possible competing explainers has been random assignment. One should note, though, that random assignment does not guarantee that the samples are equivalent. Random assignment only maximizes the probability that the samples will be equivalent. Seldom are two random samples equal on a given variable, let alone equal on all variables. Furthermore, the populations to which most researchers want to generalize are so large or entirely theoretical that a random sample of that population is physically impossible. Most often the subjects are randomly sampled from some accessible population, which may in many ways differ from the population to which the researcher wishes to generalize. Few studies in the literature employ a random sample from the population to which the researcher has generalized. These statements are not intended as a condemnation of the literature but as a condemnation of researchers relying upon random sampling as a cure-all.

On the other hand, simply arriving on the scene and measuring all the variables simultaneously does not allow causal statements either. Again, causality can only be ascertained by manipulating a variable and observing the predicted result. Variables in the real world are highly correlated, and a change in one variable, therefore, *may* change one or more other predictor variables. But just because two variables are correlated does not mean that the first variable causes the second. Both variables could be caused by one or more other variables. Again, only a tight, logical analysis can tease out the causative variables. Manipulation of the proposed causative variables is a necessary step in determining causality.

ONE-GROUP MODELS

Change hypotheses can be investigated using only one group. Such studies usually are not as conclusive as when a second (comparison) group is included, but one-group studies can be of some benefit, and the GLM approach can be used to answer those questions.

One Group: Posttest Only

The emphasis on criterion-referenced testing and on explicitly stating objectives implies that some researchers will be testing hypotheses about a single population mean. Consider a project using methods to reduce alienation. One of the objectives might be that, after six weeks of participation, the alienation mean score of the children in the project will be less than five. If the project director is interested only in how the project works for the few children in the project, all that needs to be done is to look at the sample alienation mean to see if it is less than five. But a more reasonable desire is to infer to the adequacy of the project, with the intent of adopting it in other schools. With this desire, the project director wants to infer to a population of children. The Research Hypothesis here would be:

> After six weeks of instruction, the alienation mean score in the population will be less than five.

To answer any Research Hypothesis in the GLM, a Full Model and a Restricted Model must be constructed. The Research Hypothesis dictates a Full Model that must allow the alienation mean to manifest itself:

$$\text{(Model 10.1)} \ Y \ = \ aU + E_1$$

where:

Y = alienation scores; and

a = weighting coefficient for the unit vector (which could have been represented as a_0; in either case, the weight will be the mean of the criterion scores).

Readers by now should recognize this model as the unit vector model, yielding no differential predictability ($R^2 = .00$). The one regression coefficient that must be determined is a, and this will be the sample mean. The Statistical Hypothesis implies the restriction $a = 5$. Forcing this restriction on the Full Model results in the following algebraic gyrations:

Full Model: $Y \ = \ aU + E_1$

Restriction: $a \ = \ 5$

Restricted Model: $Y = 5U + E_2$

But because $U = 1$ for all subjects, 5U is a constant, and subtracting that constant from both sides yields the final form of the Restricted Model:

$$(\text{Model } 10.2) \ (Y - 5) = E_2$$

The same general F test formula is applicable to all hypotheses, providing that the unit vector is in both the Full and Restricted Models. If this is not so, and in the present situation it is not, then an alternate formula for the F test must be used as in Equation 10.1 (Bottenberg & Ward, 1963):

$$F = \frac{(\text{ESS}_r - \text{ESS}_f)/(m1 - m2)}{(\text{ESS}_f)/(N - m1)} \tag{10.1}$$

where:

ESS$_r$ = error sum of squares in the Restricted Model;
ESS$_f$ = error sum of squares in the Full Model;
m1 = number of linearly independent vectors in the Full Model (the full number of pieces of information); and
m2 = number of linearly independent vectors in the Restricted Model (the restricted number of pieces of information).

Equation 10.1 is more general than the F test in terms of R^2, although the latter is more conceptually appealing. Notice that the degrees of freedom are calculated in exactly the same way in the two formulae. If one were to apply Equation 10.1 to any of the previous examples in this text, the identical F would result.

The sum of the squared elements in E_1 in Model 10.1 will be the ESS$_f$. The sum of the squared elements in E_2 (or Y - 5) in Model 10.2 will be the ESS$_r$. Note that the Full Model uses one piece of information (the unit vector), whereas the Restricted Model uses no information; therefore, m1 = 1 and m2 = 0. The difference between m1 and m2 is one, being equal to the number of restrictions made and also being the degrees of freedom numerator for the F test. The vector display given below contains the intermediate values for the solution of this kind of hypothesis on 24 subjects. The significance of the resultant F of 101.5 must be judged by referring to tabled values; and because this was a directional hypothesis, one must use the 90th percentile of F if alpha was .05 and the sample results are in the hypothesized direction. If the alienation sample mean was greater than five, there would have been no need to go through the statistical gyrations; it would have sufficed to report "not significant" and then suggest dropping the project.

Hypotheses about a proportion also could be tested in this manner—the criterion vector here would be a dichotomous vector rather than a continuous vector as in the alienation example. In Generalized Research Hypothesis 10.1, a hypothesis for testing a single population value is given.

Researchers having access to computing facilities can perform the required analysis quickly, as one computer run can provide all the component values of the F test. The substitution of the numerical values into Equation 10.1 must be done by hand, but that is a small price to pay for the utilization of the flexible GLM technique.

Y	=	aU1	+	E_1	E_1^2	E_2	E_2^2
1		1		-1.5	2.25	-4	16
1		1		-1.5	2.25	-4	16
1		1		-1.5	2.25	-4	16
1		1		-1.5	2.25	-4	16
4		1		1.5	2.25	-1	1
3		1		.5	.25	-2	4
3		1		.5	.25	-2	4
2		1		-.5	.25	-3	9
1		1		-1.5	2.25	-4	16
4		1		1.5	2.25	-1	1
2		1		-.5	.25	-3	9
2		1		-.5	.25	-3	9
1	= 2.5	1	+	1.5	+ 2.25	+ -4	+ 16
3		1		.5	.25	-2	4
2		1		-.5	.25	-3	9
3		1		.5	.25	-2	4
1		1		1.5	2.25	-4	16
5		1		2.5	6.25	0	0
3		1		.5	.25	-2	4
2		1		-.5	.25	-3	9
3		1		.5	.25	-2	4
3		1		.5	.25	-2	4
3		1		.5	.25	-2	4
4		1		1.5	2.25	-1	1
4		1		1.5	2.25	-1	1

$$\Sigma = 34 \qquad \Sigma = 184$$

$$F_{(m1-m2,N-m1)} = \frac{(ESS_r - ESS_f)/(m1-m2)}{(ESS_f)/(N-m1)}$$

$$F_{(1-0,24-1)} = \frac{(184-34)/1}{(34)/23}$$

$$F_{(1,23)} = 101.5$$

Applied Research Hypothesis 10.1 is provided for the reader to verify skills in testing a single-population-mean hypothesis (or single-population-proportion hypothesis, if the criterion is dichotomous).

One Group: Pretest and Posttest

The researcher is often interested in the gain from the beginning of the treatment to the end of the treatment. Therefore, not only a posttest (post) is obtained but

Generalized Research Hypothesis 10.1

Directional Research Hypothesis: For a given population, the criterion mean is greater than some specified value a.

Nondirectional Hypothesis: For a given population, the criterion mean is different from some specified value a.

Statistical Hypothesis: For a given population, the criterion mean is equal to some specified value a.

Full Model: $Y = b_1U + E_1$ (pieces full = m1 = 1)

Want (for directional Research Hypothesis): $b_1 > a$; restriction: $b_1 = a$
Want (for nondirectional Research Hypothesis): $b_1 \neq a$; restriction: $b_1 = a$

Restricted Model: $Y = aU + E_2$; or $(Y - aU) = E_2$ (pieces restricted = m2 = 0)

where:
  Y = criterion;
  U = 1 for all subjects; and
  b_1 is a least squares weighting coefficient calculated to minimize the sum of the squared values in the error vector.

Degrees of freedom numerator = (m1 - m2) = (1 - 0) = 1
Degrees of freedom denominator = $(N - m1) = (N - 1)$

where:
  N = number of subjects.

Note: Use Equation 10.1 to calculate F because the unit vector is not in the Restricted Model.

---

also a pretest (pre) on the same measure. There are three ways to analyze the data, with the last method we present being preferred.

One group: Pre and post with gain as the criterion. The pretest can be subtracted from the posttest for each subject, yielding a gain, difference, or change score. This gain score can be used as a criterion in the following Research Hypothesis: "The mean gain from pre to post is greater than zero (or any specified value)." The structure of this Research Hypothesis is the same as that in Generalized Research Hypothesis 10.1, with the pre-post difference as the criterion.

One group: Two repeated measures. Because the pretest and posttest measure the same variable, one could treat both measures as a single criterion and use a pre vector and a post vector as the two predictor variables (to see if knowledge of whether the score was pre or post helps to predict the score). A pre vector would be a vector that identifies those scores in the criterion that are pretests by containing a one if the score is a pretest, zero otherwise. A post vector identifies the posttests in the criterion in the same manner (these are the same as the time vectors—T1 and T2—in

Applied Research Hypothesis 10.1

Directional Research Hypothesis: For a given population, the population mean of X2 is greater than the national norm of 20.

Statistical Hypothesis: For a given population, the population mean of X2 is equal to the national norm of 20.

Full Model: $X2 = b_1U + E_1$ (pieces full = m1 = 1)

Want: $b_1 > 20$; restriction: $b_1 = 20$

Restricted Model: $(X2 - 20) = E_2$ (pieces restricted = m2 = 0)

alpha = .05
$ESS_f = 7,771$; $ESS_r = 8,723$, and using Equation 10.1:

$$F = \frac{(ESS_r - ESS_f) / (m1 - m2)}{(ESS_f) / (N - m1)}$$

$$F = \frac{(8,723 - 7,771) / 1}{(7,771) / 59}$$

$$F_{(1,59)} = 7.23$$

Interpretation: Since the sample mean of X2 is 23.98 (which is greater than 20), the 90th percentile of F can be referred to. The necessary F value is 2.75, and since the calculated F is larger, the Statistical Hypothesis can be rejected, and the Research Hypothesis can be accepted: "For the given population, the population mean of X2 is greater than the national norm of 20."

Generalized Research Hypothesis 10.2.) The Research Hypothesis here would be: "The posttest mean is higher than the pretest mean."

The number of scores in the criterion would be equal to twice the number of subjects because every subject would have both a pretest and a posttest. Generalized Research Hypothesis 4.1 would be used to test this Research Hypothesis, with the denominator degrees of freedom equal to $[(2N) - 1]$, where N is equal to the number of subjects.

One group: Repeated measures, person vectors. The second analysis (above) is not as powerful as it might be: Some very important information about the criterion scores has been omitted. Indeed, one assumption of statistical analysis—independence of observations—is violated when person vectors are not included. Because every person is tested twice, the researcher knows that one pretest and one posttest come from a given person. One other pretest and one other posttest come from another person, and so on. This is valuable information; individual differences are large in most areas. That is to say, if a subject scores low on the pretest, that same subject is expected to score relatively low on the posttest. Conversely, if a person scores high on the pretest, that subject is expected to score high on the posttest. The inclusion of person vectors, as illustrated in the following vector display, drains the denominator degrees of freedom, but the increase in the amount of variance accounted for is usually of such a large magnitude that it offsets the loss in degrees of freedom.

$$Y = aU + a_1T1 + a_2T2 + p_1P1 + p_2P2 + p_3P3 + p_4P4 + p_5P5 + p_6P6 + p_7P7 + E$$

	U	T1	T2	P1	P2	P3	P4	P5	P6	P7	E
Sam's Pre	1	1	0	1	0	0	0	0	0	0	?
Joe's Pre	1	1	0	0	1	0	0	0	0	0	?
Sue's Pre	1	1	0	0	0	1	0	0	0	0	?
Pat's Pre	1	1	0	0	0	0	1	0	0	0	?
Sal's Pre	1	1	0	0	0	0	0	1	0	0	?
Hal's Pre	1	1	0	0	0	0	0	0	1	0	?
Tom's Pre	1	1	0	0	0	0	0	0	0	1	?
Sam's Post	1	0	1	1	0	0	0	0	0	0	?
Joe's Post	1	0	1	0	1	0	0	0	0	0	?
Sue's Post	1	0	1	0	0	1	0	0	0	0	?
Pat's Post	1	0	1	0	0	0	1	0	0	0	?
Sal's Post	1	0	1	0	0	0	0	1	0	0	?
Hal's Post	1	0	1	0	0	0	0	0	1	0	?
Tom's Post	1	0	1	0	0	0	0	0	0	1	?

Of the three designs discussed for one-group studies, the inclusion of the person vectors best uses the information in the study. Another way of saying this is that, in most cases, the inclusion of the person vectors will yield a lower probability value than will the other two designs. This repeated-measures design is widely used and is of sufficient value to present it as Generalized Research Hypothesis 10.2.

Additional repeated measures might be obtained, necessitating more time vectors in the Generalized Research Hypothesis 10.2 Full Model. Specific hypotheses would dictate the restrictions made on that Full Model, and one can apply the multiple comparison notions discussed in chapter 4 to the present design.

TWO-GROUP MODELS

The desire in two-group studies is often to have a strict comparison group, but usually subjects cannot be randomly assigned to a treatment group and a comparison group. Therefore, researchers in the behavioral sciences often have to be satisfied with comparison groups that are somewhat similar to the treatment group but that are probably different.

Two Groups: Post Only

The introduction of a comparison group often allows the elimination of many competing explainers. The simplest design entailing both a comparison and an experimental group is the one discussed here. Only a posttest is obtained, and for change

Generalized Research Hypothesis 10.2

Directional Research Hypothesis: For a given population, the post mean is higher than the pre mean, over and above expected individual differences.

Nondirectional Research Hypothesis: For a given population, the post mean is different from the pre mean, over and above expected individual differences.

Statistical Hypothesis: For a given population, the post mean is equal to the pre mean, over and above expected individual differences.

Full Model: $Y = bT1 + cT2 + p_1P1 + p_2P2 + \ldots + p_NPN + E_1$ (pieces full $= m1 = N + 1$)

Want (for directional Research Hypothesis): $c > b$; restriction: $c = b$
Want (for nondirectional Research Hypothesis): $c \neq b$; restriction: $c = b$

Automatic unit vector analogue
Full Model: $Y = aU + bT1 + p_1P1 + p_2P2 + \ldots + p_NPN + E_1$ (pieces full $= m1 = N + 1$)
Want (for directional Research Hypothesis): $b < 0$; restriction: $b = 0$
Want (for nondirectional Research Hypothesis): $b \neq 0$; restriction: $b = 0$

Restricted Model: $Y = aU + p_1P1 + p_2P2 + \ldots + p_NPN + E_2$ (pieces restricted $= m2 = N$)

where:

Y	$=$	criterion of both pretest and posttest;
$T1$	$=$	1 if pretest, 0 otherwise;
$T2$	$=$	1 if posttest, 0 otherwise;
$P1$	$=$	1 if score from subject 1, 0 otherwise;
$P2$	$=$	1 if score from subject 2, 0 otherwise;

.
.
.

PN = 1 if score from subject N, 0 otherwise; and
$a, b, c, p_1, p_2, \ldots, p_N$ are least squares weighting coefficients calculated to minimize the sum of the squared values in the error vectors.

Degrees of freedom numerator $= (m1 - m2) = [(N + 1) - N] = 1$

Degrees of freedom denominator $= (N - m1) = [(N + N) - (N + 1)] = (N - 1)$

where:

N = number of elements in each vector—in all other Generalized Research Hypotheses, there was only one observation for each subject, and $N =$ the number of subjects. For repeated-measures analyses, there will be as many elements in each vector as there are subjects, multiplied by the number of times they were measured. In this example, N subjects times 2 measurements $= 2N$ or $(N + N)$.

to be attributable to the experimental method, the directional Research Hypothesis in Generalized Research Hypothesis 10.3 must be supported. Generalized Research Hypothesis 10.3 has exactly the same structure as Generalized Research Hypothesis 4.1, the two-group structure. Notice, though, in this context the definite importance of stating and obtaining a directional difference in favor of the experimental method.

Lack of knowledge about the standing of the subjects before treatment forces one to qualify one's interpretation. Random assignment does not guarantee that the two groups are initially equal, and a higher post score by the treatment group could be a function of that group's having been initially better. The next design allows the researcher to eliminate this possible competing explainer.

Generalized Research Hypothesis 10.3

Directional Research Hypothesis: For a given population, the post mean of the experimental method is higher than the post mean of the comparison method.

Nondirectional Research Hypothesis: For a given population, the post mean of the experimental method is not equal to the post mean of the comparison method.

Statistical Hypothesis: For a given population, the post mean of the experimental method is equal to the post mean of the comparison method.

Full Model: $Y = aG1 + bG2 + E_1$ (pieces full = m1 = 2)

Want (for directional Research Hypothesis): $a > b$; restriction: $a = b$
Want (for nondirectional Research Hypothesis): $a \neq b$; restriction: $a = b$

Automatic unit vector analogue
Full Model: $Y = aU + aG1 + E_1$ (pieces full = m1 = 2)
Want (for directional Research Hypothesis): $a > 0$; restriction: $a = 0$
Want (for nondirectional Research Hypothesis): $a \neq 0$; restriction: $a = 0$

Restricted Model: $Y = a_0U + E_2$ (pieces restricted = m2 = 1)

where:
Y = criterion;
U = 1 for all subjects;
$G1$ = 1 if subject given the experimental method, 0 otherwise;
$G2$ = 1 if subject given the comparison method, 0 otherwise; and
a_0, a, and b are least squares weighting coefficients calculated to minimize the sum of the squared values in the error vectors.

Degrees of freedom numerator = (m1 - m2) = (2 - 1) = 1

Degrees of freedom denominator = $(N - m1) = (N - 2)$

where:
N = number of subjects.

Two Groups: Pre and Post

A pretest is one way of obtaining some knowledge of initial standing. There are four different ways of analyzing such data, the first three being slightly easier to handle conceptually and operationally, the last being more powerful statistically.

Two groups: Pre and post, with gain as the criterion. The pretest can be subtracted from the posttest to yield a gain score, a difference score, or a change score, which then becomes the criterion. The Research Hypothesis again fits into the Generalized Research Hypothesis 4.1 and Generalized Research Hypothesis 10.3 framework: "The treatment method is better than the comparison method in terms of the criterion Y (gain from pretest to posttest)." Applied Research Hypothesis 10.2 gives the reader a chance to apply the two-group hypothesis to data regarding gain scores.

The comparison group may show a significant gain, whereas the experimental group shows a real gain, over and above that of the comparison group. The net difference between the two groups may be thought of as the final difference between the two groups, corrected for their initial difference. Such a correction, though, involves the assumption that each unit of difference in initial standing will produce a constant difference in final standing. This assumption can be stated statistically as a correlation of 1.0 between the pretest and the posttest. Although this assumption probably would never be valid in an applied setting, we recommend testing gain scores over

---

Applied Research Hypothesis 10.2

Directional Research Hypothesis: For a given population, X12 achieves more gain (X2 - X3) than does X13.

Statistical Hypothesis: For a given population, X12 achieves gains equal to those of X13.

Full Model: $(X2 - X3) = bX12 + cX13 + E_1$ (pieces full = m1 = 2)

Want: b > c; restriction: b = c

    Automatic unit vector analogue
    Full Model: $(X2 - X3) = aU + bX12 + E_1$ (pieces full = m1 = 2)
    Want: b > 0; restriction: b = 0

Restricted Model: $(X2 - X3) = a_0U + E_2$ (pieces restricted = m2 = 1)

alpha = .05
$R_f^2 = .0002$; $R_r^2 = .00$
$F_{(1,58)} = .009$
Computer probability = .92; directional probability = .54

Interpretation: Since the mean gains are in the direction opposite from that hypothesized, support for the Research Hypothesis has not been established, and the Statistical Hypothesis cannot be rejected.

those procedures previously discussed in this chapter. Most statisticians recommend the difference method over that of matching, mainly because of the difficulty in selecting good matched samples, as shall be discussed shortly.

Two groups: Pre and post, with pre used as a covariate. If one does not want to assume that the correlation between pretest and posttest is 1.0, then the pretest can be used as a covariate. The Research Hypothesis here would be: "The treatment method is better than the comparison method on the posttest, over and above any differences observed on the pretest." This Research Hypothesis fits the framework of Generalized Research Hypothesis 8.1, with the pretest as a covariate and the treatment and comparison dichotomous vectors as the only other vectors in the Full Model. Applied Research Hypothesis 10.3 gives one a chance to test such a hypothesis.

Two groups: Pretest(s) different from posttest. There are many research situations in which it is unwise to administer the same test before treatment as will be administered after treatment. In this situation, other behaviors related to the posttest can be assessed before the treatment and used as covariates. The researcher can assess as many behaviors as are suspected might be different between groups. The Research Hypothesis here would be: "The treatment method is better than the comparison method on the posttest Y, over and above the possible contaminating variables of A, B, C, etc." This Research Hypothesis fits the framework given previously in Generalized Research Hypothesis 8.1, with the addition of multiple covariables.

Applied Research Hypothesis 10.3

Directional Research Hypothesis: For a given population, X12 is better than X13 on the posttest (X2), over and above differences observed on the pretest (X3).

Statistical Hypothesis: For a given population, X12 is equal to X13 on the posttest (X2), over and above differences observed on the pretest (X3).

Full Model: $X2 = bX12 + cX13 + dX3 + E_1$ (pieces full = m1 = 3)

Want: b > c; restriction: b = c

> Automatic unit vector analogue
> Full Model: $X2 = a_0U + bX12 + dX3 + E_1$ (pieces full = m1 = 3)
> Want: b > 0; restriction: b = 0

Restricted Model: $X2 = a_0U + dX3 + E_2$ (pieces restricted = m2 = 2)

alpha = .05
$R_f^2 = .928; R_r^2 = .927$
$F_{(1,57)} = .81$
Computer probability = .56; directional probability = .82

Interpretation: Since the weighting coefficient for X12 in the Full Model is less than the weighting coefficient for X13, the Research Hypothesis cannot be supported, and the Statistical Hypothesis cannot be rejected.

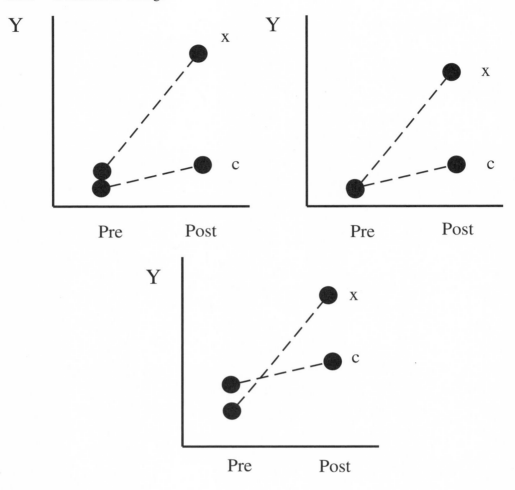

Figure 10.1.
Three possible empirical outcomes in which the treatment (X) increases more than the comparison (C).

Two groups: Pre and post—interaction between method and time, with person vectors. The long, probably confusing heading simply means that we are now going to be combining the notions of interaction and repeated measures (with directionality, of course). Figure 10.1 contains three possible empirical outcomes that would satisfy the following Research Hypothesis:

> The relative effectiveness of the treatment method as compared to the comparison method will be greater at post than at pre, over and above the expected individual differences.

Statistical Hypothesis:

> The relative effectiveness of the treatment method as compared to the comparison method will be the same at post as at pre, over and above the expected individual differences.

The Full Model reflecting the Research Hypothesis is:

(Model 10.3) $Y = a_0U + b(T1 * X) + c(T1 * C) + d(T2 * X) + e(T2 * C) + p_1P1 + p_2P2 + \ldots + p_NPN + E_3$

The restriction implied by the Research and Statistical Hypothesis is: $(b - c) = (d - e)$, resulting in the following Restricted Model:

(Model 10.4) $Y = a_0U + fT1 + gT2 + hX + iC + p_1P1 + p_2P2 + \ldots + p_NPN + E_4$

where:

Y	=	the criterion variable containing both pretest and posttest;
T1	=	1 if score from pre, 0 otherwise;
T2	=	1 if score from post, 0 otherwise;
X	=	1 if score from treatment, 0 otherwise;
C	=	1 if score from comparison, 0 otherwise;
P1	=	1 if score from person 1, 0 otherwise;
P2	=	1 if score from person 2, 0 otherwise;

.

.

.

PN = 1 if score from person N, 0 otherwise.

To find out the number of linearly independent vectors in the Restricted Model, first consider the unit vector. Because one of the time vectors can be obtained by subtracting the other time vector from the unit vector, only one of the time vectors will be linearly independent. Likewise, only one of the treatment vectors will be linearly independent. Now, if all the person vectors representing persons in the treatment group are added, the treatment vector will be obtained. Therefore, one of those person vectors is linearly dependent. Likewise, one of the comparison person vectors is linearly dependent. Therefore, only N - 2 of the person vectors are linearly independent. Adding N - 2 to the three previously determined linearly independent vectors results in $(N + 1)$ linearly independent vectors. Therefore, $df_n = (m1 - m2) = [(N + 2) - (N + 1)] = 1$; and $df_d = (N - m1) = [(N + N) - (N + 2)] = (N - 2)$.

Notice that the experimental group does not have to be equal to the comparison group on the pretest. The experimental group can be higher than the comparison group, or lower—the concern is in the *relative* effectiveness (on the posttest as compared to their relative standing on the pretest). In fact, if the groups are randomly assigned, we would expect little pretest difference, and we might only look at posttest differences. But remember: Random assignment does not guarantee that the groups will be equal. So the interaction question here seems reasonable to "take care" of the problem of initial group differences. Indeed, the only interesting question here is the interaction question. The average treatment effect will be watered down because the groups should be similar at pretest. The average time effect will be watered down because there is less change (as expected) from pretest to posttest for the comparison group. Thus, while the notion of orthogonal contrasts introduced in chapter 4 would indicate the availability of three tests of significance, only one is of any interest in the design. Applied Research Hypothesis 10.4 provides the models appropriate to the Appendix A data for this kind of hypothesis.

Applied Research Hypothesis 10.4

Directional Research Hypothesis: For a given population, the relative superiority of treatment X12 over comparison X13 is greater on the criterion (X2) at posttest (X11) than at pretest (X10), over and above any individual differences (X21 . . . X80).

Statistical Hypothesis: For a given population, differences (if any) between X12 and X13 are the same on posttest (X2) as on pretest (X3), over and above individual differences.

Note that the data are being treated here as two sets of 30 subjects. Data records 1 and 31 are listed in X16 as Subject 1, and data records 2 and 32 are listed in X16 as Subject 2, etc. The odd-numbered people are considered to be in the treatment and the even-numbered people are considered to be in the comparison. (See Appendix D-10.4 for the SAS statements.)

Full Model: $X(2 \text{ and } 3) = a_{17}X17 + a_{18}X18 + a_{19}X19 + a_{20}X20 + p_{21}X21 + \ldots + p_{80}X80 + E_1$ (pieces full $= m1 = N + 3$)

Want: $(a_{18} - a_{20}) > (a_{17} - a_{19})$; restriction: $(a_{18} - a_{20}) = (a_{17} - a_{19})$

   Automatic unit vector analogue
   Full Model: $X(2 \text{ and } 3) = a_0U + a_{17}X17 + a_{18}X18 + a_{19}X19 + p_{21}X21 + \ldots + p_{79}X79 + E_1$ (pieces full $= m1 = N + 3$)
   Want: $a_{18} > (a_{17} - a_{19})$; restriction: $a_{18} = (a_{17} - a_{19})$
   Restricted Model: $X(2 \text{ and } 3) = a_0U + a_{12}X12 + a_{10}X10 + p_{21}X21 + \ldots + p_{80}X80 + E_2$ (pieces restricted $= m2 = N + 2$)

Restricted Model: $X(2 \text{ and } 3) = a_{12}X12 + a_{13}X13 + a_{10}X10 + a_{11}X11 + p_{21}X21 + \ldots + p_{80}X80 + E_2$ (pieces restricted $= m2 = N + 2$)

where:
   $X10 = 1$ if pretest, 0 otherwise;
   $X11 = 1$ if posttest, 0 otherwise;
   $X12 = 1$ if treatment, 0 otherwise;
   $X13 = 1$ if comparison, 0 otherwise;
   $X17 = X12 * X10$;
   $X18 = X12 * X11$;
   $X19 = X13 * X10$; and
   $X20 = X13 * X11$.

alpha $= .05$
$R_f^2 = .8831$; $R_r^2 = .8829$
$F_{(1,58)} = .016$
Computer probability $= .9231$; directional probability $= .46$

Interpretation: The difference between a_{17} and a_{19} in the Full Model is less than a_{18}, therefore the directional probability can be referred to. Since the directional probability is not less than the predetermined alpha, the Statistical Hypothesis cannot be rejected, and the Research Hypothesis cannot be accepted.

MATCHED GROUPS

This and the following section are included solely to familiarize students with the process of matching more thoroughly than in chapter 8. While common in early empirical research, the pitfalls of matching have become more apparent, and the availability of computers has brought such analyses as in the previous section to the forefront.

There are two somewhat different ways to match. One may match individual pairs or match groups based on group indexes (means, standard deviations, etc.). Basically, subjects are pretested on several supposedly relevant variables and then assigned to groups such that the resultant groups are similar (not necessarily equivalent) on the several supposedly relevant variables. The sole requirement of similarity is often that of equal means. An additional requirement that is sometimes used is that of similar standard deviations. Unfortunately, only visual comparisons are usually used to ascertain similarity. A totally indefensible procedure that is sometimes used in its place is that of visually comparing the extreme scores of the various groups, and as most researchers know, extreme scores tell little about the bulk of the scores.

After the two matched groups are obtained, the format of the Research Hypothesis would be similar to that of Generalized Research Hypothesis 4.1 if there is no pretest measure; or if there is, then Generalized Research Hypothesis 10.3 would be applicable. That is, the matching variables are not always used in the analysis of the data; sometimes they are used only for the delineation of the two groups. The matching variable could be used as a *blocking variable*. That is, the matching variable could be used as a factor not of interest to the researcher but as a means of reducing the within variability, thus resulting in a smaller error component. It is somewhat cumbersome to use more than one blocking variable in this manner, but Generalized Research Hypothesis 8.1 could be used.

The two most important problems related to matching groups follow. First, the relevancy of the matching variables is often of questionable value. If there is no relationship between the matching variable and the criterion, then the results may be generalized to subjects that differ on the matching variable, although the experimenter who has needlessly matched does not know that this generalization can be made.

A second problem involves the difficulty in equating the groups on the matching variables. It is very difficult to come up with two groups having equal means on several matching variables. Of course, the more matching variables one is interested in, the more unlikely it is that one can come up with equivalent groups. This problem is further delineated in the following section concerning matched pairs of individuals.

MATCHED PAIRS OF INDIVIDUALS

A more thorough procedure for matching involves matching individuals. Thus, for every person in the treatment group, there is a like person in the comparison group. The extent to which the comparison subject is like the treatment subject is again a function of the number and relevancy of the matching variables. Gender may be an important matching variable in one study, whereas in another study it may not be relevant. Whether a variable should be considered a matching variable is an empirical question that can only be estimated ahead of time on theoretical grounds by the researcher.

Because for every person in the experimental group there is a like person in the comparison group, person vectors can be used, as in Generalized Research Hypothesis 10.2. Notice that here, as in the matching of groups, the matching variables are not used in the analysis of the data. The assumption in the matched-pairs design is that the person in the experimental group is exactly like the "like" person in the comparison group. That this assumption is not valid should be readily apparent.

There are three major problems with the technique of matched pairs. First, because the selection of one kind of individual demands a like person, the degrees of freedom are drastically reduced. For an "uncorrelated" or "independent" t test, the degrees of freedom is $N - 2$, where N is the total number of people in the two groups. For the matched-pairs analysis (referred to as a *correlated* or *dependent t test* by some), the degrees of freedom is the number of pairs minus one. Thus, if one decides to match, the gain that one obtains from matching must be enough to override the loss in degrees of freedom (exactly half as many degrees of freedom). If the matching variables are chosen judiciously, the loss in degrees of freedom will be of little consequence.

A second and more important problem with matched pairs concerns the availability of identical subjects. With individual differences as large and as variable as they typically are, it is often difficult to find an adequate match. Empirical studies using this procedure often demand an initial sampling pool of 300 or 400 to get even a crude match for 30 pairs of subjects.

The third problem with matched pairs is related to the above problem. The search for matching individuals often eliminates the extreme subjects, those who are deviant on only one measure. Conversely, those deviant matches that are found are often a function of a large error component, and on a second testing this good match may indeed turn out to be a bad match after all.

POINT CHANGE MEASURING MULTIPLE SUBJECTS

Some research questions involve the detection of a change at a certain point in time as a function of some stimulus change or input. An ideal situation would be to measure several subjects in several testing periods both before and after the stimulus change. The number of time periods before the stimulus change would not have to be equal to the number of measured time periods after the stimulus change, although the following example does have an equal number.

Figure 10.2 depicts a situation in which the mean of the last 20 time periods is much greater than the mean of the first 20 time periods. But the careful researcher surely would not want to attribute a causal influence to the stimulus input that occurred between Time Periods 20 and 21. A regression model that allows the two lines depicted in Figure 10.2 would be:

$$(\text{Model } 10.5)\, Y = aU1 + bX1 + cU2 + dX2 + E_5$$

where:

Y = the criterion to be predicted;
$U1$ = 1 if observation from Time Period 1–20, 0 otherwise;
$X1$ = the time period if observation from Time Period 1–20, 0 otherwise;
$U2$ = 1 if observation from Time Period 21–40, 0 otherwise; and
$X2$ = the time period if observation from Time Period 21–40, 0 otherwise.

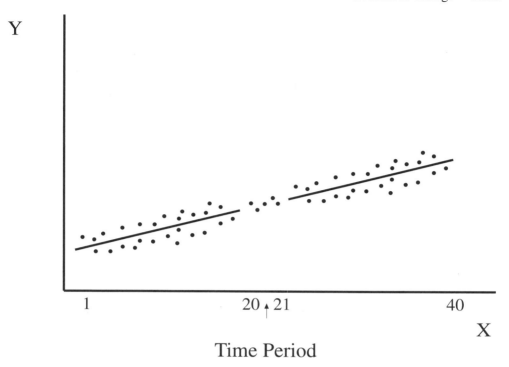

Figure 10.2.
A situation in which the stimulus input (identified by the arrow between Time Periods 20 and 21) has not caused a change.

A reasonable question to ask of the data in Figure 10.2 is, "How much error would be introduced if only one line were used to depict the data?" Another way of expressing this question is to ask if the slopes of the two lines are the same and if the Y-intercepts of the two lines are the same. In Model 10.5, the symbols b and d represent the two slopes, so one is essentially asking if it is reasonable to force b equal to d. Also in Model 10.5, the Y-intercepts are a function of a and c. If a is forced to be equal to c, then one is forcing the two lines to have the same Y-intercepts. If one simultaneously forces the two slopes to be the same (b = d, both equal to e, a common weight) and the two Y-intercepts to be the same (a = c, both equal to f, a common weight), then one is indeed asking if one line can do as adequate a job as two lines. Forcing these two restrictions on Model 10.5, with minor algebraic manipulation, one obtains:

$$Y = fU1 + eX1 + fU2 + eX2 + E_6$$

$$Y = f(U1 + U2) + e(X1 + X2) + E_6$$

$$Y = fU + e(X3) + E_6$$

$$(\text{Model } 10.6)\ Y = fU + eX3 + E_6$$

where:

$\quad$ X3 = a continuous vector of time periods.

The Y-intercept of this one line will be represented by f, and the slope will be represented by e, the coefficient of the continuous vector of time periods. If both of the restrictions are indeed true for the sample, then the elements of E_6 will be the same as the elements of E_5; the errors in prediction using a single line will be comparable to the errors in prediction using two separate lines.

If the amount of accounted-for variance in the criterion scores in the Restricted Model (Model 10.6) is significantly less than that in the Full Model, then one line does not do as good a job as do two lines, and there may be an effect due to the stimulus change. Reasons for a cautious interpretation are discussed in the next section on curvilinear relationships.

In another situation, the researcher may expect that the slopes of the lines (the increase in performance level) will stay the same before and after the stimulus change but that, following the stimulus change, there is a jump in performance level, so that the Y-intercepts are not equal, as shown in Figure 10.3. Here, the Full Model would allow for different Y-intercepts but only one slope for the two lines:

$$(Model\ 10.7)\ Y\ =\ aU1 + cU2 + eX3 + E_7$$

where:

Y = performance level;
U1 = 1 if observation from Time Period 1–20, 0 otherwise;
U2 = 1 if observation from Time Period 21–40, 0 otherwise; and
X3 = time period.

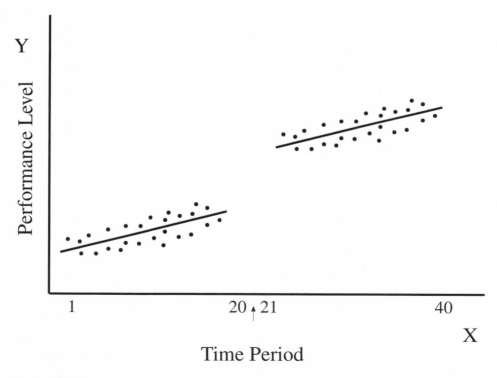

Figure 10.3.
A case in which the slope of the line after stimulus change is the same as before stimulus change.

Because the researcher expects the Y-intercept after stimulus change to be higher than the one before stimulus change, the Statistical Hypothesis would state that the Y-intercepts are equal (a = c). Forcing the restriction onto the Full Model would yield:

$$Y = aU1 + aU2 + eX3 + E_8$$

Gathering vectors together that have the same weighting coefficient yields:

$$Y = a(U1 + U2) + eX3 + E_8$$

The Restricted Model thus would be:

(Model 10.8) $Y = aU + eX3 + E_8$

If the restriction (a = c) is not viable and the magnitude of c is greater than that of a, then one can attribute a causal heightening effect to the stimulus change. Obviously, if a > c, then the stimulus change has decreased the performance level.

CURVILINEAR RELATIONSHIPS THAT ACCOUNT FOR THE DATA BETTER THAN STIMULUS CHANGE

The interpretations in the above discussions have been somewhat cautious because the models were assuming and acting as if there were rectilinear relationships. We, in this text, suspect that there are quite a few curvilinear relationships in the real world, but these relationships have not been discovered because researchers have not looked for them. If there is an underlying curvilinear relationship, one may be making a costly mistake by trying to map the data with two straight lines.

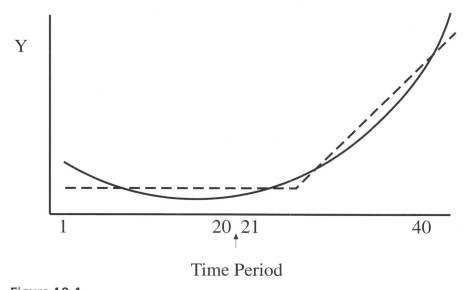

Figure 10.4.
A case in which a single second-degree curved line depicts the data as well as do two second-degree curved lines, and better than two straight lines.

Figure 10.4 depicts a set of data in which the performance scores after the stimulus change continue to follow the curvilinear trend that started before the stimulus change. Visually, it appears that a single second-degree curved line can account for the data, rather than two second-degree curved lines, or two straight lines.

The two dashed straight lines in Figure 10.4 might lead to the conclusion that the stimulus change after Time Period 20 had an effect (i.e., raising the rate of change in Y3). However, if a single second-degree curve fits all the data well (yielding a high R^2), then what was happening systematically before the stimulus change is still happening after the stimulus change. A researcher would be hard-pressed to argue that the stimulus change in Figure 10.4 upset the normal course of events.

The regression approach to verifying the visual impression would be as follows. Just as in the case with rectilinear lines, one needs to generate a model containing two separate second-degree curved lines.

$$(\text{Model } 10.9)\ Y = a_0 U + a_1 Z1 + b_2 Z2 + s_3 Z3 + a_4 Z4 + b_5 Z5 + s_6 Z6 + E_9$$

where:

Y = the criterion performance score;
$Z1$ = 1 if observation is from Time Period 1–20, 0 otherwise;
$Z2$ = the time period if observation is from Time Period 1–20, 0 otherwise;
$Z3$ = the squared value of the element in $Z2$;
$Z4$ = 1 if observation is from Time Period 21–40, 0 otherwise;
$Z5$ = the time period if observation is from Time Period 21–40, 0 otherwise; and
$Z6$ = the squared value of the element in $Z5$.

To test the effectiveness of a single line, all that one needs to do is equate the corresponding components of the two separate lines. Because a_1 and a_4 are the two Y-intercepts, b_2 and b_5 are weights for the two linear components, and s_3 and s_6 are weights for the two second-degree components, the restrictions are: $a_1 = a_4$, both equal to a_7, a common weight; $b_2 = b_5$, both equal to b_8, a common weight; $s_3 = s_6$, both equal to s_9, a common weight. Inserting these restrictions into Model 10.9 yields:

$$Y = a_0 U + a_7 Z1 + a_7 Z4 + b_8 Z2 + b_8 Z5 + s_9 Z3 + s_9 Z6 + E_{10}$$

$$Y = a_0 U + a_7 (Z1 + Z4) + b_8 (Z2 + Z5) + s_9 (Z3 + Z6) + E_{10}$$

$$Y = (a_0 + a_7)(U) + b_8 (Z7) + s_9 (Z8) + E_{10}$$

$$(\text{Model } 10.10)\ Y = a_{10} U + b_8 Z7 + s_9 Z8 + E_{10}$$

where:

$Z7$ = the continuous vector of time periods; and
$Z8$ = the squared value of the corresponding element in $Z7$.

The unit vector allows for a nonzero intercept; the $Z7$ vector allows for a linear component; and the squared elements of vector $Z7$ appearing in vector $Z8$ allow for a second-degree curvilinear component. Model 10.10 thus allows for a single second-degree curve. With reference to data such as in Figure 10.4, Model 10.10 will yield as good a fit as will Model 10.9. Thus, the kind of relationship between time and the criterion that was occurring before the stimulus change is also occurring after the stimulus change. Therefore, the stimulus change had no impact on what was occurring.

Whether the researcher chooses a curvilinear or rectilinear fit to the data is a function of both theoretical notions and empirical data. A statistician cannot tell a researcher when to investigate curvilinearity. What we have done is to show the applicability of the GLM procedure to curvilinear types of problems (and to encourage such application whenever the researcher thinks that such might be fruitful).

FUNCTIONAL CHANGE IN INDIVIDUAL-ORGANISM RESEARCH

The discussion in the preceding section has many ramifications for individual-organism research. Kelly, McNeil, and Newman (1973) have documented a more extensive application of the GLM approach to repeated measures of a single organism. The following paragraphs should only serve to whet the appetites of empirically oriented single-organism researchers.

Eyeballing the data that emanate from standard individual-organism designs can be bolstered with probability statements of empirical reproducibility. Suppose that some functional relationship is established under a set of conditions, A. Then a new set of conditions, B, is introduced to the organism. Finally, the A conditions are reintroduced, A' (A prime). Data points depict the functional relationship between time period and performance level for each condition. Figure 10.5 illustrates one possible outcome.

As with any collection of data, several research questions can be tested. For instance, one might ask if the A baseline function has been achieved during the A' condition. This question is answered by testing Model 10.11 against Model 10.12 on only the data from A and A'.

$$(\text{Model 10.11}) \; Y \; = \; a_1 U1 + a_2 X2 + a_3 U3 + a_4 X4 + E_{11}$$

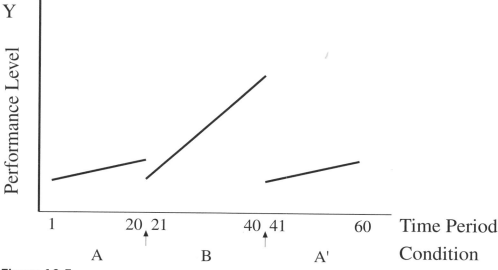

Figure 10.5.
One possible data set emanating from single-organism research.

where:

Y = criterion of performance level;

$U1$ = 1 if criterion from Time Period 1–20, 0 otherwise (Condition A);

$X2$ = time period if observation from Time Period 1–20, 0 otherwise;

$U3$ = 1 if observation from Time Period 41–60, 0 otherwise (Condition A'); and

$X4$ = time period minus 40 if observation from Time Period 41–60, 0 otherwise.

The restrictions implied by the Research Hypothesis are: $a_1 = a_3$, both equal to a_5, a common weight; and $a_2 = a_4$, both equal to a_6, a common weight. Forcing these restrictions onto the above Full Model results in the following Restricted Model:

$$(\text{Model } 10.12)\ Y\ =\ a_5 U + a_6 X5 + E_{12}$$

where:

$X5$ = the vector containing the (relative) time at which performance level was observed.

Note that the adjustment made on the time scores in vector $X4$ has the effect of superimposing the data obtained during the A' condition onto the data obtained during the A condition. If there is a significant decrease in the amount of criterion variance accounted for by Model 10.12, then one cannot assume that the same (linear) functional relationship exists under the A and A' conditions. If there is no significant decrease in the amount of variance accounted for, then the Restricted Model is as valid as the Full Model, implying that the difference in the two functional relationships is no more than what would be expected through random fluctuation. If a nonlinear baseline is expected under A, such a condition could be tested by including in the model the appropriate nonlinear terms.

A possibly more interesting change question involves the comparison of the functional relationship under condition B with the (average) functional relationship under conditions A and A'. If the B condition is effective, then the rate of increase ought to be faster than under baseline conditions (A and A'). The superimposed starting point for all conditions might reasonably be expected to be the same (an empirically testable situation, of course). The Research Hypothesis of interest could be: "Assuming a common starting point for all conditions, the B condition yields a faster rate of increase on the criterion Y than does the average of conditions A and A'." The Full Model to test this Research Hypothesis would be:

$$(\text{Model } 10.13)\ Y\ =\ a_0 U + a_2 X2 + a_4 X4 + E_{13}$$

where:

Y = criterion of performance level;

$X2$ = time period if observation from Condition A, time period minus 40 if observation from Condition A', 0 otherwise; and

$X4$ = time period minus 20 if observation from Condition B, 0 otherwise.

The restriction implied by the Research Hypothesis is: $a_2 = a_4$, both equal to a_8, a common weight. Forcing this restriction onto the above Full Model yields the following Restricted Model:

$$(\text{Model } 10.14)\ Y\ =\ a_0 U + a_8 X5 + E_{14}$$

where:

X5 is a vector containing the relative time at which performance level was observed.

The Research Hypothesis is supported only if $a_4 > a_2$ in the Full Model and if the directional probability is less than the predetermined alpha. The careful reader will begin to notice the underlying similarities in the testing of various hypotheses.

Kelly et al. (1973) illustrate some ways the GLM can be used to answer meaningful questions with single-organism research designs. Although some of the material is redundant, repeating earlier material, the entire paper is incorporated here (following) so as to retain the flow. Researchers should remember that these are only a few examples and that the Research Hypothesis must be stated first. Once the Research Hypothesis is stated, the models used to test that hypothesis are easily constructed.

SUGGESTED INFERENTIAL STATISTICAL MODELS FOR RESEARCH IN BEHAVIOR MODIFICATION

For several reasons, research in operant psychology usually has been summarized using descriptive statistics. We suspect the chief reason inferential statistics has been given little attention is that early laboratory research yielded such clear-cut distinctions that one did not have to resort to tests of statistical significance. Another reason also might be attributed to poor advice from statisticians regarding limitations of single-subject data.

Operant psychological techniques have diffused to applied behavior modification, where complex behaviors are manipulated in settings where control of extraneous variables is difficult to achieve. Consequently, data often fail to exhibit the clear magnitude of effects observed in the data derived from laboratory manipulation. Under these circumstances, often the data no longer hit the reader between the eyes, though the expected trend seems present.

We believe researchers who use behavioral modification procedures should consider wider use of inferential statistics, especially when some doubt exists regarding the outcome of an experimental manipulation. Given this belief, we present several inferential statistical models that may aid the operant investigator in presenting data. None of the procedures presented are particularly new nor startling, yet we believe their use has not been fully appreciated as research tools. The following designs are illustrated:

1. single-organism curve fitting of response latency
2. multiple-organism curve fitting
3. evaluating A-B-A'-B' observed mean differences
4. evaluating A-B-A'-B' observed curve differences

Example 1: Single-Organism Curve Fitting of Response Latency

In a study of reinforcement schedules (see Figures 10.6 and 10.7), if the experimenter was interested in the latency of the response (criterion variable) over sets

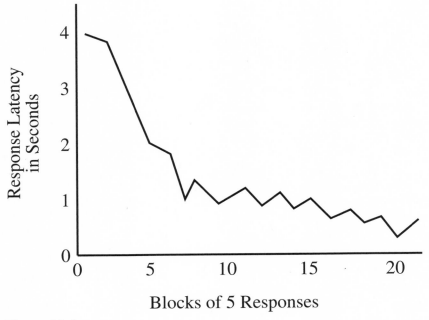

Figure 10.6.
An observed response latency relationship on a fixed-ratio 100 schedule for Day 1.

of five response blocks within the schedule (independent variable) the following are only a few of the many research questions that can be asked regarding the design:

1. What kind of functional relationship between response blocks and response latency exists on Day 1?
2. Is there a different functional relationship between response blocks and response latency for Day 1 and Day 2?
3. Is there a difference between the initial response latency on Day 1 and the initial response latency on Day 2?

If one expects a second-degree functional relationship in Question 1, above, the following models can test this expectation.

$$\text{(Model 10.15) } Y = a_0 U + b_1 X + c_1 Z + E_{15}$$

$$\text{(Model 10.16) } Y = a_0 U + b_1 X + E_{16}$$

where:
Y = the criterion latency score;
U = 1 for all sessions;
X = the number of the response block within the schedule;
Z = the squared values of X; and
a_0, b_1, and c_1 are least squares weighting coefficients computed to minimize the squared components in the error vectors E_{15} and E_{16}.

The proportion of variance (R^2) accounted for by the above two models can be compared by using the generalized F test used in all regression analyses. If the F test

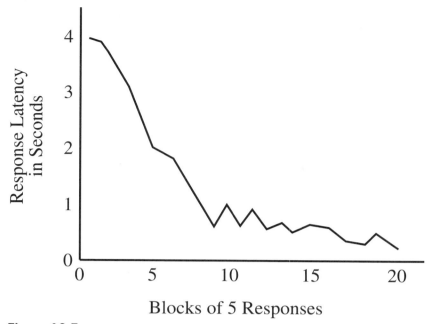

Figure 10.7.
An observed response latency relationship on a fixed-ratio 100 schedule for Day 2.

is significant at the specified alpha level, then the Full Model (Model 10.15) is pre-ferred over the Restricted Model (Model 10.16), showing that there is a second-degree curvature to the data. If the F test is not significant, then the Restricted Model is preferred over the Full Model because the Restricted Model is a more parsimoni-ous accounting of the criterion. Other kinds of curves could be investigated, such as third-degree, fourth-degree, and logarithmic curves.

Question 2, "Is there a different functional relationship between response blocks and latency for Day 1 and Day 2?," can be tested statistically by determining if the graphs for the two days have the same Y-intercepts, the same linear component, and the same second-degree curvature. A Full Model must be developed that will allow for differences on the three parameters, and then this model must be compared with a model that forces a common value for the three parameters. Models 10.17 and 10.18 below are those two models:

$$\text{(Model 10.17) } Y = a_1 U1 + b_1 X1 + c_1 Z1 + a_2 U2 + b_2 X2 + c_2 Z2 + E_{17}$$

$$\text{(Model 10.18) } Y = a_0 U + b_3 X3 + c_3 Z3 + E_{18}$$

where:

Y = the criterion latency score;

$U1$ = 1 if the criterion from Day 1, 0 otherwise;

$U2$ = 1 if the criterion from Day 2, 0 otherwise;

$X1$ = number of the response block within the schedule if criterion is from Day 1, 0 if criterion from Day 2;

$X2$ = number of the response block within the schedule if criterion is from Day 2; 0 if criterion from Day 1;

$Z1 = X1 * X1 =$ the squared elements of X1;

$Z2 = X2 * X2 =$ the squared elements of X2;

$U = 1$ for all observations;

$X3 =$ number of the response block within the schedule (disregarding day);

$Z3 =$ the squared number of response block within the schedule (disregarding day); and

$a_0, a_1, a_2, b_1, b_2, b_3, c_1, c_2$, and c_3 are least squares weighting coefficients determined to minimize the squared elements in the error vectors E_{17} and E_{18}.

The reader should note that Model 10.17 allows reflection of two second-degree lines, one for Day 1 and one for Day 2. Model 10.18 forces the line of best fit for Day 1 and Day 2 to be the same. That is, the two days have common intercepts, linear components, and second-degree components.

The R^2 of the above two models can again be tested with the generalized F test. If the R^2 of Model 10.17 is found to be *not* significantly different from the R^2 of Model 10.18, then one can conclude that the functional relationships for Day 1 and Day 2 are the same. Indeed, a stable state can be defined where the functional relationships among several contiguous days are *not* statistically different. In the example given above, one may say a stable state is observed when *two* contiguous days can be reflected by one line. If stability for three or four sessions is desired to specify a stable state, then one need only expand Model 10.18 to provide aU, bX, and cZ components for each day under consideration. A comparison of this new model to Model 10.18, using the F statistic, could provide an answer for the stable state. If the observed F is nonsignificant, then a stable function can be said to exist. To find out whether that function required a second-degree component would necessitate a comparison, such as Model 10.15 versus Model 10.16.

If the R^2 of Model 10.17 is found to be significantly larger than the R^2 of Model 10.18, then the two second-degree curves of best fit are different. Stated in another fashion, it could be concluded that the functional relationship is not the same for the two days. But because three parameters were simultaneously investigated, one could not, from the above analysis, conclude that the three parameters were different. If one were interested in specifically how the two days differed, then any one of the following three restrictions could be individually made and tested: $a_1 = a_2$ (replaced by a_0, a common Y-intercept); $b_1 = b_2$ (replaced by b_3, a common linear component); $c_1 = c_2$ (replaced by c_3, a common second-degree component), resulting, respectively, in the following models:

$$(\text{Model 10.19}) \ Y = a_0U + b_1X1 + b_2X2 + c_1Z1 + c_2Z2 + E_{19}$$

$$(\text{Model 10.20}) \ Y = a_1U1 + a_2U2 + b_3X3 + c_1Z1 + c_2Z2 + E_{20}$$

$$(\text{Model 10.21}) \ Y = a_1U1 + a_2U2 + b_1X1 + b_2X2 + c_3Z3 + E_{21}$$

By testing Model 10.17 against Model 10.19, one can determine if the model that allows for, among other things, different Y-intercepts accounts for a greater proportion of variance than does a model that forces the Y-intercepts to be the same. If Model 10.17 does have a significantly larger R^2 than Model 10.19, then the data for the two days do have different Y-intercepts. Such a finding would imply that there is a significantly different initial response on Days 1 and 2. If there is *not* a significant difference between Models 10.17 and 10.19, then the Y-intercepts for the two days cannot be considered different. Note that the above conclusion does not say anything at all about the linear and second-degree slopes.

By testing Model 10.17 against Model 10.20, one can determine if the first-degree function of Day 1 is different from the first-degree function of Day 2. A question similar to traditional statistics that one might ask of these data is: "Is there a linear interaction between day and trial blocks?" Model 10.22, allowing for interaction, would be compared against Model 10.23, which depicts a state of affairs of no interaction.

$$(\text{Model } 10.22)\ Y\ =\ a_1U1 + a_2U2 + b_1X1 + b_2X2 + E_{22}$$

$$(\text{Model } 10.23)\ Y\ =\ a_1U1 + a_2U2 + b_3X3 + E_{23}$$

where:
   $X3 = X1 + X2.$

This test is investigating linear interaction, and if one suspected a second-degree interaction (above and beyond the linear one) the appropriate models would be Models 10.17 and 10.21, which we have already compared.

The F tests just presented relate to inferences that can be made regarding one organism's population of responses under a particular schedule. The above analyses considered sample estimators of two potentially different populations (i.e., a_1 and a_2; b_1 and b_2; and c_1 and c_2). If the observed differences between the two sample estimators of population values are likely due to sampling variation (a nonsignificant F), then one can assume that the two samples come from the same within-organism response population. One now has a good idea as to what this one organism will do under this one schedule.

Example 2: Multiple-Organism Curve Fitting

Once a stable curve is determined within one organism, one may wish to compare this functional relationship with another organism with a similar history.

Suppose the curved relationship depicted in Figures 10.6 and 10.7 were found to be the same and all the regression weights were nonzero, then Model 10.18 ($Y = a_0U + b_3X3 + c_3Z3 + E_{18}$) would reflect the relationship for that particular organism.

Given another organism with a similar history, one may use the procedures in Example 1 to express the functional relationship for this individual.

A legitimate question may be: "Is the functional relationship within Organism 1 different from the functional relationship within Organism 2?"

To answer this question, a model (Model 10.24) can be constructed to incorporate the two separate equations and then restricted down to another model (Model 10.25) that imposes common weights for the two organisms.

$$(\text{Model } 10.24)\ Y\ =\ a_1U1 + b_1X1 + c_1Z1 + a_2U2 + b_2X2 + c_2Z2 + E_{24}$$

where:
   Y = the criterion latency score;
   $U1$ = 1 if the criterion is from Organism 1, 0 otherwise;
   $X1$ = number of the response block within the schedule if the criterion is from Organism 1, 0 otherwise;

Z1 = the squared elements in X1;
U2 = 1 if the criterion is from Organism 2, 0 otherwise;
X2 = number of the response block within the schedule if the criterion is
     from Organism 2, 0 otherwise; and
Z2 = the squared elements in X2.

$$\text{(Model 10.25) } Y = a_0 U + b_3 X3 + c_3 Z3 + E_{25}$$

Model 10.25 is different from Model 10.18 only in that data for two organisms are being considered.

If the differences between Model 10.24 and Model 10.25 are likely to be due to sampling variation (nonsignificant F test), then one can conclude with some degree of confidence that, for the schedule under consideration, the response characteristics for the two organisms reflect a population that does not differ significantly.

If the differences between Model 10.24 and Model 10.25 are not likely to be due to sampling variations (significant F test), then one must figure out which (one, two, or three) of the three population estimators (a_0, b_3, and c_3) are not equal for the two organisms. A highly probable expectation might be that the shape of the curve is stable over organisms, but the initial response latency varies over organisms (some just get to work sooner). Model 10.26 can be cast and tested against Model 10.24 to ascertain this expectation.

$$\text{(Model 10.26) } Y = a_1 U1 + a_2 U2 + b_3 X3 + c_3 Z3 + E_{26}$$

Note that Model 10.26 differs from Model 10.25 only in the elements that reflect the intercept ($a_1 \neq a_2$). Given a nonsignificant difference between Model 10.24 and Model 10.26, one can conclude with some degree of confidence that the population estimators of the linear and second-degree slope are common for the two organisms. A significant difference between Model 10.26 and Model 10.25 would then indicate that the two Y-intercepts (initial response latency) were different.

Example 3: Evaluating A–B–A' Observed Mean Differences

A third type of example is the traditional problem of evaluating mean differences in the A–B–A' design. The discussion here can be applied to single subjects or means of a group of subjects; care must only be exercised to be sure the generalization is to the appropriate populations (within subject or across subjects).

Some of the questions one may be interested in answering are:

1. Was there an increase in response rate during Condition B over Condition A?
2. Was there a decrease in response rate under the A' condition as compared to the B condition?
3. Are A and A' response rates different?

Each of these questions could be tested by comparing the scores in one condition to the scores in another via the t test. We prefer to use the data in the third condition to help obtain a more stable estimate of the expected variability of scores in the two conditions being tested. This approach assumes that the variability within all three conditions is somewhat similar. (The fourth example will discuss what may be done

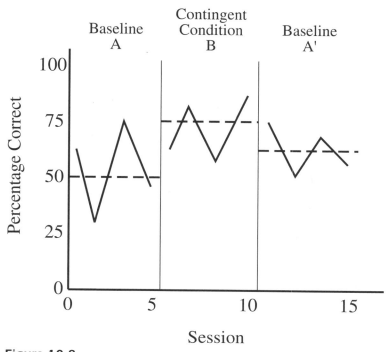

Figure 10.8.

Percentage correct for 15 sessions under three different conditions in A–B–A' design (dashed lines represent means for each condition).

if the variability within conditions is not similar.) Figure 10.8 is one typical example where the above questions would be appropriate.

All of the questions stated above can be answered by making specific restrictions on a model that allows for three mean response rates, Model 10.27.

$$(\text{Model 10.27}) \; Y \; = \; a_1 U1 + a_2 U2 + a_3 U3 + E_{27}$$

where:

    Y   = response rate;
    U1 = 1 if the criterion from Condition A, 0 otherwise;
    U2 = 1 if the criterion from Condition B, 0 otherwise; and
    U3 = 1 if the criterion from Condition A', 0 otherwise.

Model 10.27 allows for three separate means, one for each condition. Since the three Research Hypotheses are concerned with means and tacitly assume, by omission, that slopes are of no concern, slope functions are not specified in the model. We shall return to questions about slopes in Example 4.

To answer Question 1, "Was there an increase in response rate during Condition B over Condition A?," would require restrictions on the weighting coefficients in those two conditions: $a_1 = a_2$, both equal to a common value of a_4. This single restriction would yield Model 10.28:

$$(\text{Model 10.28}) \; Y \; = \; a_4 U4 + a_3 U3 + E_{28}$$

where:

U4 $=$ U1 + U2; and a_4 is the common mean for Conditions A and B.

Note should be made that the above question is directional (B increase over A, rather than just a change from A to B) and, as such, an adjustment on probability associated with the *F* value is required.

If the weight a_2 is greater than a_1 (in Model 10.27) and the comparison of Model 10.27 to Model 10.28 yields a significant *F* value (directional test), one can conclude with some degree of confidence (one minus alpha) that response rate under Condition B is greater than under Condition A.

Question 2, "Was there a decrease in response rate under the A' condition as compared to the B condition?," can be answered by restricting the weighting coefficients in these two conditions: $a_2 = a_3$, both equal to a_5, a common weighting coefficient. This single restriction requires Model 10.29 to be:

$$(\text{Model } 10.29)\ Y\ =\ a_1 U1 + a_5 U5 + E_{29}$$

where:

U5 $=$ U2 + U3.

If the *F* test yielded an adjusted probability value that was less than the set alpha level, then one can conclude the decrease was significant.

The last question, associated with recovery of the baseline, Question 3, "Are A and A' response rates different?," can be answered by comparing Model 10.30 and Model 10.27:

$$(\text{Model } 10.30)\ Y\ =\ a_2 U2 + a_6 U6 + E_{30}$$

where:

U6 $=$ U1 + U3.

If the *F* value (nondirectional because A' can be significantly lower or higher than A) is nonsignificant, then one can conclude with some degree of confidence that the baselines are *not* significantly different. Here one would want to use a somewhat high alpha level (say .60) because acceptance of the no-difference hypothesis is here desired. One would also want to have many observations, otherwise a low number of observations could lead to lack of significance.

Inspection of Figure 10.8 might be instructive of the value of the preceding hypothesis testing. Condition B does seem to yield an increase in response rate and may be interpreted to be an obvious difference. But what about the decrease in response rate in A' from B? The eyeball detects a difference, yet not really an unequivocal difference. Likewise, A and A' look a little different; is this a likely sample variation? We think these last two questions show the need for an inferential test in all instances.

Example 4: Evaluating A–B–A'–B' Observed Curve Differences

A fourth example we have encountered is an A–B–A'–B' study that involves a situation where a functional relationship exists between treatment condition, day in the treatment, and response rate. Consider Figure 10.9. A group of children is ob-

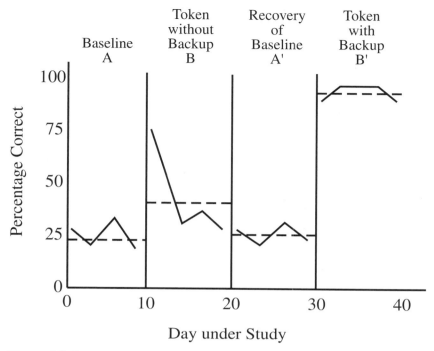

Figure 10.9.

Percentage correct by days within four experimental conditions (dashed lines represent means for each condition).

served for 10 days under a typical condition (A), then tokens are instituted for correct response on work material for a 10-day period (B), at which time tokens are withdrawn for 10 days (A'), and then there is a return to tokens, with an added stipulation that the tokens can be cashed in for preselected toys or other items that the children "desire."

For illustrative purposes, we specify a few questions that may seem reasonable. In the simple test of means, as provided in Example 3, we would ignore the negative slope under Condition B. The similarity of variability that we assumed and probably observed in Example 3 does not hold true in Figure 10.9. The variability about the Condition B mean is much greater than in the other three conditions.

A set of questions different from the ones asked below or the same questions asked in a different sequence would result in different models. One may ask the following sequence of questions:

1. Has percentage-correct baseline been achieved in Condition A? This can be operationally defined as a slope in Condition A that is not significantly different from 0.
2. Is there a negative slope relating day to percentage correct in Condition B?
3. Has percentage-correct baseline been recovered under A' condition? This can be operationally defined as a nonsignificant difference in the means of A and A'.
4. Is percentage correct higher under Condition B' than under Conditions A and A'?

These four questions can be answered by placing the appropriate restrictions on Model 10.31:

(Model 10.31) $Y = a_1U1 + b_1X1 + a_2U2 + b_2X2 + a_3U3 + b_3X3 + a_4U4 + b_4X4 + E_{31}$

where:

Y = percentage correct;
$U1$ = 1 if Y is from Condition A, 0 otherwise;
$X1$ = day (1–10) associated with Y under Condition A, 0 otherwise;
$U2$ = 1 if Y is from Condition B, 0 otherwise;
$X2$ = day (1–10) associated with Y under Condition B, 0 otherwise;
$U3$ = 1 if Y is from Condition A', 0 otherwise;
$X3$ = day (1–10) associated with Y under Condition A', 0 otherwise;
$U4$ = 1 if Y is from Condition B', 0 otherwise; and
$X4$ = day (1–10) associated with Condition B', 0 otherwise.

The reader will note that Model 10.31 is simply an equation that allows for a slope and intercept for each of the four conditions. $Y = a_1U1 + b_1X1$, for example, is $Y = a + bX$ for Condition A; $Y = a_2U2 + b_2X2$ is $Y = a + bX$ for Condition B, and so on.

To answer Question 1, the weight b_1 can be set to 0; that is, we force the line under Condition A to have a 0 slope. Model 10.32 reflects this restriction.

(Model 10.32) $Y = a_1U1 + a_2U2 + b_2X2 + a_3U3 + b_3X3 + a_4U4 + b_4X4 + E_{32}$

We can test Model 10.32 (the Restricted Model) against Model 10.31 (the Full Model) and if there is *no* significant loss in the R^2 value, then we can conclude with some degree of confidence (nondirectional *F* test) that a baseline is achieved, and Model 10.32 may be used as a new Full Model because it is the more parsimonious model. (Let us assume b_1 does equal 0, and therefore we may continue the sequence of testing.)

Question 2, "Is there a negative slope relating day to percentage correct in Condition B?," can be answered by restricting Model 10.32 by setting b_2 (the slope coefficient for Condition B) equal to zero. Model 10.33 represents this restriction.

(Model 10.33) $Y = a_1U1 + a_2U2 + a_3U3 + b_3X3 + a_4U4 + b_4X4 + E_{33}$

Note that both b_1 and b_2 are deleted from this model: b_1 because we found it equal to zero, and b_2 because we wish to test the slope for Condition B. In the case pictured in Figure 10.9, we should find a significant loss in R^2 from Model 10.32 to Model 10.33 because a distinct negative slope exists. This would be a directional test because we asked a directional question (negative slope). Given this situation, we retain Model 10.32 as the Full Model because it best describes the data.

The meaning of this finding is that it is ridiculous to ask, "Is the mean of B greater than the mean of A?" There may be a significant difference between these means, but the observed state of affairs is that initially tokens increase percentage correct, but over time the effect dissipates toward a return to the baseline.

Model 10.32 has a mean for Condition A and an intercept and a slope for the other three conditions. An inspection of Figure 10.9 should lead one to conclude that the means of A and A' are about equal and that the slopes are equal to zero. To answer Question 3, "Has percentage-correct baseline been recovered under A' condition?," we must set $a_1 = a_3$, both equal to a_5, a common weight, and $b_3 = 0$. Note that two restrictions are being made (equal means and no slope for Condition A'). Model 10.34 reflects these restrictions.

$$(\text{Model } 10.34)\ Y\ =\ a_5 U5 + a_2 U2 + b_2 X2 + a_4 U4 + b_4 X4 + E_{34}$$

where:

U5 = 1 if percentage correct (Y) is from Condition A or A', 0 otherwise.

If the R^2 for Model 10.34 is not significantly different (nondirectional test) from the R^2 of Model 10.32, then one can conclude that the baseline was not significantly different from the original baseline (mean of A = mean of A'). Let us assume this is the case, which means that we may continue our sequence of hypothesis testing.

Question 4, "Is percentage correct higher under Condition B' than under Conditions A and A'?," implies that there is a zero slope for Condition B'; however, Model 10.34 includes a weighting coefficient for a slope (b_4). Before we answer Question 4, we must cast Model 10.35 to reflect an assumed slope of zero in Condition B'.

$$(\text{Model } 10.35)\ Y\ =\ a_5 U5 + a_2 U2 + b_2 X2 + a_4 U4 + E_{35}$$

Then we can answer Question 4 by setting in Model 10.35 $a_4 = a_5$, both equal to a_6, a common mean that yields:

$$(\text{Model } 10.36)\ Y\ =\ a_6 U6 + a_2 U2 + b_2 X2 + E_{36}$$

where:

U6 = 1 if Y is from Condition A, A', or B', 0 otherwise; and
a_6 = the mean of Conditions A, A', and B'.

Given the data in Figure 10.9, we would expect the R^2 of Model 10.36 to be significantly smaller (directional test) than the R^2 of Model 10.35; therefore, we could conclude that percentage correct under Condition B' is significantly higher than under baseline conditions.

These four questions were asked and tested using inferential statistics. The particular set of questions was a limited set of those questions that could have been asked; they were, however, the four questions the researcher was interested in investigating.

The regression models have been presented to illustrate clearly and simply how the GLM procedure can help researchers who are interested in behavior modification designs. The models that have been presented so far, specifically for the questions under Examples 3 and 4, do reflect the Research Hypothesis asked in as simple a manner as possible so that the concepts can be most easily understood. However, more precise models (repeated measures, for example) can be written to answer these questions, but they have not been presented until this point because they tend to be a little more complex, though the concepts are similar.

The following is a brief and oversimplified introduction to the concept of repeated measures. For Question 1 in Example 3, which was originally answered by testing Model 10.27 against Model 10.28, a repeated-measures model will be written to more precisely answer that question. Another repeated-measures model will be written to answer Question 1 in Example 4.

When experiments are designed so that all subjects (S) are used in all treatments (K), then one has S * K observations. This situation constitutes a repeated-measures design. It is unlikely that, when using this design, there are S * K uncorrelated (independent) observations. Because the same subjects are in several treatments, there is likely to be a correlation between treatments due to subject or subject-by-treatment interaction. This problem can be considerably reduced by employing *person vectors* in the GLM.

A *person vector* (P) is a vector that contains information about which person was responsible for that particular criterion score. By using such vectors, one can control for the variance in each of the K treatments due to individual differences (person differences). One thus can determine the amount of variance due to the treatments, independent of individual differences.

The following is a repeated-measures model to test Question 1 in Example 3: "Was there an increase in response rate in Condition B over Condition A?" (assuming that the same subjects are used in each condition). This question can be tested with more precision by using Models 10.37 and 10.38 than previously done using Models 10.27 and 10.28. Models 10.37 and 10.38 assume that there are three subjects, and each is in all treatments.

$$(\text{Model } 10.37)\ Y = a_1 U1 + a_2 U2 + a_3 U3 + a_4 P1 + a_5 P2 + a_6 P3 + E_{37}$$

$$(\text{Model } 10.38)\ Y = a_7 U7 + a_3 U3 + a_4 P1 + a_5 P2 + a_6 P3 + E_{38}$$

where:

Y = response rate;
U1 = 1 if the criterion is from Condition A, 0 otherwise;
U2 = 1 if the criterion is from Condition B, 0 otherwise;
U3 = 1 if the criterion is from Condition A', 0 otherwise;
U7 = U1 + U2;
P1 = 1 if the score on the criterion was made by Person 1, 0 otherwise;
P2 = 1 if the score on the criterion was made by Person 2, 0 otherwise; and
P3 = 1 if the score on the criterion was made by Person 3, 0 otherwise.

Because the Research Hypothesis is dealing with means, the Full Model (Model 10.37) allows each treatment to have its own mean. The Research Hypothesis did not require information about slopes; therefore, no vectors dealing with slopes are included.

By making the restriction $a_1 = a_2$, both equal to a common weight a_7, the means of Conditions A and B are being forced to be equal. Any differences between the amount of variance accounted for by the two models would have to be due to this restriction. If $a_2 > a_1$, and if there is a significant difference between Models 10.37 and 10.38, using a directional test, then one can say that the mean of B is significantly greater than the mean of A at some alpha level above and beyond any variability likely to be caused by individual differences. The part of the statement dealing with individual differences is what differentiates the results of testing Models 10.27 and 10.28 from testing Models 10.37 and 10.38. The testing of Models 10.37 and 10.38 is more precise because it considers variance that can be accounted for by individual differences.

The following is a repeated-measures design to test Question 1 in Example 4: "Has percentage-correct baseline been achieved in Condition A?" If we operationally define and conceptually accept the idea that, if the slope of the baseline data is not significantly different from zero (a zero slope would indicate a stable percentage correct), then one could statistically test this question with more precision than one could using Models 10.31 and 10.32 by using Models 10.39 and 10.40—assuming that the same subjects are being used in all treatments, and in this case we assume three subjects.

(Model 10.39) $Y = a_1U1 + b_1X1 + a_2U2 + b_2X2 + a_3U3 + b_3X3 + a_4U4 + b_4X4 + p_1P1 + p_2P2 + p_3P3 + E_{39}$

(Model 10.40) $Y = a_1U1 + a_2U2 + b_2X2 + a_3U3 + b_3X3 + a_4U4 + b_4X4 + p_1P1 + p_2P2 + p_3P3 + E_{40}$

where:

Y	=	percentage correct;
U1	=	1 if Y is from Condition A, 0 otherwise;
U2	=	1 if Y is from Condition B, 0 otherwise;
U3	=	1 if Y is from Condition A', 0 otherwise;
U4	=	1 if Y is from Condition B', 0 otherwise;
X1	=	day (1–10) associated with Y under Condition A, 0 otherwise;
X2	=	day (1–10) associated with Y under Condition B, 0 otherwise;
X3	=	day (1–10) associated with Y under Condition A', 0 otherwise;
X4	=	day (1–10) associated with Y under Condition B', 0 otherwise;
P1	=	1 if the score on the criterion was made by Person 1, 0 otherwise;
P2	=	1 if the score on the criterion was made by Person 2, 0 otherwise; and
P3	=	1 if the score on the criterion was made by Person 3, 0 otherwise.

The inclusion of the person vectors in Models 10.39 and 10.40 increases the possibility of finding significance if indeed baseline has not been achieved. Certainly one would not want to act as if baseline had been achieved when in fact it had not. We strongly believe that the regression procedures as illustrated represent the proper case in which the Research Hypothesis dictates the statistical procedure.

A Note Regarding Time Series

A statistical analysis of time measures often has been a special and troublesome problem. Traditional methods, when applied to time designs, can yield results that are incorrect. The basic reason for this is that such data as time measures tend to fluctuate. If it is possible to assume that the variability is uniform over the whole criterion, there is no longer a statistical problem (Boneau, 1960; Kerlinger & Pedhazur, 1973). In operant conditioning designs, the legitimacy of making the assumption that the variability is stable over the area of interest is more likely to be correct, owing to the procedures used in calculating the baseline. The procedures generally require collection of enough data so that one can determine if the responses are stable over the area of interest. How many observations are necessary depends upon the stability and adequacy of the function being fit. At the lower limit, only one more observation than predictor pieces of information is needed to calculate the statistical index. In essence, one needs only a sufficient number of observations to provide a stable indicator of the functional relationship, and this is primarily a nonstatistical decision.

11

MISCELLANEOUS QUESTIONS ABOUT RESEARCH THAT REGRESSION HELPS ANSWER

Using an interview format, we cover in this chapter several somewhat-unrelated topics. Many questions in this chapter can best be answered through logic and a strong theoretical background in one's research subject area, but a good grasp of the GLM provides some added insight. Many researchers have come to us through the years with these very questions, and it is to their research zeal that the character named "Budding Researcher" is dedicated.

SAMPLES AND POPULATIONS

Budding Researcher: When I am reporting my research, what do I need to say about the population to which my results apply?

Dr. G. L. Model: The population to which you want to infer should be clearly indicated. The scope and meaningfulness of the population is entirely up to you as the researcher and to the constraints under which you perform your research. But it is up to you to justify that the sample you investigate represents that population. If the population is a theoretical one, then you must justify that there is no reason to believe that the sample and those subjects in the theoretical population are discrepant.

There are no indexes to indicate when a representative sample has not been obtained, although the sample and its selection must pass the scrutiny of your colleagues. The ultimate test is, of course, the repeated finding of the same results in samples from that population.

Bud: Can I achieve a representative sample by random sampling?

Dr. GLM: Random sampling does not guarantee that the sample will be representative, only that the chances that the (random) sample will be highly unrepresentative are minimized. Any single random sample could be quite unrepresentative of the population.

Random sampling demands that every entity in the population have an equal opportunity to appear in the sample. Very few pieces of research that appear in the behavioral science journals strictly meet this criterion. It is more often the case that only a limited number of entities are available for testing, only a certain percentage of questionnaires are returned, or the research is performed at a certain university in a certain class. Researchers must remember that the observed findings (unless replicated) may be a function of the particular sample and may not be generalizable to the (different) population; and collecting data on more and more subjects will not necessarily overcome this difficulty. If the sample is unrepresentative in the first place, then measuring more (unrepresentative) subjects simply solidifies the discrepancy. Similarly, striving for a 100% return on questionnaires through second and third mailings (and other associated harassment) will probably generate responses different from what the first questionnaire would have obtained if it had been responded to voluntarily.

Bud: How large a sample should I use?

Dr. GLM: The desirability of large sample sizes has been an illusion for many years. As you have already learned, the number of subjects in a study enters into the calculation of the degrees of freedom denominator (N minus pieces of information in the Full Model). The F test of significance is adjusted for the number of subjects in the sense that, with a fixed R_f^2 and R_r^2, it is harder to get statistical significance with a smaller number of subjects. With reference to the F denominator degrees of freedom, the minimum sample size is one more subject than predictor variables in the Full Model. If there were only as many subjects as predictor variables in the Full Model, the denominator degrees of freedom would be zero (a researcher's bad dream). And if you were analyzing fewer subjects than predictor variables, the denominator degrees of freedom would be negative (a researcher's nightmare). If only as many subjects are used as predictor variables, the R^2 would always be 1.0, exemplifying the notion of producing a spurious result by overfitting the data.

I'm not saying you should *always* use a small number of subjects. What I'm trying to communicate is that you need only one more subject than the number of predictor variables to perform the F test of significance. Now, if a researcher has a good grasp of the functional relationship being studied, one needs only a few subjects. But I suspect that most relationships being studied in the behavioral sciences right now and in the near future will *not* yield extremely high R^2 values; thus larger sample sizes are needed to be fairly certain that one is getting a good estimate of the relationship in the population. Fortunately, there is no recognized rule of thumb about the number of subjects to use. The less tight a grasp the researcher has on the functional relationship, the larger the recommended sample size. The wider the range of values on the continuous predictor variables, the larger the recommended sample size. The more reliance on data snooping rather than on theory and past research, the larger the recommended sample size. On the other hand, you must never let a small sample size eliminate an analysis. You must realize that small sample sizes reduce statistical power—the probability of finding significance if the Research Hypothesis is true in the population. Indeed, increasing the sample size will lower the probability value, while change in sample size has no systematic effect on the R^2 value (after the effects of overfitting have been eliminated—say, 20 subjects per predictor variable). Increasing the sample size merely results in the sample R^2 being a better estimate of the population R^2. That is to say, increasing the sample size tends to yield a sample R^2 that is *not* closer to one or to zero but to the population R^2, whatever it is (Figure 13.4b illustrates this). This seems a compelling reason to rely upon R^2 for decisions rather than solely on the probability value. (You might want to reread the section on power in chapter 8.)

All findings require replication, and unexpected findings on small samples simply indicate a reduced chance of being replicated. The proof is always in the prediction, and whether a finding from a small sample has merit rests in empirical replication.

CRITERION AND PREDICTOR VARIABLES

Bud: What variable do I use as my criterion?

Dr. GLM: Strangely enough, that question is often asked of us statisticians, but it can never be specifically answered by anyone but the researcher. The criterion variable is the variable that measures the construct of interest to the researcher. The researcher is intrigued that subjects vary on this variable and sets out to account for variability on this variable. Often the construct is measurable by more than one variable, although the GLM approach as discussed in this text can handle only one criterion variable at a time. You could use the same predictor model to predict two different criterion variables separately. A friend of mine named Spaner played around with that idea and published his thoughts on it in 1970.

Another notion about variables is that a particular variable may be a criterion variable for one researcher and a predictor variable for another researcher. This is often the case in developmental studies and studies on academic success.

Bud: How do I know what variables to use as predictors?

Dr. GLM: Most important, the nature of the predictor variables you use should emanate from a combination of past research and your theoretical structure. Given this as a premise, what specifically do you want to know about predictors?

Bud: How should I measure the constructs I've decided to use as predictors?

Dr. GLM: Two researchers may agree on using a given construct as a predictor, but they might use two quite different measures of that construct. You must remember that the R^2 and resulting interpretations are the result of the actual data and not of the construct that you hope the variable is measuring. The scores that are obtained are, one hopes, good approximations of the construct to which you would like to infer. But, for instance, you must always keep in mind that the measure of IQ that is used in a statistical analysis is most likely not synonymous with the construct of IQ. Until someone is able to "see" the IQ construct, we must act as if the operational measure is at best a (good) approximation. The goodness of that approximation is, in measurement terms, referred to as *validity*. You also should be aware that the measures need not be of an interval scale, as demonstrated by my friends McNeil and Kelly in 1970. Many authors of statistics texts state that only interval data can be analyzed by the GLM. But the bottom line is accounting for variance, and they presented several examples in which noninterval data resulted in an R^2 close to 1.0.

I would like to stress that the researcher does not need to follow any of the "guidelines" about using "recognized" tests or measurements. The predictor variables can be measured in any way the researcher sees fit. If a high R^2 is obtained, more credit to the researcher. There is nothing sacred about any of the prevalent measures in any of the behavioral sciences. Indeed, the many years of attention paid to certain well-known tests may have stagnated the growth of knowledge about relationships in that field. On the other hand, low R^2 values may not indicate that the criterion cannot be accounted for by the constructs measured in the predictor set but that the measures of the constructs were poor ones and need to be rescaled or replaced by other measures of the construct.

Bud: Should all my chosen predictor variables be uncorrelated?

Dr. GLM: That is another of those guidelines I would rather you ignored. Most multiple correlation literature encourages the use of uncorrelated predictor variables. ANOVA designs are set up such that the predictor variables are uncorrelated. But the real world is never fashioned that way. More important, the ultimate measure of the value of a predictor variable is its effect on the R^2. If the R^2 is significantly increased by the inclusion of a correlated predictor variable, then by definition that variable is a good variable (in that set of predictor variables).

When you use both linear and polynomial terms in a regression model, those variables are very highly correlated, but you have come to learn how valuable and even necessary they may be. Additional group membership variables, beyond those for which you hope to establish differences, also may increase the R^2, though they also will be correlated. If you want to get further into that, my fellow workers McNeil and Spaner wrote an article in 1971 that you could read.

Lewis-Beck (1980) provides four recommendations if you want to avoid highly correlated predictors (referred to as *multicolinearity*). First, one can obtain a larger sample to see if the correlation is an artifact of the particular sample. Second, one could combine the correlated predictors into a more encompassing predictor. Third, one could restrict the use of the regression equation to that of prediction and not try to extend the model to that of interpretation. Fourth, one could discard one of the "offending variables," but that would be a "willful commission of specification error"—that is, omitting a predictor that you originally considered relevant.

There is a special predictor variable—one that is uncorrelated with the criterion but highly correlated with other predictor variables—and it has been given the name *suppressor variable*. This kind of variable (whatever name you want to give it) is a predictor variable that was of no value in the bivariate way of looking at the criterion but is indeed valuable in the multivariate model. Because most past research is of a bivariate nature, you might want to have less faith that past research will give you many insights into productive multivariate hypotheses. Another of my friends stated the results of multivariate studies quite succinctly: "Multivariate statistical procedures have the darnedest habit of doing what they are designed to do (considering context in finding optimal linear combinations), rather than simply confirming what we think we know from examining univariate statistics taken out of context, one variable at a time" (Harris, 1992, p. 11).

Some suppressor-variable situations will merely be rescaling problems. Others will depict intricate and inseparable relationships between distinct constructs. These will be the variables that "complexly" measure the "complex behavior" so often discussed in the behavioral sciences.

Bud: How many predictor variables should I use in my analysis?

Dr. GLM: You must use as many predictor variables as it takes to result in an R^2 that satisfies you. And I hope you will not be satisfied until you get an R^2 close to 1.0. You have an infinite number of predictor variables at your disposal. You choose to measure a limited number of variables from an infinite number of possibilities. And once those limited number have been measured, the possibilities of polynomials and interactions are unlimited. Because you are the researcher, it is up to you to choose from the unlimited set of transformed variables those that are to be used in the analysis. This decision is a very difficult one, for the value of a variable is in its ability to increase R^2 not in whether it is a linear interaction term or a third-degree term, and so forth. But if you choose to defend the inclusion of predictor variables on some grounds other than practical utility, remember that the originally measured variable (the linear term) must be defended just as much as the second-degree term, or as much as any other term for that matter. And, as humans are probably too frail to be able to posit reasons for including one variable out of an infinite set, the final decision to include a predictor variable thus should be deferred to pragmatic reasoning—whether or not the variable increases R^2.

When I first introduce curvilinear models to eager young researchers like yourself, most of these researchers excitedly include the second-degree term in their prediction models. Then when third-degree models are introduced, the researchers excitedly include third-degree terms. When it becomes apparent that any power can be used, then they want to know, "How high a power should I include in my model?" I cannot be emphatic enough about this: Previous research and theoretical views determine the Research Hypothesis and therefore also the predictor variables. Lacking solid previous research or a tight theoretical view, predictor variables should be used that will increase the R^2.

Some researchers have been led to believe that if a third-degree power is used, the second-degree power and linear term also must be used. But I am saying that the researcher does not have to include lower-powered terms. Likewise, the inclusion of an interaction term does not demand that the variables being interacted also appear as separate terms. The inclusion or omission of a variable should always be thought out by the researcher. One must remember that interpretation of a variable's effects is always within the context of the other variables in the regression model. If a variable is not included in a model, one is assuming that the effects of that variable, within the context of the other variables, is negligible. These notions are quite important when one considers the drain on the degrees of freedom by including these sometimes undesired variables and the additional cost involved in measuring them and analyzing them.

THE DOLLAR COST OF PREDICTION

Bud: You have been encouraging the inclusion of many predictor variables, but each additional predictor I use costs money, time, and effort.

Dr. GLM: Of course. The cost may be in the collection of the data, or it may be in the computer time to transform the data or to obtain the results. The F test you use to compare models places the same cost on each predictor variable. In reality, some variables cost much more than others, while some may be essentially free. Measures that require much testing time or that require much intensive follow-up would cost a great deal. Some variables are readily available in school files or in published data banks (ICPSR, 1993). All of the transformed variables cost essentially nothing except the small amount of computer time needed to make the actual transformation.

Researchers often implicitly take these concerns into consideration in determining what variables to measure. You should not overlook a potentially important variable just because it may cost a great deal—you should have more interest in accounting for your criterion! On the other hand, you should pay more attention to transformed variables for two reasons. First, the transformed variables may increase the R^2. Second, again, the cost of transforming variables is negligible.

The business of considering the cost of predictor variables is a difficult and complex problem. A method of assigning cost figures to the predictor variables and then using this information in determining the "best" model was proposed by Pohlmann in 1973. Until cost values can be more validly assessed, Pohlmann's method remains ahead of its time.

ATTENUATION, REPLICATION, AND CROSS-VALIDATION

Bud: I have another concern about my criterion variable. I have read that an unreliable criterion causes something called *attenuation*.

Dr. GLM: Attenuation is the reduction in predictability due to the unreliability of the criterion. Formulae exist for "correction" due to attenuation and are often used; but it seems a little ironic to boast about the degree of predictability you would have if you had a perfectly reliable criterion. Often the researcher does not have perfectly reliable criteria; and furthermore, when you try to predict a criterion, you must have some faith in the measurement of that criterion. Prediction of a criterion is an exercise in validity, and increasing the "statistical" reliability of the criterion may decrease the validity of the functional relationship.

The R^2 you get from your sample is most likely an overestimation of the population R^2 because the weighting coefficients in a regression model are a function of those sampled subjects. Several adjustment formulae exist for reducing the R^2 by considering the number of subjects and the number of pieces of information in the model (Klein & Newman, 1974; Newman, 1973b). But these formulae, like the one that SAS employs, are average statistical adjustments. I prefer to *replicate* my findings (perform the same study on new subjects to see if I get significance again) or to cross-validate by applying the weighting coefficients from my original sample to a new sample of subjects from the same population. Suppose that for a given sample, the following results were obtained:

(Model 11.1) $Y = (6 * U) + (1.3 * X1) + (2.6 * X2) + (3.3 * X3) + (-66.1 * X4) + E_1$

The weighting coefficients in this model are unique to the sample, but if the R^2 is relatively high, they should be applicable to other samples from the same population with little loss in R^2. To check this out, you would have to obtain another sample of subjects, measure them on the same criterion and predictor variables, and then multiply the original sample weights and the appropriate predictor variables in the new sample to obtain a predicted criterion for each subject in the new sample. The correlation between the actual criterion and the predicted criterion would give you an idea about the "goodness" of those initially discovered weighting coefficients. In this process, called *cross-validation*, you are investigating the replicability of the *magnitude* as well as the *direction* of the weighting coefficients.

Now, if you wish to practice this notion to get a better understanding of it, here is something you might try on the data in Appendix A. Take the first 30 subjects and find the regression weights (using the computer, of course) for this model:

(Model 11.2) $X2 = a_{12}X12 + a_{13}X13 + a_{14}X14 + E_2$

If you follow the SAS setup in Appendix D-11.1, you will get a solution like this one:

$X2 = (-69.82533 * U) + (4.62768 * X12) + (0.0 * X13) + (.92690 * X14) + E_2$

Then take the last 30 subjects and generate a new variable, which would be a predicted criterion, by using the weights you obtained on the first 30 subjects. Then obtain the correlation between the criterion (X2) and the predicted criterion (X15).

That correlation (.99) indicates the degree of stability of the function as expressed by the variables and their weighting coefficients (see Appendix D-11.2).

Cross-validation on a study that is based on a strong theoretical background should yield equally high R^2 values for the original and later samples; but cross-validation on a study based on data-snooping procedures can be expected to yield a lower R^2 on the second sample than on the first. Small samples will even cross-validate well if the "true" functional relationship has been found. For instance, with a very small sample, you could cross-validate Einstein's theory ($e = mc^2$) or Newton's law ($d = 1/2gt^2$), as illustrated in chapter 13.

TRADITIONAL SEQUENCING OF HYPOTHESES

Whenever a researcher collects data, there are often reasons for several inter-related hypotheses to be tested. Sometimes the order of testing makes a difference, such as when making sure that certain assumptions are tenable. There are other times when the sequence depends on the design itself. At other times, the sequence depends on the desired conclusions. Therefore, there is no one "correct" way to sequence one's hypotheses. Indeed we have seen research studies that use many different sequences and use them very appropriately. The important point is that a Research Hypothesis should dictate an analysis, as well as the sequence.

Two-Way ANOVA

The two-way ANOVA (discussed in chapter 4) usually follows this sequence: First, test for interaction between the A factor and the B factor. If the interaction is significant, then stop. If the interaction is *not* significant, then test both the A main effect and the B main effect. There may be times, though, when the interaction question is not of interest, say, when two curricula are being compared and the blocking variable IQ is used. Although there may be an IQ-by-curriculum interaction, a given school district may not have the capacity for offering two different curricula—the district cannot afford to buy two sets of curricula and to train teachers in each of the two curricula. The purpose for including IQ as the blocking variable is to reduce the error variance in the analysis. Here, though, the researcher would not test for either an IQ-by-curriculum interaction or the IQ main effect. The IQ main effect is almost a given, and what could the researcher do, change subjects' IQ levels?

ANCOVA

As discussed in chapter 8, ANCOVA assumes that there is no interaction between treatment and the covariate. One could choose to test this assumption and, if significant, decide either to "proceed with caution" or to stop the sequence of analyses and report the significant interaction results. Again, there is no correct way, it depends on the particular research application.

Curvilinear Hypotheses

As discussed in chapter 9, a nonlinear function might fit the criterion better than a linear function. The degree of curvilinearity investigated is dictated by the research design or theory. Just because there is not a linear fit to the data does not mean that there is not a higher-order fit, as the inverted U example demonstrated. Likewise, just because an investigator cannot find a linear or second-degree fit does not mean that a third-degree or fourth-degree or fifth-degree fit is not inherent in the data. Where does one start? How far does one go? Does one start by looking for the fifth-degree fit and then testing that model against simpler models? Or does one start with the simplest model and go to increasingly more complicated models?

Ideally, one starts with *Truth* being mapped by either the Full Model or the Restricted Model. Unfortunately, Truth is an elusive reality and therefore must be pursued with the utmost determination. It would be easiest to pursue Truth from a Full Model that has an R^2 close to 1.00. Subsequent models could be investigated that would yield the same R^2 but in a more parsimonious fashion. How does one find a Full Model with such a high R^2? Theory, past research, and data snooping can all assist in that endeavor.

Hypotheses and Data Snooping

How one sequences one's analyses depends upon one's theory, design, and sequence of questions. Most of the questions will, one hopes, be posited before the data are collected, while some data-snooping hypotheses may be entertained. The answers to the a priori hypotheses are definitive for the population from which the subjects were sampled. The answers to the data-snooping investigations are not definitive but are the bases for future research.

The story of Archimedes is appropriate here. When he discovered the relationship between volume and displacement while taking his bath, he was so excited that he went running through the streets naked shouting "Eureka! Eureka!" Because he had discerned a law, which by definition has an R^2 of 1.00, a law that was extremely parsimonious and quite replicable, he had every right to run naked. Present-day researchers should have as much excitement about their research and as much investment in the results that they would be willing to run through the streets of their hometown crying "Eureka!" Once one contemplates this act, one realizes that one would certainly want to be correct in his or her euphoria. If one were wrong, it would certainly be embarrassing—probably more so than would be a retraction in a future journal issue. But such contemplation should lead to more serious thought about the choice of variables measuring the construct, the design of treatments or choice of intact groups, the sequencing of hypotheses, and the a priori specification of these Research Hypotheses. One would definitely *not* want to run naked if something were discovered from a data-snooping analysis that had not been expected. Modesty would require a replication of unexpected findings before such a public performance.

SEQUENCING OF HYPOTHESES

Bud: Must I state and test just one hypothesis at a time, or is there any way to combine the testing of several hypotheses?

Dr. GLM: The GLM approach makes it quite easy to posit Research Hypotheses and to build models and F tests to test those hypotheses. Most regression computer programs allow for many hypotheses to be tested on one computer run. The problem then becomes finding the most efficient and logically defensible sequence of testing those hypotheses. There are several ways of approaching this problem—from statistical, computer, and researcher points of view.

Bud: What sequencing has been developed by statisticians?

Dr. GLM: The statistician's major methods are *interaction* and *multiple comparisons*. Some statisticians admonish researchers to test for interaction before investigating main effects; then, if interaction is found, look only at simple effects. John Williams (1974b) indicated two concerns he had about this procedure. First, a test of significance is being performed in hopes of *not* finding significance. And second, the Type I error rate—the chance of accepting one of the Research Hypotheses falsely—is greater than the stated alpha. The stated alpha is appropriate for *each* hypothesis, but if alpha equals, say, .05 for each of three hypotheses, the probability of finding at least one hypothesis significant is .14 (considerably higher than .05). Furthermore, a researcher may be interested in the main effects even if interaction exists. The point is that not all researchers need to take the same route.

Most multiple-comparison procedures demand that the researcher compute the one-way F first and then investigate an orthogonal set of hypotheses. But the researcher may be interested in only one comparison or a subset of the available nonorthogonal comparisons. If the comparisons have been hypothesized before the data are analyzed, then the researcher is not capitalizing on vagaries in the data. Furthermore, there is no need for all the (orthogonal) multiple comparisons to be calculated and interpreted by the researcher. If one blindly tests all multiple comparisons, one is getting answers to questions unasked. (These concepts were introduced in chapters 4 and 5.)

If all orthogonal comparisons are desired, then a more efficient way to define vectors than 1,0 is available, the orthogonal coding as discussed in chapter 4. Kerlinger and Pedhazur (1973) and Williams (1987) present the various alternatives. My position is that one is not interested in those kinds of questions often enough to justify learning those procedures. As Kerlinger and Pedhazur (1973) indicate, the same F values will be obtained under the various systems.

Bud: If I do not need to follow the multiple-comparison route, what is the procedure developed from the computer point of view?

Dr. GLM: I do not recommend this procedure for hypothesis-testing purposes, but you should be aware of it for hypothesis-generating purposes. Most computer installations have a *stepwise regression program*. A priori hypotheses are not tested with such a program. The hypothesis-testing sequence is controlled only very slightly by the researcher, although some stepwise programs allow for an "importance" indicator for each variable. But specific models generally cannot be constructed and tested with stepwise programs.

The researcher reads into the stepwise program the criterion variable and all the predictor variables of interest. The idea is to find which variables form the "best" predictor model for that criterion. Forward selection procedures find the one variable in the predictor set that is most highly correlated with the criterion. This variable, with the unit vector, is referred to as the best model at Step 1. Then the program searches the variables to find which predictor, in combination with the two already in the system, will yield the highest R^2. Once found, these three variables become the best model at Step 2. The process continues until all variables are in the system or until no variable can be found such that the increase in R^2 is significant at a level specified either by the program or by the user.

The stepwise procedure was developed for continuous variables. Few applications have involved interactions or nonlinear terms. Few have involved dichotomous variables for two reasons. First, the inclusion of linearly dependent vectors will result in either an error signal or the program will abort. Second, if a variable is to be represented by three mutually exclusive dichotomous vectors, what interpretation would one place on the situation in which one of those dichotomous vectors was in the system but the other two were not?

That friend of mine, John Williams, whom I just spoke of, with his friend Lindem, developed in 1974 a procedure, referred to as *setwise regression*. Instead of using single variables, the researcher is allowed to define *sets* of variables. While sets can be defined on a logical basis, the use of a setwise procedure seems mandatory when binary-coded variables are included and there are more than two categories involved. The setwise procedure includes one set at a time in a stepwise fashion.

Some problems arise with the stepwise approach. There are many hypotheses being tested, none of which has been specified by the researcher. The resulting "best" model will most likely be quite drastically overfit, and therefore replication of that model is unlikely. Also, the stepwise procedure may stop too soon, in that two variables considered simultaneously might significantly increase the R^2, whereas neither one of them may separately significantly increase the R^2. Some stepwise versions allow variables already in the system to be deleted if they are not making a contribution in a much larger model. Versions that do not allow for this additional flexibility may end with an inferior best model.

Some of these problems with forward stepwise regression are resolved with backward stepwise regression. In this approach, all variables are considered the best model at Step 1. Then the variables (or sets, as in setwise regression) are evaluated one at a time to find which one will, when omitted from the system, reduce the R^2 the least. That variable is then omitted, and the model using all the variables except that one is the best model at Step 2. The process is repeated for Step 3 and so on, either until all variables have been omitted or until omitting any of the variables would result in a significant loss in predictability. The backward procedure is preferable, though many hypotheses are tested (resulting in much computer time and a large probability of making at least one Type I error). Thus, the resulting model may not replicate because of overfitting. At any rate, the particular hypothesis you want to test very likely will not appear—and if it does appear, you may have difficulty recognizing it.

Applications of the stepwise procedure have typically ignored curvilinear variables, as well as interaction variables. Also, too much attention has been placed on the sequence in which variables either enter or leave the system. This sequencing depends on all the intercorrelations between the predictors, as well as how highly the predictors correlate with the criterion. Small changes in one or two correlations could quite drastically change the sequencing.

In essence, stepwise regression programs perform a hypothesis formulation function, whereas the regression procedures discussed in this text are concerned with testing those formed hypotheses. Thayer (1990) reviewed all of the stepwise options and provided recommendations for obtaining the best model for *explanatory* purposes. He describes in excellent detail the use of various setwise procedures. He admonishes,

> When variables are selected for a regression model, the stepwise method can be helpful if the initial choice of variables is chosen as much as possible using theory, the defaults of the computer program used are not used automatically, more than one computer run is done using different variable selection methods, and the final model is chosen through an intelligent process, not automatically using the final model generated by the computer program. When the

model is described, all subjective decisions made in the model selection process should be reported. (p. 67)

Another SAS option, RSQUARE, can assist in data snooping. This procedure provides the R^2 for each possible combination of predictor variables. Appendix D-11.3 contains the SAS setup for backward, forward, and RSQUARE.

Bud: What method for sequencing hypothesis testing can you recommend from the researcher's point of view?

Dr. GLM: Build your hypotheses upon *past research* (yours or someone else's) *and your theoretical views*. If you have enough information and enough confidence in your theory to state directional Research Hypotheses, then state them in the order of your interest and expectations, and test them in that order. You can expect statistical support for such hypotheses, and you should be able to progress toward causal interpretation.

Bud: What about those areas of my theory where past research has given me no clues about what functional relationships to expect? I cannot state directional Research Hypotheses about those relationships, can I?

Dr. GLM: You make a good point. I am emphatic about the researcher stating the Research Hypothesis. The Research Hypothesis should be a statement of expected outcome based upon past relevant research, theoretical relationships, and synthesis of relevant constructs. But most past research has been bivariate in nature; that is, only one predictor variable was used to account for criterion variance, and the amount of criterion variance accounted for has usually been quite small. Therefore, theory development has been held back, and so there is little knowledge from which one can synthesize. Data snooping is one way of finding functional relationships. Because one constantly remembers to replicate the findings on a new sample of data, I see nothing wrong with "seeing what works." Using a stepwise regression program is one way to discover predictor variables; accidental (serendipitous) findings are another way. Guidelines about how to obtain serendipitous findings are nonexistent, but you should be willing to cross-validate a variable that is valuable in a pilot study. Many important variables in the sciences have been discovered accidentally, although their value was repeatedly tested (cross-validated) before they were widely accepted. Mosteller and Tukey (1968) aptly summarized how the researcher should react to results found through data snooping when they said, "Here we must stop our calculations with indications and be careful to think of our results only as hints about what to study next, rather than as established results."

To give you an idea of what data snooping is all about, I give you a "data-snooping analogy," in which the bread is analogous to a meaningful discovery, and the telling of friends about the news is analogous to publishing one's results for the benefit of the profession.

Suppose that you have been asked by your significant other to go down to the milk discount store to buy some milk. Now, if you do exactly as you are asked, you will go directly to the milk counter, pick up the milk, pay for the milk, and return home.

Let us suppose that you snoop around instead, and you accidentally notice that this store has bread at an exceptionally low price. In fact, at the price listed, you might label this "the best bread buy in town." If you are the least bit economically minded, you will take note of this price and purchase a loaf or two, especially if you are aware of procedures by which bread can be stored for a long period.

You are not "sure" that this low price will prevail for long, but you would consider informing some of your friends about the "best bread buy in town." You probably would indicate your reservations by saying something like, "Yesterday I accidentally discovered that the milk discount store that I trade at had a good price on bread. You might want to go down there and see if they still have this good bargain; I really do not know if they still have that good buy." That is, you are somewhat reluctant to yell too loudly about the bargain until the bargain can be verified. The longer the store retains this low price the louder and more frequently you will announce your discovery to more of your friends. If the bargain remains long enough and enough people are able to verify your finding, then the "best bread buy in town" becomes a lawful fact.

That lawful fact might not have been discovered if you had not snooped around while at the milk discount store looking for milk. An auxiliary bargain discovered while looking for something completely different must be verified on subsequent trips before much faith can be placed in the stability of that bargain.

As a researcher, please take the analogy to heart. Snooping around in your data is not antithetical to good research. If you find a meaningful relationship while snooping around, then you are obligated to replicate your results before you say too much about them. Your Research Hypotheses should be based upon theory, which is often loosely based on past research. Often there is little past research upon which to base your hypotheses. Furthermore, the past research is often of poor quality and bivariate in nature and therefore possibly uninformative or even misleading. When one snoops around in data, that study must be considered to be a *pilot* (past research) upon which hypotheses are developed for the replication sample.

It could be argued that researchers have drastically held back scientific advancement by not snooping around in their data. It seems unfortunate to collect large amounts of data and then to look at those data with blinders. We should take those blinders off—but simultaneously remember to leave our replicators on!

Bud: Can I use multiple regression procedures to help in my data snooping?

Dr. GLM: Data snooping is easily done within the GLM approach. All your measured variables (about which you have hypotheses, or hunches, or you just want to see what is going on) can be included in a Full Model. If many predictor variables are being used, many subjects should be used, so that the resulting functional relationship is not due entirely to the idiosyncrasies of the sample at hand.

The R^2 for the Full Model using all measured predictor variables could be compared against the unit vector model to see if any significant criterion variance is being accounted for. If you want to limit your predictor set to, say, one-half the number of predictors you started out with, and you want to find the best ones, you might use one of the stepwise regression approaches.

Data snooping also can be thought of in terms of trying to increase R^2 over the value you have obtained based on your theoretical testing. (This snooping will then lead to further hypothesis testing.) Suppose a researcher would like to obtain 90% predictability. The R^2 of the Full Model is only .70. What can be done now? More snooping is called for, but where? By now you should be thinking of possible interactions and polynomial forms. One could use not only X, X^2, X^3, X^4, one could even give the predictor the fifth degree. One could use not only $X * W$ but also $X * W^2$ and $X^3 * W^2$. One could use not only X^2 and W^3 but also $X^{1.3} * W^{3.3469}$, ad infinitum. Yes, there is an infinite number of transformed (e.g., interaction and nonlinear) variables that you could get from any set of measured variables. But an infinite set of predictors cannot be used because you cannot have the required number of subjects (infinite plus one). Ideally, theory and past research should have guided us and assisted us in selecting the "appropriate" set to use. The primary guideline, though, should be pragmatic predictability. When R^2 is up to .90, you might be happy. (Astronauts going to the moon desired an R^2 around .95, and those desiring to return to Earth wanted a substantially higher R^2.) The actual minimum R^2 value is up to the individual researcher and will depend upon the state of the art in the research area and the amount of impact one desires to make in that area.

Snooping may be a somewhat costly approach. Computer time is needed to generate the various transformed variables. Furthermore, the amount of computer time required to solve for a regression model increases drastically when the number of predictor variables increases. Some of this computer time can be saved by splitting the sample in half and snooping only on one half—saving the second half for the required cross-validation (as discussed earlier). Only those transformations that were found useful in the snooping sample would have to be computed in the cross-validation sample. The cross-validation sample would require additional computer time, but cross-validation is a research step that one should take anyway.

Some statisticians have suggested that factor analysis could be used first to reduce the set of predictor variables to a small set of uncorrelated predictors (Connett, Houston, & Shaw, 1972). This procedure would have the advantage of reducing the number of predictors, but the possibility of obtaining the desired level of R^2 might be lost (Newman, 1972, identified other problems). Most interesting criterion variables are, by the admonition of the researchers investigating them, complexly caused. It is too naive to believe that real-world causers are both rectilinear and uncorrelated.

What does the researcher have after snooping and subsequently cross-validating? One has a highly predictable and highly reliable way of accounting for variability in the criterion of interest. One may likely not understand why the predictors are combined in the way they are. Indeed it could be argued that scientific advances occur only when data hit you between the eyes and say, "Hey, we're something you didn't expect to find. We're reliable and predictable. Use our functional relationships because we work."

Data snooping may lead to serendipitous findings, and when I realize how many aspects of our world (e.g., pasteurization, vulcanization, popcorn) have been found serendipitously, I am amazed. The concepts of transformed variables and cross-validation within the data-snooping framework discussed above should give more researchers the opportunity to advance science.

INTERPRETATION OF WEIGHTS

Bud: How do I interpret the weighting coefficients?

Dr. GLM: You must remember that the tests of significance are about population values, and these weights are sample values. As such, the weights will usually fluctuate from sample to sample. Some statisticians (Harris, 1992, 1993; Lunneborg, 1992) encourage researchers to interpret the weights. I take a cautionary stance on the issue, as do my friends McNeil (1990, 1991b, 1992, 1993) and Brown (1992). McNeil (1993) identified three conditions that should be satisfied before one can interpret the weights: (a) the R^2 should be very close to 1.00; (b) the results (including magnitude of weights) should have been replicated; and (c) manipulations of manipulatable variables should have occurred.

MISSING DATA

Bud: What do I do if some subjects are missing data?

Dr. GLM: The problem of some subjects having missing data is highlighted by the GLM procedure because of the inclusion of multiple predictor variables. The possibility of any one subject missing a score is geometrically increased when one considers each additional predictor variable. Some computerized regression procedures would interpret the blank (missing) scores as zero and would produce a valid-appearing printout, but the results would not be valid, as one certainly did not want each of the missing data points to be expressed as zero. Other computer programs require specification of what code is to be interpreted as missing data.

The easiest solution to the missing data problem is to try to get complete data, thereby avoiding the missing data problem. This is more likely to be the situation if the study is carefully designed, the questionnaire is carefully constructed, the treatments are carefully developed and applied, and so on. Researchers should be cautioned, though, that forcing complete data often generates bad data. For example, mailing a second or third or fourth questionnaire, continually knocking on someone's door, and demanding that subjects appear for an experimental situation will often generate negative attitudes in those subjects and therefore invalid data. In these cases it would be better to have missing data than to have bad data.

Another solution to the missing data problem is to exclude those subjects who have any missing data (which is what SAS does). But eliminating subjects because of missing data most likely redefines the population from which one has sampled and, therefore, to which one can generalize. Students in school for whom the researcher is missing some data during a two-week period may be the unhealthy kids or the truant kids. Therefore, if one continually uses only complete data, the population to which one may legitimately generalize might be only healthy and nontruant students. Furthermore, the Full and Restricted Models will have different subjects, and if the two models are both generated, the R^2 of the Restricted Model may in fact be larger than the R^2 of the Full Model (Haupt, 1993).

Another solution to the missing data problem would be to insert the mean value of the predictor variable. Insertion of a mean value assumes that the person with missing data is like the average subject with data. Another problem with inserting mean values is the reduction of the variance of the predictor variable, resulting in a variable with lowered predictive value. The extreme case is that in which there is only one subject who has a score on the predictor variable. All other subjects are missing that predictor score. Insertion of the mean (the score of the one subject) would result in all subjects having the same score; and therefore, the variance of the variable would be zero, resulting in a vector with no predictive value. One way of getting around the problem of reduced variance is to insert a random score from the vector into the missing data locations. However, if the persons who are missing data are different from those with data, then the procedure of inserting random scores is, on the average, decreasing the relationship between that predictor variable and any other variable.

The best solution to this missing data problem may be to test directly to see if subjects missing data are different on the criterion from those who do not have missing data, using the procedures outlined in Generalized Research Hypothesis 4.1. The Full Model would contain a 1,0 vector indicating persons who have missing predictor data. This Full Model would be compared to a Restricted Model that does not contain the missing data vector. A significant difference would indicate that those subjects who are missing data are different on the criterion from those subjects who have complete data.

Another procedure would be to predict the value of the missing score from the other predictor variables; that is, use as a criterion the predictor variable of concern and include only those subjects who have complete data on all the predictor variables. Then find the weighting coefficients for the functional relationship between the criterion (predictor variable of concern) and the remaining predictor variables. These weighting coefficients can then be applied to the subjects who have missing data on that particular predictor variable. This last procedure seems of benefit when there are many variables relative to the (small) number of subjects. This procedure, though, does lead to a more systematic relationship between the predictors and the criterion than really exists. Replication is therefore an important corollary of this procedure.

OUTLIERS

Bud: What are outliers and what can I do about them?

Dr. GLM: *Outliers* are data points that are different from the rest. They stick out from the crowd. If these outliers are erroneous, then the data should be corrected. Often, though, outliers are real data points that simply are not modeled well by the data.

Say a researcher correlates nonverbal IQ with math achievement and finds a substantial negative correlation (say, -.50). This correlation is completely opposite to the literature. Suppose that, upon inspection of the 140 subjects, it is discovered that one subject's IQ had been entered as 555 instead of 55. This one outlier, and it is an extreme outlier, forces the line of best fit to go through it, resulting in the negative correlation. Once the error has been corrected, the correlation turns out to be .80, more in line with the literature. One error out of 140 subjects can have a drastic effect on the correlation! Never forget how literal the computer is—it will always analyze the data you give it, never questioning their veracity.

Levine and Stephenson (1988) present an interesting example of outliers that were not erroneous, just "different." They report a reanalysis of data from an original study that reported a "strong, positive correlation of .47 between the percent of over-age students in elementary schools and the subsequent dropout rate of their graduates." Levine and Stephenson (1988) determined that the high correlation was a function of only a few schools (7) out of the 300 schools. These few schools were all vocational schools, and once they were removed from the analysis, there was only a very small correlation.

Tracz and Leitner (1989) reanalyzed the metanalysis of the relationship between class size and learning originally conducted by Glass and Smith (1979). The original study had concluded that smaller classes resulted in more effective learning. Tracz and Leitner (1989) omitted some outliers (classes of five or fewer students) and found that there was no relationship between class size and learning. They argued that these class sizes should be omitted because they lay outside the normal possible class size.

Lewis-Beck (1980) identified five actions that a researcher can take with respect to outliers: (a) check to make sure they are valid data; (b) exclude them from the analysis; (c) report different models for the outliers and the nonoutliers; (d) transform the data, so that the outliers are now more in line with the rest of the data; (e) gather more subjects or more variables to see if there are other outliers or if the outliers can be modeled with the nonoutliers.

HOW TO BEAT THE HORSES

We conclude this chapter with some insight about how one might earn some money at the racetrack. It might be considered a reward for sticking with the GLM.

When a bet is made at the racetrack, the bettor, unless he enjoys throwing his money away, acts as if he is making a prediction. He is putting not only faith but also money behind some kinds of information that he has mixed in some fashion. Most bettors cannot verbalize their prediction model. Some bettors do not even have a model, picking their choice randomly or through some kind of superstitious behavior. Some bettors have models that they apply differentially; for example, they let past earnings and track conditions play a large part in the model, unless the horse ran badly the last time. If the horse did not place the last time out, then the only variables looked at are jockey weight and starting position. But give the horse one of the top jockeys in the country and that becomes the sole determiner—a winner for sure.

A little reflection should indicate that very seldom will a single variable, or even two variables, be of enough predictive value to be useful. What one needs to do is develop a regression model with (probably) many predictor variables. This model then should be applied uniformly to all horses.

Lest one become too excited, such a scheme has already been developed by someone who calls himself "The Wizard" and has been written up by Morgan in *Look* (June 1, 1971). Although he investigated 120 possible predictor variables, The Wizard evidently did not use interactions and polynomial terms. Incorporating such variables in the prediction model would most likely increase predictability. If the criterion (winning a horse race) is a complex phenomenon, then a complex prediction model is needed to account for that complex phenomenon.

Attempting to use a prediction model at the racetrack assumes that winning a horse race is a predictable behavior. Random events, such as the randomly assigned starting position and potholes, may lower predictability but not eliminate the possibility of obtaining a somewhat reasonable R^2. Doping the horses, making illegal "deals," and giving "slowdown" instructions to jockeys also may be important variables (although usually not available to the average regression person).

If winning a horse race is a predictable behavior (*lawful behavior* might be a better term here), there are still other factors that automatically reduce the predictability. The track does not want to lose money, so it has built-in variables that force the R^2 toward zero. Good horses are required to carry extra weight, and if a horse has been consistently winning, that horse is forced to race with a higher level of competition. If these efforts do not reduce meaningful predictability, then the odds come into play. For if there is clearly a probable winner in the field, most bets will be placed on that horse, reducing the odds and therefore the payoff if the horse does in fact win. But if you know how to use interactions and nonlinear functions (and you do now), you should be able to make more accurate predictions than those who do not. Keep regressing, and see you at the track!

12

APPLICATION TO EVALUATION

In this chapter we discuss various notions about evaluation within the context of the GLM. We first discuss three compensatory education evaluation models in terms of the GLM. We then discuss a variety of applications in various evaluation settings. We conclude the chapter with an interesting application to the identification of decision-making policies.

EVALUATION OF COMPENSATORY EDUCATION PROGRAMS

The federal government has been the major impetus behind compensatory education program evaluation. The emphasis is on wanting to make sure that the states and the local education agencies (LEAs) use federal dollars in the way intended and on wanting to determine if those dollars have made any difference. The major push for program evaluation came in the funding for compensatory education, first known as Title I and then, after the legislation was changed, as Chapter 1. These funds are earmarked for the remediation of low-achieving students who reside in low socio-economic areas.

RMC Research Corporation was awarded a contract to develop and describe evaluation models that would be appropriate for the evaluation of compensatory programs at the local level. These models had to detect the added benefit of the federal dollars. RMC described three evaluation models: Model A was the norm-referenced model; Model B was the comparison-group model; and Model C was the regression model.

Each of the models relies on norm-referenced testing and on making an estimate of what achievement level the compensatory students would have achieved had they not received the federal compensatory dollars (Tallmadge & Wood, 1981). The observed performance at the end of the project is referred to as the *postproject* performance, and the estimate of how well students would have performed without the project is referred to as the *no-project* expectation. The difference between the observed postproject performance and the no-project expectation defines the effect of the project on the participating students as shown in Equation 12.1:

$$\begin{array}{ccccc} & & \text{OBSERVED} & & \\ \text{PROJECT IMPACT} & = & \text{POSTPROJECT} & - & \text{NO-PROJECT} \\ & & \text{PERFORMANCE} & & \text{EXPECTATION} \end{array} \quad (12.1)$$

The models are similar in that they all use posttest scores to measure students' achievement at the end of the project. They differ in the manner in which the no-project expectation is generated. Model A uses national norms to estimate the no-project expectation; Model B uses a comparison group that is similar to the compensatory group; and Model C uses a statistical relationship of a specially selected group of higher-scoring students (Tallmadge & Wood, 1981). Although each model has a criterion-referenced counterpart, those models will not be discussed, as the data analysis is essentially the same.

Owing to the concern some statisticians have about analyzing percentile scores, RMC refined an interval-level transformation of the percentile scale by anchoring the new "normal-curve equivalent" (NCE) scale to the percentile scale at the first and 99th percentiles and then arbitrarily making interval NCEs. Thus NCEs of 1, 50, and 99 are equivalent to percentiles of 1, 50, and 99, respectively, but all other NCEs and percentiles are different, as indicated in Figure 12.1.

Although detailed hand calculation formulae and worksheets were developed by RMC, they made no references to tests of significance. As is demonstrated in the following sections, all three models are amenable to being analyzed by the GLM.

Model A: The Norm-Referenced Model

In Model A, test norms are used in place of a comparison group. The no-project expectation is based on the assumption that students, without a special additional

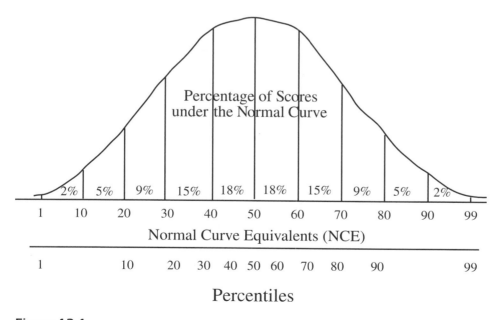

Figure 12.1.
Relationship between percentile and NCE.

curriculum, will maintain their respective percentile standings from pretest to posttest. Project impact is defined as shown in Equation 12.2:

	PROJECT GROUP		PROJECT GROUP	
PROJECT IMPACT =	POSTTEST	-	PRETEST	
	SCORE		SCORE	(12.2)

The first term to the right of the equal sign corresponds to the observed postproject performance in Equation 12.1 and the second term to the no-project expectation in Equation 12.1. While Model A does not have an identifiable comparison group (that requires testing), it does make an assumption that the students being tested are receiving a regular curriculum that is, on the average, as effective as that received by the norming students.

The GLM formulation of the model is as follows:

Full Model: $(Post - Pre) = a_0 U + E_1$

Want: $a_0 > 0$; restriction: $a_0 = 0$

Restricted Model: $(Post - Pre) = E_2$

Note that the above regression models are the same as for testing a single-population mean (see Generalized Research Hypothesis 10.1). If the F is significant, and if the results are in the direction hypothesized, then the compensatory program has been of some effect.

Model B: The Comparison-Group Model

In Model B, the posttest status of the compensatory students is compared to that of a similar group that did not receive the compensatory program. This model should be recognized as the two-group model described in chapter 4 and again in Generalized Research Hypothesis 10.3. Although providing an accurate assessment of the impact of the compensatory funding, few LEAs were able to implement the model as it required withholding funds from students in the comparison group—neither a politically wise nor a legally allowable action. In Model B, the no-project expectation is the mean posttest score of the comparison group, and project impact is defined as in Equation 12.3:

	PROJECT GROUP		COMPARISON GROUP	
PROJECT IMPACT =	POSTTEST	-	POSTTEST	
	SCORE		SCORE	(12.3)

Such an expectation can be analyzed with GLM as follows:

Full Model: $Post = a_1 COMPENSATORY + a_2 COMPARISON + E_3$

Want $a_1 > a_2$; restriction: $a_1 = a_2$

Restricted Model: $Post = a_0 U + E_4$

If the *F* is significant, and if the results are in the direction hypothesized, then the compensatory program has been more effective than the comparison.

Model C: The Regression Model

In Model C, all students in the LEA are pretested on a selection test; those scoring below a cutoff receive both the regular and compensatory programs. Those who are at or above the cutoff receive just the regular program. Once the students take the posttest and the relationship between the selection test and the posttest is determined for the noncompensatory students, one can use that prediction equation for the compensatory students. This prediction indicates how the compensatory students would have scored on the posttest had they participated only in the regular program. This no-project expectation is then subtracted from the compensatory students' actual posttest scores to yield the measure of project impact. For Model C, Equation 12.1 becomes:

$$\text{PROJECT IMPACT} = \begin{array}{l} \text{PROJECT GROUP} \\ \text{POSTTEST} \\ \text{SCORE} \end{array} - \begin{array}{l} \text{POSTTEST SCORE} \\ \text{ESTIMATED FROM} \\ \text{COMPARISON DATA} \end{array} \quad (12.4)$$

Such an expectation can be analyzed with the GLM as follows:

Full Model: Post $= a_1\text{PROJECT} + a_2\text{COMPARISON} + a_3\text{SELECTION} + E_5$

Want: $a_1 > a_2$; restriction: $a_1 = a_2$

Restricted Model: Post $= a_0\text{U} + a_3\text{SELECTION} + E_6$

If the *F* is significant, and if the results are in the direction hypothesized, then the compensatory program has been of some effect.

Model C has many advantages over the other models. First, Model C reduces the amount of testing by using the selection test in the analysis. Model A requires a separate selection and pretest, as well as the posttest. Model C does not require the assumption of an average curriculum, as both groups receive the regular curriculum of the LEA. Finally, Model B requires a comparison group similar to the compensatory group of students, whereas Model C uses students in the same LEA who are different (on the selection test). In Model C, the major assumption of concern to evaluators is that the relationship between selection test and posttest is assumed to be the same in the comparison group and the compensatory group. Additional information on the GLM approach to Model C can be found in McNeil and Findlay (1980) and Trochin and Stanley (1991).

The Full Model described above allows for two parallel lines, each over a different section of the pretest. Figure 12.2 depicts the desired results and how the lines would fit the two ellipses of the data. Chapter 1 students perform better at posttest than predicted by the relationship between pretest and posttest for the regular students (the dashed line).

Another distinct advantage of Model C is that one can look for interaction and nonlinear effects. Such findings might have implications for how the compensatory program is designed and working. For instance, results such as in Figure 12.3 indi-

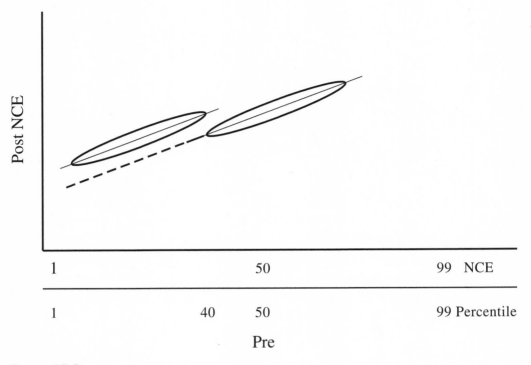

Figure 12.2.
Results indicating Chapter 1 students below the 40th percentile performing better at posttest than comparison students (40th percentile and above).

cate that the program is not working for those students most in need. The other two models do not allow for such investigation. Such findings would lead one to be concerned about how the program was initially conceived or how it is being implemented. Clearly, some of the very neediest, lowest-scoring students are not gaining from the program.

Smith, McNeil, and Mitchell (1986) discovered results such as those depicted in Figure 12.3, and they recommended that the cutoff (in the LEA studied) be lowered to the 30th percentile, as there was no apparent improvement for those students above the 30th percentile. An alternate recommendation would have been to revamp the compensatory curriculum, but the current curriculum was effective with the majority of students (who needed help the most).

The interaction expectation can be analyzed with the GLM as follows:

Full Model: Post = a_1PROJECT + a_2COMPARISON + a_3(PROJECT * SELECTION) + a_4(COMPARISON * SELECTION) + E_5

Want: $a_3 \neq a_2$; restriction: $a_1 = a_2$

Restricted Model: Post = a_1PROJECT + a_2COMPARISON + a_3SELECTION + E_6

Notice that this Restricted Model is the same as the Model C Full Model. If the F is significant, then the results should be plotted to determine in what way the interac-

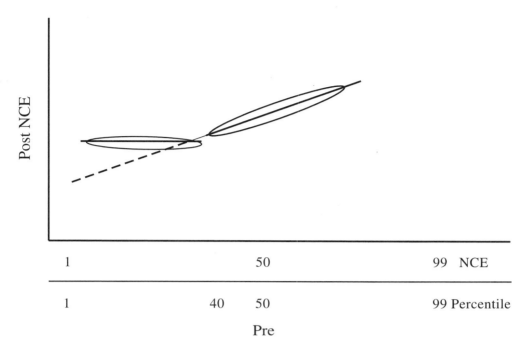

Figure 12.3.
Results indicating that the Chapter 1 program is not helping students above the 30th percentile.

tion is occurring. Specific recommendations would rest on how that interaction is occurring, along with the magnitude of the treatment effect.

Smith et al. (1986) also discovered a curvilinear relationship in the regular group, so this second-degree "line of best fit" was extended into the comparison group. Again, the other two evaluation models are not capable of determining such relationships. We should note that the students who scored above the 80th percentile were eliminated from the analysis because many of them were in a very different "regular" curriculum—the talented and gifted program. While it is generally a good idea to analyze as many subjects as possible, subjects who are outliers, who do not meet the definition for inclusion into the group, or who do not meet other requirements (such as in this case, receiving the regular curriculum), should be omitted from the analysis.

The Research Hypothesis of two different second-degree lines can be analyzed with the GLM as follows:

Full Model: Post $= a_1$PROJECT $+ a_2$COMPARISON $+ a_3$(PROJECT $*$ SELECTION) $+$ a_4(COMPARISON $*$ SELECTION) $+ a_5$(PROJECT $*$ SELECTION)$^2 +$ a_6(COMPARISON $*$ SELECTION)$^2 + E_7$

Want: $a_1 \neq a_2$; $a_3 \neq a_4$; $a_5 \neq a_6$; restriction: $a_1 = a_2$; $a_3 = a_4$; $a_5 = a_6$

Restricted Model: Post $= a_1$U $+ a_3$SELECTION $+ a_5$SELECTION$^2 + E_8$

APPLICATIONS TO OTHER EVALUATION SITUATIONS

An increased emphasis has been placed on evaluation in recent years. This is a welcomed trend, as too often new techniques or procedures have been adopted without a critical check on their effectiveness. In this section the notions of evaluation are related to the designs of previous chapters, as are some notions of cost efficiency, restricting weighting coefficients to values other than zero, teacher effectiveness, and the prediction of under- and overachieving students. We also present some ideas about the place of regression analysis in the context of one evaluation model.

There is not total agreement that evaluation and research represent the same activities. The following paragraphs are presented not with the intent of settling the question but with the idea that the designs in chapters 4 through 10 can be used for evaluation purposes.

Criterion-Referenced Testing

The lack of either adequate or available control groups in many practical evaluation situations often has led to the specification of criterion levels to be reached. When the value of the technique or procedure is to be generalized beyond the specific sample, inferential techniques are called for. Only one group of subjects is tested, but some rationale exists to posit a criterion level that they should have reached. Testing a single population mean, Generalized Research Hypothesis 10.1 would be applicable. Whether that rationale is logical may be questioned, but at least the criterion level is specified. That the technique or procedure caused the subjects to reach criterion is usually equivocal and can only be ascertained through comparisons to a control group.

Control Groups

Control groups provide a tighter control over some competing explainers of the data but are not always feasible or available in practical applications. When the researcher has authority over the selection of the control group, designs such as Generalized Research Hypotheses 4.1, 4.2, and 4.3 can be used. Often, though, control groups are "given" to the researcher, and therefore similarity cannot always be assumed. In these cases, statistically controlling for the pretest or several entering behaviors is a reasonable solution and can be accomplished using the covariance notions discussed in chapter 8.

Interaction Between Pretest and Treatment

Treating possible contaminating pre variables as covariates may not always be sufficient. The traditional ANCOVA assumption is that the relationship between the covariate and the criterion is the same for all treatments. This is the assumption of no interaction (parallel slopes). The reader should be familiar enough and comfortable enough with regression by now to realize that one could test this assumption. The discovery of interaction should not be treated as a negative finding but as an important finding by itself. With reference to Figure 12.4, the insight method is better, overall, than the rote method. But clearly the low-IQ subjects do better with the rote method. Assuming the two methods to be of equal cost, the recommendation from

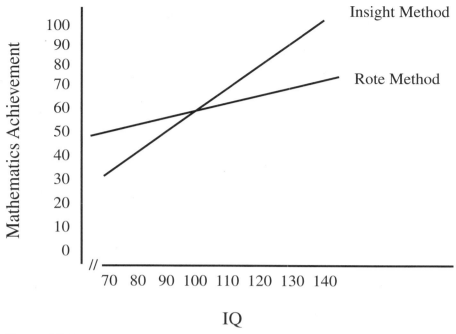

Figure 12.4.
Interaction between type of method and IQ upon the criterion of math achievement.

the evaluator would be to use the rote method with subjects who have an IQ of below 100, whereas the insight method should be used for subjects who have an IQ of 100 or higher.

Now assume that no subjects with an IQ of less than 100 were placed in the rote method. The best guess through extrapolation is the line in Figure 12.4 for the rote method. But not having observed subjects with an IQ less than 100, one cannot say for sure what their resultant math achievement would be. One must always be careful when extrapolating outside the range of observed values. Obtaining a predicted score for unobserved values within the observed range is a much safer procedure, especially if one has obtained a somewhat high R^2.

Novick and Jackson (1970) proposed that Bayesian statistics could provide better solutions to prediction problems because they did not have to use common Y-intercepts and common slopes. Although many applied regression analyses (and all covariance analyses) in the past used common slopes, one must realize that common slopes *do not* have to be used. The Bayesian prediction models may be more fruitful than the regression approach, but as of yet the Bayesian approach has not been compared to regression models using varying slopes (see Newman, Lewis, & McNeil, 1973, or Newman, Lewis, McNeil, & Fugita, 1978, for a more complete discussion).

Evaluating Programs That Have Cost Differences

Very seldom will two programs being compared have identical costs. Cost here is meant to include not only the announced dollar cost but all associated costs, such as teacher in-service training, teacher rejection, changes in janitorial policy, increased use of library books, and increased vandalism by dissident students. It is

Generalized Research Hypothesis 12.1

Directional Research Hypothesis: For some population, Method A is more than c units better than Method B on the criterion Y.

Nondirectional Research Hypothesis: For some population, Method A is other than c units better than Method B on the criterion Y.

Statistical Hypothesis: For some population, Method A is c units better than Method B on the criterion Y.

Full Model: $Y = aA + bB + E_1$ (pieces full = m1 = 2)

Want (for directional Research Hypothesis): $a > (b + c)$; restriction: $a = (b + c)$
Want (for nondirectional Research Hypothesis): $a \neq (b + c)$; restriction: $a = (b + c)$

Automatic unit vector analogue
Full Model: $Y = aU + bB + E_1$ (pieces full = m1 = 2)
Want (for directional Research Hypothesis): $b < -c$; restriction: $b = -c$
Want (for nondirectional Research Hypothesis): $b \neq -c$; restriction: $b = -c$

Restricted Model:
$$Y = (b + c)A + bB + E_2$$
$$Y = b(A + B) + cA + E_2$$
$$Y = bU + cA + E_2$$

But: Since c is a known quantity, cA is also a known quantity, and it can be subtracted from both sides, resulting in:

Restricted Model: $(Y - cA) = bU + E_2$ (pieces restricted = m2 = 1)

where:
Y = criterion;
U = 1 for all subjects;
A = 1 if subject in Method A, 0 otherwise;
B = 1 if subject in Method B, 0 otherwise; and
a and b are least squares weighting coefficients calculated to minimize the sum of the squared values in the error vectors.

Degrees of freedom numerator = (pieces full - pieces restricted) = (2 - 1) = 1

Degrees of freedom denominator = $(N$ - pieces full) = $(N - 2)$

where:
N = number of subjects.

Note: To use the general F test, the criterion must be the same in both the Full and Restricted Models, which is not so here. The test of significance must use the formula incorporating the error sum of squares of the two models, Equation 10.1.

difficult to list and measure all possible costs and extremely difficult to change these costs into a unit, such as dollars. The ultimate decision of adopting a given program should, though, consider these costs.

One way to consider the differential cost of two programs in an evaluation plan is to demand more from the program that is more costly. If most of the important costs have been considered, one could arbitrarily demand that the more costly program be more than, say, three units better than the cheaper program before adoption of the more costly program. In times of austerity, a criterion of five units better might be set. In times of unlimited monetary resources from parents who want the best for their children, the most effective program (anything more than zero units better) would be the program adopted. Though the required criterion number of units is arbitrarily set, the units are decided before the data are obtained, making the final decision determined by the data and not by any unique and unanticipated impressions from the data. Adoption of more stringent alpha levels would have a similar effect, but stating a particular unit value allows one to make a more conclusive statement if significance is found.

Generalized Research Hypothesis 12.1 presents the "c-unit question," and Applied Research Hypothesis 12.1 presents a "three-unit question." Appendix D-12.1 contains the SAS setup for Applied Research Hypothesis 12.1.

Nonzero restrictions also can be made on a continuous predictor variable as indicated in Figure 12.5. Suppose a community has indicated a willingness to pay more taxes if it can be reasonably predicted that an increase for each one dollar spent

Applied Research Hypothesis 12.1

Directional Research Hypothesis: For some population, X6 is more than 3 units better than X7 on the criterion X2.

Statistical Hypothesis: For some population, X6 is 3 units better than X7 on the criterion X2.

Full Model: $X2 = a_6X6 + a_7X7 + E_1$ (pieces full = m1 = 2)
Want: $a_6 > (a_7 + 3)$; restriction: $a_6 = (a_7 + 3)$

Automatic unit vector analogue
Full Model: $X2 = a_0U + a_6X6 + E_1$ (pieces full = m1 = 2)
Want: $a_6 > 3$; restriction: $a_6 = 3$

Restricted Model: $(X2 - 3X6) = a_0U + E_2$ (pieces restricted = m2 = 1)

alpha = .05
$ESS_f = 7656.166$; $ESS_r = 7656.983$
$F = .006$
Computer probability = .93; directional probability = .53

Interpretation: Since the weighting coefficients are not in the hypothesized direction, the directional probability is not lower than alpha; therefore, fail to reject the Statistical Hypothesis.

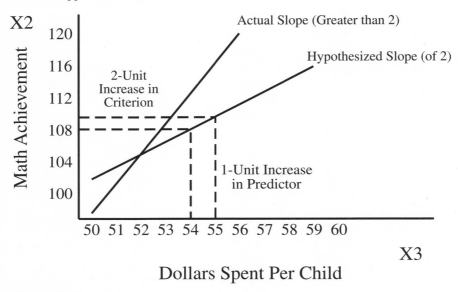

Figure 12.5.
Data that yield a slope greater than 2.

Applied Research Hypothesis 12.2

Directional Research Hypothesis: For some population, for each additional unit increase in X3, the increase in the criterion X2 will be more than 2 units.

Statistical Hypothesis: For some population, for each additional unit increase in X3, the increase in the criterion X2 will be equal to 2 units.

Full Model: $X2 = a_0U + a_3X3 + E_1$ (pieces full = m1 = 2)

Want: $a_3 > 2$; restriction: $a_3 = 2$

Restricted Model: $X2 = a_0U + 2X3 + E_2$

But: Since 2X3 is a known quantity, it must be subtracted from both sides of the equation, resulting in:

Restricted Model: $(X2 - 2X3) = a_0U + E_2$ (pieces restricted = m2 = 1)

alpha = .05
$ESS_f = 559.878$; $ESS_r = 566.178$
$F = .65$
Directional probability $\approx .20$

Interpretation: Although the weighting coefficient a_3 in the Full Model is in the hypothesized direction, the directional probability is not lower than alpha; therefore, fail to reject the Statistical Hypothesis.

on each child in math class will result in more than a two-unit mean increase on the math test. Assuming that data are available from other communities concerning dollars spent per child in math class (X3) and resultant scores on the math test (X2), and assuming a linear increase over the range of dollars spent, Applied Research Hypothesis 12.2 might be appropriate. (Appendix D-12.2 contains the SAS setup.)

Restricting weighting coefficients to values other than zero is important enough to present Generalized Research Hypothesis 12.2. Any of the previously discussed Generalized Research Hypotheses in any of the previous chapters could have had a weight restricted to some constant value other than zero. Often zero is a meaningful restriction, although we feel that more often a weight might have been restricted to some nonzero value if the researcher had felt comfortable in doing so. Restricting a weighting coefficient to zero is only one of an infinite number of ways of restricting

Generalized Research Hypothesis 12.2

Directional Research Hypothesis: For some population, for each additional unit increase in the predictor X, there will be more than k units increase in the criterion Y.

Nondirectional Research Hypothesis: For some population, for each additional unit increase in the predictor X, there will be other than k units increase in the criterion Y.

Statistical Hypothesis: For some population, for each additional unit increase in the predictor X, there will be k units increase in the criterion Y.

Full Model: $Y = a_0U + a_1X + E_1$ (pieces full = m1 = 2)

Want (for directional Research Hypothesis): $a_1 > k$; restriction: $a_1 = k$
Want (for nondirectional Research Hypothesis): $a_1 \neq k$; restriction: $a_1 = k$

Restricted Model: $(Y - kX) = a_0U + E_2$ (pieces restricted = m2 = 1)

where:
    Y = criterion;
    U = 1 for all subjects;
    X = predictor score; and
    a_0 and a_1 are least squares weighting coefficients calculated to minimize the sum of the squared values in the error vectors.

Degrees of freedom numerator = (m1 - m2) = (2 - 1) = 1
Degrees of freedom denominator = (N - pieces full) = (N - 2)

where:
    N = number of subjects.

The F test using error sum of squares would have to be used with this hypothesis, as with any that specifies a nonzero weight.

that weighting coefficient. *It is important to have a rationale for restricting a weighting coefficient to some value, but the rationale must be as thoroughly thought out and as defensible for restricting the coefficient to zero as to any other numerical value* (McNeil, 1991a).

If all the costs of programs can be assessed, then one might want to incorporate the actual cost into the analysis. The *gain per dollar cost* could be used as a criterion when evaluating the relative effectiveness of two programs. That is, the gain or criterion worth of the program could be divided by the dollar outlay or cost. The two "program vectors" would be the predictor vectors, resulting in a model structure exactly like that of Generalized Research Hypothesis 4.1. The decision about which program to implement would be an administrative decision based upon the amount of money available as compared to the amount of educational gain demanded (or loss tolerated) by the community.

Teacher and Student Evaluation

Accountability has entered the classroom, and many school systems are interested in determining the relative effectiveness of teachers within the same grade level. Students are not always randomly placed into classrooms; therefore, teachers are faced with students who have different entering characteristics from students in other classrooms. Covariates are a necessity. Perhaps previous achievement, motivation, IQ, and a host of other variables should be considered possible contaminating variables. The criterion variable(s) should be ones that are relevant to the objectives of that particular school system. The possible contaminating variables and the criterion should be proposed and discussed by all the teachers. A list of the agreed upon contaminating variables and criteria should then be passed on to the administration for their consideration. Once the variables have been established by the administration, then each teacher knows the basis upon which an evaluation will be made. Once the criterion variable and contaminating covariables are agreed upon, then the covariance notions incorporated in chapter 8 would be appropriate. The Dallas, Texas, school district has implemented these notions for several years (Webster & Mendro, 1992).

A slightly different route might be useful in the evaluation of individual student achievement. One can use the standard error of estimate to ascertain whether the student is achieving "above that of similar students" (the overachiever) or whether the student is achieving "below that of similar students" (the so-called underachiever).

Figure 12.6 depicts a simple example of these notions. Those students who are *above the top dashed line* have achieved more than other students who had the same entering IQ. Those students *below the bottom dashed line* are achieving less than other students who had the same entering IQ. Note that overachievers can be both high- and low-IQ students. Likewise, underachievers can be both high- and low-IQ students. The distance between the dashed line and the solid line of best fit is dependent upon how many students the particular school system wants to identify and either reward (the overachievers) or remediate (the underachievers). If the achievement scores are normally distributed about the line of best fit, then the functional relationship being used is a good one. (If the previous sentence seems cryptic, re-read the discussion in chapter 8 on homogeneity of variance.) The unit normal curve can be referred to after the percentage of identified students has been set. If 5% of the students are to be rewarded and 5% remediated, then the two dashed lines would be 1.64 standard deviation units away from the line of best fit. The standard deviation of the discrepancies between the actual achievement and the predicted achieve-

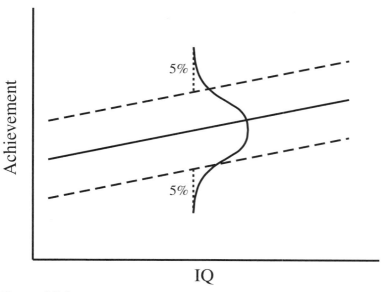

Figure 12.6.
Identification of overachievers and underachievers.

ment is the standard deviation of the errors in prediction. This standard error of estimate can be found from the following equation:

$$S_{est} = S_y\sqrt{(1 - R^2)} \qquad (12.5)$$

R^2 is the proportion of criterion variance accounted for; $(1 - R^2)$ will then be the proportion of criterion variance unaccounted for. The square root of variance is the standard deviation, so $(1 - R^2)^{1/2}$ is the proportion of criterion standard deviation that is not accounted for, or the proportion of criterion standard deviation that is error. Multiplying this proportion by the actual criterion standard deviation (S_y) gives the desired quantity, the standard deviation of the error in predicting the criterion from the predictor variable(s) (the standard error of estimate).

Note that if R^2 is low, say zero, then $S_{est} = S_y$, or all the criterion variance is error variance. The larger the R^2, the smaller S_{est} becomes in relation to S_y. When $R^2 = 1$, all the criterion variance is accounted for, all the errors in prediction equal zero, the standard deviation of these errors would be zero, and $S_{est} = 0$.

SAS can print the criterion and the error in prediction for all students as in Appendix D-12.3. The statements in Appendix D-12.3 print the identification numbers (ID) of the students, the actual criterion (X2), the predictor (X3), the error in prediction (RESIDUAL), and the cutoff indicating over- or underachievement (one standard deviation in this application). The scores of those students who are within ±2.52 should not be attended to, as these discrepancies between the actual criterion and the predicted criterion could simply be random fluctuation rather than something systematic that needs rewarding or remediating. Note that 18 out of 60 students (30%) are identified as over- or underachievers in this application, using one standard error of estimate.

The GLM procedure allows for a more numerous set of predictors than that depicted in Figure 12.6. The notions are the same though, although figures cannot be drawn to depict, say, five continuous predictors. The identified underachiever using five continuous predictors is achieving much below students with the same profile of those five entering scores. Although another student may not exist in the school exactly like each one on these five entering variables, the use of continuous variables allows the inferences made in the preceding sentence.

A few precautionary remarks are in order here. Students may be labeled over-achievers or underachievers when in fact they are not. That is, chance may have placed them in one of the two extremes, or one bad or good test score might have influenced their placement. Erroneous placement is especially crucial for identified underachievers. Therefore, it is suggested that all students' folders be extensively examined before attempting any remediation. Statistical and computer techniques can save time identifying potential students to work with, but humans should not abdicate all their decision making and responsibility. One mistake during data entry can make Johnny look like an underachiever rather than the overachiever that he really is. More important, all the relevant variables may not have been taken into consideration. Who would ever have predicted that Johnny would not be doing well this year because Sally no longer is his girl?

The presentation here employs the standard error of estimate for the identification of individuals. Some statistical authors use similar procedures and concepts to state confidence intervals around the observed sample statistic, whether it be the mean, the differences between the means, or the line of best fit. The emphasis in this text has been on the testing of hypotheses and the increasing of R^2, rather than on the identification of population values.

Regression Models for One Evaluation Model

The place of regression analysis in the evaluation setting has been well doc-umented by Webster and Eichelberger (1972). They present the context, input, process, and product (CIPP) model and then develop regression models for an actual applica-tion in the education setting. The CIPP model delineates four kinds of evaluation.

> Context evaluation serves planning decisions by identifying unmet needs, un-used opportunities, and underlying problems that prevent the meeting of needs or the use of opportunities; input evaluation serves structuring decisions by projecting and analyzing alternate procedural designs; process evaluation serves implementing decisions by monitoring project operations; and product evalu-ation serves recycling decisions by determining the degree to which objectives have been achieved and by providing insights about the causes of the obtained results. (p. 91)

Each of these four kinds of evaluation is discussed in more detail by Webster and Eichelberger, and then they present regression models with alternate lines of se-quencing of hypotheses, depending upon whether there is previous significance or lack of significance. The possible inferences that can be drawn from results are discussed, and decisions that might be made as a result of these inferences are delineated.

The important aspect of their article is that it indicates the close connections among research, evaluation, and statistical procedures (particularly regression analy-sis). There must be arbitrary administrative decision making in the real world—the point that Webster and Eichelberger are making is that before the administrative decision could be "yes," statistical significance must be established. But statistical significance may not be enough evidence of the administrators' decision. Statistical significance identifies the likelihood of random occurrence, but an administrator might be justified in not taking action until at least, say, 80% of the variance in the criterion is accounted for ($R^2 > .80$). As in all applications, statistical significance is a necessary, but not sufficient, condition neither for practical significance nor for decisions in the real world.

POLICY CAPTURING

Evaluations are made, for instance, when people apply for jobs, when they are considered for advancement, and when salary increases are considered. These evaluations are usually made based on several criteria, but the actual combination of the criteria may be unspecified. A committee is often convened to review the information and make the decision, but such a review requires much time, and reviewers frequently discover that they do not agree on how the criteria should be combined. Some reviewers may place more weight on certain pieces of information, while others may be swayed by information that is irrelevant to the task at hand or that was not obtained for all of the subjects. In these cases, the reviewers have different policies, and some policies may be based on irrelevant information. The GLM can be employed (a) to identify the individual policies; (b) to determine the degree of similarity in the various policies; and (c) to evaluate new subjects in a more economical and consistent fashion.

Humans are not very good at storing a great deal of information, nor are they good at using that infrmation consistently. Computers can be easily trained to do just that. Such is the essence of policy capturing (Christal, 1968; Houston, 1987).

An Applied Policy-Capturing Example

Suppose that 200 students applied to enter a teacher education program. The program announcement would have specified the nature of the program and the minimum requirements. It is now the task of a committee to review the applicants and decide which ones to accept. Let us assume that there is room for only 100 students.

Each committee member reviews the applications and makes an "accept" or "not accept" decision for each applicant. Was each committee member internally consistent? That is, did similar applicants evoke the same decision by a given committee member? And were the decisions similar across committee members? That is, did Applicant A receive the same decision from all the committee members? If the answer is "no" to either question, then there is a problem. If a member looks at two similar applicants but makes a different decision for each, then that decision is worthless at best and litigious at worst. If two members look at the same applicant but make different decisions, then at least one of them is not upholding the integrity of the organization.

Why Does a Reviewer Make Different Decisions?

Reviewers usually do not establish a set policy. They feel that they are experts, that they "know" what kind of applicant will be successful. If a reviewer does develop a somewhat systematic policy as he or she proceeds through the task, only the later applicants will be judged by that policy. The policy is usually continuously modified throughout the task, often influenced by information provided, or not provided, by the previous applicant.

Even if a policy is established, the human mind has limits regarding how much information can be processed and how complicated a policy can be applied. Most reviewers simply cannot attend to all the relevant information. In addition, applicants may have presented the same information in very different ways. The reviewer may simply miss a relevant piece of information or may focus on an irrelevant piece of information, such as church affiliation or place of birth.

All of the above factors are exacerbated when the review task takes a long time because of the large number of applicants or a lengthy application. A reviewer's attention span is limited and can be influenced both by outside forces, such as noise or progress by other reviewers, and by internal forces, such as hunger or daydreaming. Usually the last applicants are reviewed more quickly than the first applicants— the reduction in time may be because the reviewers have become adept at the task or may be because there are other pressing priorities.

Why Do Reviewers Make Different Decisions?

All of the above factors will contribute not only to differences within a reviewer but also to differences between reviewers. In addition, reviewers may undertake the task with very different understandings or desires about the important variables. One may have the impression that rural applicants should be given special consideration, while another may "understand" that the university's mission is to enroll more foreign students. If the mission is to enroll more foreign students, some reviewers may feel incompetent in making those decisions and defer them to other reviewers.

Some reviewers may have hidden agendas that are contrary to the program. One reviewer may be biased toward math and science applicants while another toward applicants from private schools. If these hidden agendas become known to other reviewers, the other reviewers may develop compensating policies to counteract those "hidden agendas." Clearly these reviewers will rate applicants differently.

The GLM Approach to Policy Capturing

Once all the information is quantified and the decision is made, the GLM can be used to identify, or *capture*, the policy. The value of this procedure is in determining (a) how consistent a reviewer is; (b) the weights and the relevant information to see if this policy is in line with the organization's guidelines; and (c) whether the policy is consistent across the various reviewers.

The outcomes above can be determined through a stepwise regression approach. The "best model" would provide the weights for each piece of information, as well as the overall R^2. The degree of consistency is indicated by the R^2. One would want a significant R^2, greater than, say, .60. The relevant information would be those variables that are in the "best model." Other information is either irrelevant or superfluous, as judged by that reviewer.

Assuming there is a consistent policy for a reviewer, that policy can be cross-validated with the other reviewers. One hopes that all reviewers are using the same policy. What might emerge, however, is that two or more policies are being implemented, requiring a conference to decide *the* policy.

Getting the Data

Obtaining the data for a policy-capturing study requires that the reviewers make their decisions and that the information made available to the reviewers then be quantified. The GLM results will indicate the degree of consistency (R^2), and how the variables were weighted (the policy). If the policy and the consistency in applying that policy are not acceptable, then the reviewers can be informed of the problem and trained to be consistent. (Appendix H contains a policy-capturing simulation.) The steps are as follows:

1. Have the reviewer make a decision about each applicant (either dichotomous: 1 = accept, 0 = not accept; or continuous—say on a 5-point scale).
2. Determine relevant information (through program announcement or by asking all the raters individually).
3. Quantify the relevant information (dichotomously or continuously).
   Past achievement: = high school GPA
   Number of teacher education courses taken: = #
   Quality of reference letters:

   5 = great
   4 = good
   3 = medium
   2 = poor
   1 = bad

   Breadth of references:

   2 = includes education professors
   1 = includes noneducation professors
   0 = no university professors
   -1 = only friends

   Dedication to education on personal statement:

   3 = clearly dedicated
   2 = probably dedicated
   1 = wants a job
   0 = unsure of life and self

   Quality of portfolio-objective:

   4 = clearly defined educational objective
   3 = somewhat clear
   2 = somewhat vague
   1 = vague
   0 = no objective

   Breadth of portfolio:

   2 = more than one
   1 = limited to one
   0 = very idiosyncratic

Notice that some information is easier to quantify than other information. The numerical values and descriptions will be difficult to determine (but no more difficult than "determined" by the reviewers).

4. Perform a stepwise regression, using the "relevant information" to predict the decisions.

5. Determine if the R^2 of the "best model" meets the minimum criteria for this application (minimum will vary depending upon how crucial the decision is).
6. Cross-validate the results from each reviewer with the other reviewers. If a cross-validated R^2 does not meet the minimum, then those two reviewers are implementing different policies.

Advantages of Policy Capturing

The policy-capturing process identifies whether the decision process is consistent. Given that it is, the computer can be used to obtain decisions about future applicants, thus saving time and avoiding the inclusion of irrelevant information. An organization could even consider publishing their policy, so that applicants would know (a) what the relevant information was and how it is used in the policy; (b) how they currently stand on the policy scale; and (c) most important, what can be done to improve one's chances.

Disadvantages of Policy Capturing

The determination of a policy requires that the same information be obtained for all subjects. Any additional information of note for an applicant cannot be considered. The decision is taken out of the hands of humans, and some may see this as a loss of information. But if certain information is considered valuable, it should be obtained from all applicants. One compromising approach is to use the computer model (based on the reviewer's decisions) to identify a pool of the best three applicants, who are then screened in person. But such in-person screening is subject to all kinds of vagaries. Are the same questions asked of each applicant? Is the attire of each applicant the same? And is the attire viewed by each interviewer in the same way? Are applicants provided the same information? Are applicants interviewed the same time of day? Is in-person interviewing relevant to the degree program?

Most of these questions will be answered in the negative, suggesting an unsystematic information-gathering process and therefore an unsystematic policy. Humans just cannot be systematic for any length of time—we can say this because we have been unsystematic ourselves. When unsystematic behavior affects only one's own life, it is that person's decision, and it may be tolerable. But when unsystematic behavior affects other peoples' lives, it is not tolerable—nor probably legal.

13

THE STRATEGY OF RESEARCH AS VIEWED FROM THE GLM APPROACH

MEETING THE GOALS OF RESEARCH WITH THE GLM

We discuss the GLM in this chapter as it relates to several goals of research: predictability, parsimony, replicability, and validity generalization. These goals are presented with the development of a well-established physical law. Our emphasis is upon the percentage of variance that can be accounted for in the criterion under investigation rather than on statistical significance from random events. Additional remarks concerning curvilinear relationships and data snooping are also presented.

The material that follows in the next section represents one way in which Sir Isaac Newton might have developed the law of gravity. The material is adapted from McNeil (1970a) and is presented with the intent of showing the advantage of using the GLM technique as one's statistical tool. Think of it as a narrative interview with Sir Isaac Newton.

NEWTON STATES THE PROBLEM

It seems that for years some of our most competent physicists have been looking for the functional relationship between the amount of time an object has been falling in space and the distance the object has fallen. Stating the problem symbolically, $d = f(t)$, and verbally, "Distance is what function of time?" I believe that I have finally discovered this functional relationship between time and distance. This means that, if I know the amount of time an object has been falling, I can tell you how far it has fallen.

One of the surprising findings that my research strategy has led me to is that I need know *nothing* about the object, if I can assume that the object is falling in a vacuum, where there is no resistance to its fall. The only additional variable that I need to know is what I call the "gravitational constant"—computed from the forces being exerted by the earth, the sun, the moon, and other heavenly bodies. For any one time and place, this variable has the same value for any object.

Goal 1: Predictability

To give a little background, let me review previous research. Galileo published (1632) much data on falling objects, but he only investigated the linear relationship between time and distance. Galileo and, before him, Aristotle used the GLM procedure in testing their hypotheses, but these great thinkers did not realize the value and the flexibility of this procedure. They only used a very restricted form of the technique, the form that ascertains the rectilinear relationship between the two variables.

Galileo used the following model:

$$(\text{Model } 13.1)\ D\ =\ aU + bT + E_1$$

where:

  D = the vector containing the distances the objects have traveled;
  T = the vector containing the amount of time the objects have traveled;
  U = the unit vector that allows the regression constant to be nonzero;
  E_1 = the vector containing the differences between the actual distance and the predicted distance (that predicted from the pool of predictor variables on the right-hand side of the equation); and
 a and b are weighting coefficients of best fit, determined from the sample data. "Best" is here defined in the least squares sense, minimizing the sum of the squared values in E_1.

Model 13.1 produced a significant fit of the data, but the R^2 value was discouragingly low ($R^2 = .40$). This R^2 value is the squared value of the correlation between the observed and predicted values and can be interpreted as the proportion of variance in the distance-measures criterion predicted, or accounted for, by the predictor variables in the sample (in Model 13.1, the regression constant and the continuous variable of time).

A Linear Model Allowing for a Curvilinear Relationship

Upon looking at a bivariate plot of the data in Figure 13.1, it becomes obvious that there is not simply a linear relationship—there is a curvilinear relationship between time and distance. And, because the rate of acceleration of the curve continuously increases, there must be a second-degree curvilinear relationship. The GLM model that allows a second-degree relationship to exist is:

$$(\text{Model } 13.2)\ D\ =\ aU + bT + cT^2 + E_2$$

All the variables in Model 13.1 appear in Model 13.2, with the addition of T^2. Each element of T^2 is the squared value of the corresponding value in T (thus, if the object took 4 seconds to fall the designated distance, then T^2 would be 4^2 or 16).

The variable T^2 allows the second-degree curve in the data to manifest itself, if in fact there is a second-degree curve in the data. I fit Model 13.2 to the data for a 100-pound stone. What surprised me was not that I obtained a significant fit of the data but that the R^2 value was .99. Distance could be almost exactly predicted from knowledge of the regression constant, time, and time squared.

Goal 2: Parsimony

Parsimony means to explain something in the simplest way possible. With respect to the problem at hand, Model 13.2 incorporates three pieces of information in predicting the criterion of distance. The notion of parsimony demands that we inves-

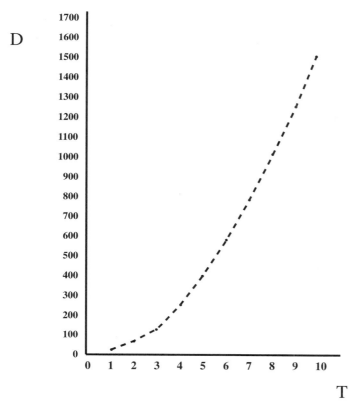

Figure 13.1.
Bivariate plot of data suggesting a possible curvilinear relationship between time (T) and distance (D).

tigate the predictability of the criterion by dropping out T, T^2, U, or some combination of these three vectors. If T^2 is dropped out of the model (actually I am hypothesizing that the weight c associated with T^2 is equal to zero), I end with the model used by Galileo. I tested the predictability (R^2) of Galileo's model (Model 13.1) against the R^2 of Model 13.2 and found a significant decrease ($p < .01$) in the amount of variance being accounted for in Galileo's model. Model 13.2 was accounting for 99% of the variance, while Galileo's model was accounting for some 40%. I would like to add parenthetically here that many researchers had interpreted Galileo's results incorrectly. Galileo had tested the predictability of his model against one that contained no differential information:

$$(\text{Model } 13.3) \ D \ = \ aU + E_3$$

The test showed that there was a significant decrease in predictability and, in particular, indicated that the rectilinear relationship between time and distance was something other than zero. As the R^2 of .99 in Model 13.2 indicated, a model can be built that more accurately predicts the criterion of distance. The point is that we should not rest on our haunches until we get close to accounting for 100% of the variance.

A second model more parsimonious than Model 13.2 involves hypothesizing that the weight associated with T is equal to zero. Again, this reduces to dropping that particular variable out of the pool of predictor variables:

$$(\text{Model } 13.4)\ D\ =\ aU + cT^2 + E_4$$

This last model does not contain the time variable at all, except in its squared form, and the surprising thing is that Model 13.4 was as predictive as the original Full Model, Model 13.2. In fact, the R^2 of Model 13.4 was also .99, exactly equal to that of Model 13.2. What was also true, but fully anticipated, was that the weight for the unit vector was equal to zero, (i.e., the Y-intercept is 0). This, in effect, substantiates our impression that the distance covered by a falling stone in zero minutes is exactly zero feet. The final values for the more parsimonious model (Model 13.4) were: $D = (0 * U) + (16 * T^2)$, or $a = 0$ and $c = 16$. The goal of parsimony has been satisfied in that the criterion has been predicted with a small amount of predictor information: Here, only one bit of predictor information is needed, that of the square of time.

Goal 3: Replicability

As indicated previously, the regression weights were optimum weights for those particular sample data. I am concerned, of course, about predicting the distance measure for not just this one sample of stones but for any stone from a clearly defined population of stones. My first concern was to check the replicability of the obtained regression weights for another set of data on 100-pound stones.

One process of checking replicability simply involves taking the obtained regression weights (from Model 13.4) and applying them to the new data. This process produces predicted distances, and the predicted distances can be then correlated with the observed distances. In this instance, every new time score was simply squared and then correlated with the associated distance score. The resulting (Pearson product moment) correlation was .98, yielding an R^2 of .96, which again is quite satisfactory. The regression weights found for Model 13.4 were checked for replicability on several other samples of data with 100-pound stones, and all R^2 values were above .95.

Goal 4: Validity Generalization

There are several aspects to the replication problem. I just discussed how I replicated the functional relationship in a single population. My students were interested in generalizing this functional relationship to other populations, such as populations of lighter stones. I discovered that the stones' weights were unimportant. Here is how I obtained evidence to make this statement. I wondered about the generalizability of the results on the 100-pound stones to lighter stones. I obtained data on 50-pound stones and proceeded to apply the regression weights from the 100-pound stones in Model 13.4. To my surprise, the R^2 resulting from this replicability study was comparable to the aforementioned replicability studies. That is, the functional relationship between time and distance was the same for both 100-pound stones and 50-pound stones.

It was not a great intellectual leap to investigate the possibility that the functional relationship was similar for stones of all weights. To check this hypothesis, I collected distance and time data on stones of various weights and then added pound measures to the predictor pool of variables in Model 13.4:

$$(\text{Model } 13.5)\ D\ =\ aU + cT^2 + hP + E_5$$

where:

P is the weight in pounds of the stone being measured.

Of course, to make this model viable, I had to measure objects of various weights. When Model 13.5 was applied to a set of data on stones of differing pounds, the regression weight h turned out to have a numerical value of zero. Thus, when the predictive efficiency of Model 13.5 was compared against the predictive efficiency of Model 13.4, no predictive information was lost. The hypothesis—that the functional relationship between time and distance was similar for objects of any weight (more specifically, for the range of weight values I used in the analysis—was supported. To put it another way, weight is not needed in defining the functional relationship between distance and time. The functional relationship has been generalized across all weights.

A Return to the Gravitational Constant

I still have not discussed the gravitational constant. This is the most difficult discussion of all, but it should be included because it increases the generalizability of my findings and also illustrates the flexibility of the GLM approach.

All the data that have been presented so far were obtained in my experimental labs. But objects move in places other than my lab. In particular, it had been known for quite some time that objects move faster when they are at lower altitudes. Also, the functional relationship that I found did not replicate on observations made by Galileo in outer space. Therefore, the functional relationship between time and distance may be modified by the gravitational field in which the object is measured; that is, various observations provided clues about what kinds of variables might be important to investigate.

I shall not go into the actual calculation of this gravitational constant except to indicate that it can be quite accurately measured. The gravitational constant will be symbolized as G. I subjected data that included objects measured within various gravitational fields to Model 13.6 (an extension of Model 13.4) to check the functional relationship within the various gravitational fields:

$$(\text{Model 13.6}) \quad D = aU + cT^2 + iG + E_6$$

Model 13.6 yielded an R^2 value of .50, indicating that 50% of the criterion variance was predicted by the predictor set. To be sure, this is a large amount but not as high as I had been accustomed to. I was careful not to make the same mistake that Galileo had made years before, that of stopping with an extremely parsimonious model that does only a fair job of accounting for the criterion variance. Generally, one reduces the amount of predictability when the variable pool is reduced. Conversely, the inclusion of another predictor variable will generally increase the predictability, the question being: Is the increase in predictability significant?

Curvilinear Interaction

The bivariate plot (Figure 13.2) between time and distance for the various G levels was visually inspected, and it became obvious that there was likely an interaction between the gravitational constant (G) and the square of time (T^2) (Figure 13.3).

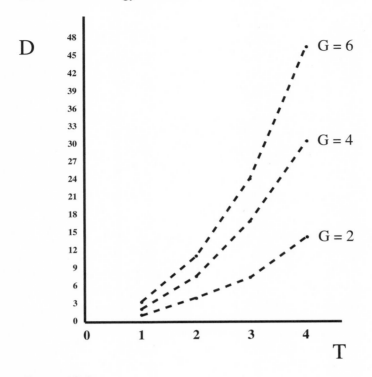

Figure 13.2.
Bivariate plot of data indicating the curvilinear interaction between time and gravity (G) in the prediction of distance.

A model was constructed that allowed this hypothesized interaction to exist, in which the vector $(G * T^2)$ represents a single number that is the product of the gravitational field and the square of time:

$$\text{(Model 13.7)} \quad D = aU + cT^2 + iG + j(G * T^2) + E_7$$

In accordance with the anticipated results, R^2 was .99, indicating that Model 13.7 was indeed a good reflection of the functional relationship. I carefully noted that a, c, and i all had numerical values extremely close to zero, indicating that the variables associated with these regression weights were all somewhat useless in obtaining the high degree of predictability. Thus, the hypothesis that the parameter value of these weights is, in fact, equal to zero was tested. By setting a, c, and i in Model 13.7 equal to zero, the following model is derived:

$$\text{(Model 13.8)} \quad D = j(G * T^2) + E_8$$

When the R^2 of Model 13.8 was tested against the R^2 of Model 13.7, a significant decrease in the amount of variance being accounted for was *not* found. In fact, the R^2 of Model 13.8 was .98, well within the decrease expected from the measurement errors in our data.

The goal of predictability was met: The interaction model predicted a high percentage of the criterion variance, although a good linear relationship did not exist in the prediction of distance. The "simplified" interaction model (Model 13.8) pre-

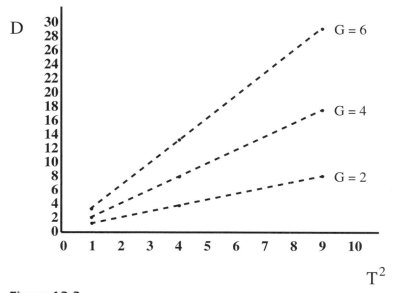

Figure 13.3.
Bivariate plot of data that led to the notion that G and T^2 interact in the prediction of D.

dicted just as satisfactorily as Model 13.7; therefore, the simplified interaction model was accepted as a better model (because it simultaneously met the goals of predictability and parsimony).

After running successful replications on other sample data, Model 13.8 was deemed to be "lawful" for the phenomenon being studied (because it met the goals of predictability, parsimony, replicability, and to some extent validity generalization). The functional relationship between time and distance (for stones of various weights dropped from various altitudes) is thus that function indicated by Model 13.8.

From my GLM analysis of Model 13.8, I note that the numerical value of j is .5. Verbally then, D is equal to one-half the product of G times T^2; symbolically, $D = \frac{1}{2}(G * T^2)$. If we call distance d, the gravitational constant g, and time t, we have the following symbolic equation: $d = \frac{1}{2}gt^2$. Because it always yields an R^2 of close to 1.00. and I found it, I will call it *Newton's law*.

Our Remarks

After reading Newton's discussion, we hope that behavioral science researchers will pay attention to the methodology that he presented. It seems inappropriate for these researchers to argue continually that the behaviors they look at are inherently complex but then to use only simplex models; the behaviors may indeed be complex, but how many researchers have even attempted to investigate curvilinear interaction as Newton did? It is time for us to realize that complex behavior demands complex methodology, as well as complex theory.

There is a problem about what variables one should choose, but that is a content question that cannot be answered by a methodologist. Transformations and intricate relationships should be sought rather than discarded. Interaction has traditionally been viewed as a bad effect rather than as an effect that will help in explaining the phenomenon at hand.

The notion of a curvilinear relationship is not new to methodology, but we must admit that few investigators make a habit of checking the curvilinear predictability. Indeed, some statistical books discourage the inclusion of highly correlated predictor vectors in regression models (all polynomials are highly correlated with the original vector). Whether a researcher should include a vector that allows for curvilinearity is partly a function of past research findings and partly a function of the nature of the data. Though the addition of another predictor variable cannot decrease the R^2, it is not so that such a vector will always statistically increase the R^2 value. Whether such investigations prove rewarding is an empirical question.

Inspecting the sample data is considered a poor research strategy by many researchers. But if the goal of research is to develop generalizable scientific findings, then it can be argued that a researcher is essentially committing a crime when one does not "squeeze" the data for all they are worth. When such squeezing is done, though, the researcher is committed to replicating findings on an independent sample. The stability, and therefore value, of the finding must rest on the replicated study rather than on the original study alone.

THE MOST GENERAL REGRESSION MODEL

The hypotheses and their associated models presented since chapter 1 have been of a rather restricted nature when compared with the research notions of chapter 1. In that first chapter, we argued that behavior (whether of humans, university administrations, plots of land, etc.) is complex and that many predictor variables may be needed to account for that behavior. In accounting for such behavior, we have shown that categorical information could be used (chapter 4), that continuous variables could be used (chapters 5 and 6), and that transformations of originally scaled variables should be considered (chapter 9). We must emphasize that the data and inherent functional relationships determine the variables necessary to obtain the desired R^2. Both dichotomous and continuous variables can be used in the same regression model (as shown in chapters 7 and 8).

Generalized Research Hypotheses were presented in the previous chapters in part to show that traditional least squares solutions are all computational simplifications for the general least squares procedure. The regression formulation was shown for such traditional solutions as the Pearson correlation (Generalized Research Hypothesis 5.1); the phi correlation (Generalized Research Hypothesis 4.4); the t test to show that the point biserial correlation equals zero (also called the t test for the difference between two independent means, as in Generalized Research Hypothesis 4.1); the t test for the difference between two dependent means (Generalized Research Hypothesis 10.2); the t test for a single population mean (Generalized Research Hypothesis 10.1); the one-way ANOVA (Generalized Research Hypothesis 4.2); the two-way ANOVA (chapter 7); and the ANCOVA (chapter 8). Every one was computed using the same least squares procedure. To aid in the calculation of these solutions by hand, simplified formulae have been developed by statisticians over the years to fit special cases; and then these solutions became grouped according to their computational similarities (correlation, t tests, F tests, etc.). Now that the computer is available to perform these calculations, statistics should be put in their conceptual place rather than their computational place. Any hypothesis can be tested by the regression approach (as long as the researcher is using the least squares approach with a single criterion). The specific Research Hypothesis dictates the Full Model and the Restricted Model and the testing of those models. Rather than spending time searching for the "appropriate" simplified statistical test that will fit the design, the researcher should spend time explicitly stating the Research Hypothesis, so that the Full Model and the Restricted Model can be generated and tested.

As a way of organizing one's approach to stating Research Hypotheses, it may be helpful to point out that all of the Research Hypotheses discussed in the previous chapters may be stated as "over and above" hypotheses. This is not to suggest that stating it that way is the best way to state *any* hypothesis but to emphasize that each is a form of the most generalized hypothesis.

All hypotheses can be either phrased or rephrased into a single structure. That structure contains a criterion variable and one or more variables that are being tested, and it may contain one or more covariables. (The unit vector is nearly always a covariable, but it could be a variable being tested, a covariable, or neither. This is more fully discussed later in this chapter.) The single structure for expressing all Research Hypotheses is:

Is knowledge of variables X1, X2, . . ., Xk valuable, over and above knowledge of variables Z1, Z2, . . ., Zg, in the prediction of the criterion Y?

One regression model is specified by the Research Hypothesis as having full knowledge of the mentioned predictor variables:

(Model 13.9) $Y = a_1 X1 + a_2 X2 + \ldots + a_k Xk + c_1 Z1 + c_2 Z2 + \ldots + c_g Zg + E_9$

where:

Y = the criterion variable;
$X1, X2, \ldots, Xk$ = the variables being tested;
$Z1, Z2, \ldots, Zg$ = the over and above variables; and
$a_1, a_2, \ldots, a_k, c_1, c_2, \ldots, c_g$ are least squares weighting coefficients calculated
so as to minimize the squared elements of the error vector E_9.

The Full Model must be compared to a Restricted Model that has as predictors only the over and above information. Each k variable of interest is restricted from the Full Model by setting its weighting coefficient equal to some numerical value. There are an infinite number of numerical values that one could use. The actual restriction(s) need to be a function of past theory, empirical findings, and expectations of the researcher. To simplify the following discussion, each of the k variables of interest is set equal to zero (making k restrictions).

The restrictions would be:

$a_1 = 0; a_2 = 0; \ldots; a_k = 0$

Forcing those restrictions onto the Full Model would result in the following Restricted Model:

(Model 13.10) $Y = c_1 Z1 + c_2 Z2 + \ldots + c_g Zg + E_{10}$

The increase in R^2 in going from the Restricted Model to the Full Model would be due to the predictive information in the X1, X2, . . ., Xk variables that is over and above the predictive information of the Z1, Z2, . . ., Zg variables in the Restricted Model.

The Research Hypothesis also could be stated as:

Are variables X1, X2, . . ., Xk predictive of Y, holding constant the variables Z1, Z2, . . ., Zg?,

or,

Are variables X1, X2, . . ., Xk predictive of Y, while covarying the effects of variables Z1, Z2, . . ., Zg?

And if the variables appearing in the Full Model but not in the Restricted Model are all mutually exclusive group membership vectors, then the Research Hypothesis could be stated as:

Are the k groups different on Y, after adjusting for differences on the variables Z1, Z2, . . ., Zg?

Because there are k variables to be assessed and g covariables, m1 = k + g. (Note that it is being assumed that all linearly dependent vectors have been omitted from the Full and Restricted Models.) The number of predictor pieces of information in the Restricted Model is g; therefore, m2 = g. The numerator degrees of freedom for the F test then is:

$$[N - m1] \ = \ [(k + g) - (g)] \ = \ k$$

The R^2 of the Full Model minus the R^2 of the Restricted Model is the proportion of criterion variance accounted for by the additional k variables. Indeed, this average increase in R^2 becomes the numerator for the F test:

$$F_{(m1 - m2, N - m1)} \ = \ \frac{(R_f^2 - R_r^2) / (m1 - m2)}{(1 - R_f^2) / (N - m1)}$$

For the F test (in terms of R^2) to be computed on regression models, three requirements must be met:

1. The same criterion must appear in both models.
2. All predictor variables in the Restricted Model must appear in the Full Model, either directly or as a linear combination of the Full Model predictor variables.
3. The unit vector must appear in both models.

A quick glance at Models 13.9 and 13.10 will verify that the first two requirements are met. The same criterion appears in both models, and the Z1, Z2, . . ., Zg predictor vectors in the Restricted Model appear in the Full Model. For most applications, researchers will want to include the unit vector in both models (as one of the over and above variables designated by Z) as a way of taking care of the arbitrary scaling of variables. Indeed, the unit vector so commonly appears in both models that most computerized regression programs provide it automatically. (Again, SAS includes the unit vector automatically but allows the user to omit it through the

NOUNIT option in the model statement.) The unit vector has been included as a co-variable so often that now it is assumed to be a covariable, unless otherwise indicated. It must be remembered, though, that the unit vector also can be tested (i.e., be one of the test variables designated by X) as in the single population mean question in Generalized Research Hypothesis 10.1. Furthermore, the unit vector need not appear in either model—the variables that appear in a given model are always a function of the question the researcher is asking. (Whenever the unit vector is not in the Full Model or is being tested and is not in the Restricted Model, one must not use the general F test with R^2 values but use Equation 10.1 with error sum of squares instead.)

Any number of Xi variables may be investigated, and any number of Zi variables may be controlled (including none if that is desired). Furthermore, the variables may be dichotomous, or they may be continuous. The criterion variable in multiple regression is limited to only one variable, but it may be either dichotomous or continuous. Combinations of each of these options result in specific kinds of Research Hypotheses, most of which were discussed in previous chapters. To reemphasize the point made earlier, though all hypotheses can be reworded in the over and above fashion, many hypotheses make more sense stated otherwise. For instance, "For a given population, Treatment A is better than Treatment B on the criterion" seems more communicative than does "For a given population, Treatment A is better than Treatment B on the criterion, over and above the overall mean."

Where is this discussion leading? Suppose Researcher A has theorized that variables X1, X2, X3, and X4 account for the variability in the criterion of interest. An obtained R^2 of .90 is used as evidence of that statement. Researcher B, though, realizes that not all of the criterion variance has been accounted for and proposes that, in addition, variable X6 is necessary. Two researchers have posited two different states of affairs, and as one is a restricted case of the other, the argument can be tested through the regression approach. Because Researcher B is proposing a new, untried variable, and because Researcher B is using more information than Researcher A, Researcher B is obligated to have a higher R^2. The difference in R^2 could be agreed to before data collection, or the two researchers could agree to use a test of significance for the difference in R^2, as in Generalized Research Hypothesis 13.1. With this Generalized Research Hypothesis, the text has finally arrived at the degree of complexity that was outlined in chapter 1.

THE GOAL OF CONTROL IN THE BEHAVIORAL SCIENCES

All behavioral science research is ultimately concerned with cause and effect—which is to say, *control*. Many researchers initially react negatively to this statement, but when hard-pressed will ultimately agree that cause and effect relationships are what they are looking for. Statisticians have traditionally indicated that cause and effect relationships cannot be ascertained from correlational studies but only from ANOVA designs applied to experimental data. Familiarity with the GLM approach makes it clear that correlation and ANOVA are both subsets of the GLM and that interpretations of cause and effect can only be made on logical grounds. If the logic is not defensible, then the cause and effect relationship is not tenable.

Unfortunately, many causal interpretations are based on significance levels rather than on the amount of variance accounted for (Byrne, 1974). Results that are highly significant (due to factors other than chance) may account for, say, only 1% of the variance in the criterion. A not too unusual bivariate outcome results in a corre-

---

Generalized Research Hypothesis 13.1

Directional Research Hypothesis: For a given population, X6 is positively predictive of the criterion Y, over and above X1, X2, X3, and X4.

Nondirectional Research Hypothesis: For a given population, X6 is predictive of the criterion Y, over and above X1, X2, X3, and X4.

Statistical Hypothesis: For a given population, X6 is not predictive of the criterion Y, over and above X1, X2, X3, and X4.

Full Model: $Y = a_0U + a_1X1 + a_2X2 + a_3X3 + a_4X4 + a_6X6 + E_1$ (pieces full = m1 = 6—assuming all linearly dependent variables removed)

Want (for directional Research Hypothesis): $a_6 > 0$; restriction: $a_6 = 0$
Want (for nondirectional Research Hypothesis): $a_6 \neq 0$; restriction: $a_6 = 0$

Restricted Model: $Y = a_0U + a_1X1 + a_2X2 + a_3X3 + a_4X4 + E_2$ (pieces restricted = m2 = 5)

where:

   Y = the criterion;
   X1, X2, X3, X4, X6 = continuous or categorical information; and
   a_0, a_1, a_2, a_3, a_4, and a_6 are least squares weighting coefficients calculated to minimize the sum of the squared values in the error vectors.

Degrees of freedom numerator = (m1 - m2) = (6 - 5) = 1

Degrees of freedom denominator = $(N - m1) = (N - 6)$

Note: If all predictors are continuous, m1 = 6. If some reflect categorical information, the linear dependencies must be ascertained.

---

lation of .44 and an R^2 less than .20. Of course, some phenomena are more highly accounted for than others, but few studies in the literature report results leading to an R^2 greater than .50. (It would be very interesting to force all researchers to report their R^2 values and not let them hide behind statistical significance inflated by large sample sizes. All other factors being equal, as the sample size increases, the results will become more "significant," whereas the R^2 will simply more closely approximate the population R^2. Thus, increasing the number of subjects does not artificially inflate the R^2 value, although it does deflate the probability value.) Figure 13.4 depicts these relationships.

   In the past, cause and effect has been interpreted on data that, although highly significant, yielded small R^2 values. One of our positions in this text is that one should spend one's time more fruitfully by finding variables that will increase the R^2 rather than trying to understand how that small proportion of criterion variance has been predicted.

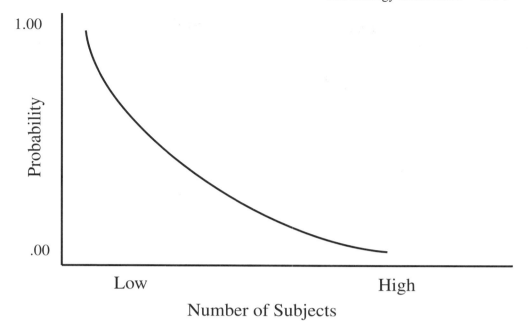

Figure 13.4a

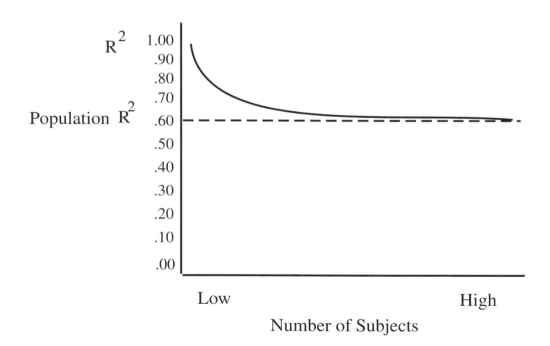

Figure 13.4b

Figure 13.4.
Relationship between number of subjects, probability, and R^2.

But while adding more predictor variables will increase the R^2, the question is, has a stable R^2 been obtained? Figure 13.5 contains the relationship between R^2 and the number of predictors. As illustrated, the R^2 can be increased by adding additional predictors—until there are as many predictors as there are subjects, resulting in an "overfit" model with an associated R^2 of 1.00. As Constas and Francis (1992) illustrated, with a figure similar to Figure 13.6, the interest should be on the adjusted R^2.

Of additional importance is the adequacy of the overlap of the criterion with the construct one wishes to consider, as discussed at the end of chapter 9. Too often the criterion is equated with the construct, when in reality one must realize that the criterion is only an approximation of the construct—that is, there is only a partial overlap between the criterion and the construct.

If the unaccounted-for criterion variance overlaps the construct, then one would do well to continue trying to build up the R^2. Several consternating situations arise when the criterion only partly overlaps the construct. It may be that the criterion variance accounted for is not overlapping the construct variance. Any interpretation regarding causality in this situation would be appropriate for the criterion but not for the construct.

On the other hand, all the common variance between the criterion and the construct may be accounted for. To attempt to build up the predictability of the criterion here would be fruitless. Interpretation of causality may be of some value here, but one must note that complete control (bringing about desired changes or upsetting the prediction) of the construct would not be possible because of the lack of an isomorphic fit between criterion and construct. This situation calls for the selection of other criterion variables, either instead of the criterion or in addition to the criterion.

Once the R^2 has approached 1.0 and the researcher has some faith in the criterion as a good measure of the construct, then one can begin considering how to

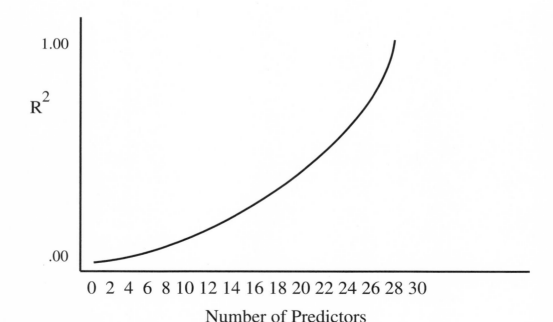

Figure 13.5.
The relationship between R^2 and the number of predictors (with 32 subjects and population $R^2 = .80$).

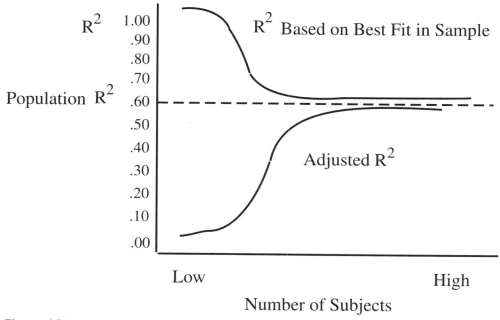

Figure 13.6.
Relationship between adjusted R^2 and R^2 (with 25 predictors and population $R^2 = .60$).

control the construct. This implies that variables must be included in the model that can be manipulated. Nonmanipulatable variables that must remain in the predictive model simply indicate differences in functional relationships for various populations. That is, if gender must remain in the model, then the functional relationship is different for males than it is for females. But if all the predictor variables are nonmanipulatable, then the model is simply describing a state of affairs rather than indicating some possible ways of upsetting the state of affairs.

Given a choice between greatly overlapping predictor variables, some of which are more manipulatable than others, one would choose to include the more manipulatable variables in the model. Some variables are easier to manipulate than are others; but, unfortunately, current statistical procedures do not allow us to incorporate these differences. These decisions must be made on subjective bases rather than on objective bases.

It may be the case that even though the R^2 is extremely high, the predictor variables may not be the causers of the criterion. The criterion and predictor variables may both be caused by some additional variable or set of variables. The technique of *path analysis* has been introduced by statisticians as a possible means of assessing the causal sequence. Those developments have not been entirely successful, but the intent is appropriate (see Pedazur, 1982).

One real-world advantage of the GLM, that of allowing the use of correlated predictor variables, turns out also to be one of the problems in detecting control. When one predictor variable is changed, one would expect that the values of the other (correlated) variables also might change. This is another reason for not interpreting regression weights. Indeed, the whole swarm could change as a result of changing only one variable. If the criterion is increased, then this might well be a desired goal. But when control is being considered, one usually wants to know how much effect there will be on the criterion, given a specified change in a predictor.

The GLM may give one insight about what variable to change, but it does not generally indicate what effects to expect. If the subjects have been randomly assigned to the independent variables, then these effects are implied by the GLM approach. But these kinds of designs do not adequately reflect the real world. When subjects are sampled from the real world, the predictor variables will be correlated, and therefore the effects cannot be directly known. The effects of changing predictor variables must wait for subsequent empirical manipulation. We have already indicated that some variables are more easily manipulated than others; it also might be the case that some kinds of subjects are more easily manipulated than others. So, the assessment of control must wait for empirical verification. Ascertaining causality is always a tentative proposition. Empirical manipulation helps rule out other competing explainers.

As is shown in the next two sections, the GLM can help in the various stages of one's research. Ultimately, causers will be identified. But no statistical technique will provide an understanding of why that variable is a cause. Knowledge of one's field, theory, and common sense, combined with statistical results, nurture explanation and understanding.

A PROPOSED RESEARCH STRATEGY

Various research strategies were discussed at the end of chapter 11, although all of them were concerned with the sequencing of hypotheses. In the present section, we put the processes of data snooping, hypothesis testing, replication, and manipulation into the perspective of an encompassing research strategy.

The strategy involves (a) the use of *data snooping* to generate hypotheses (preferably directional hypotheses); (b) the *testing of hypotheses* on static variables; (c) the *replication* of the hypothesis testing; (d) the development and testing of hypotheses about the effects of *manipulation* of some previously static variables; and (e) the *replication* of the manipulated-variable hypotheses. Table 13.1 displays the strategy schematically.

Table 13.1

Stages of a Proposed Research Strategy and Their Relative Emphases on Probability and R^2

Stage	Emphasis on Low Probability	Emphasis on High R^2
(Static Variables Stages 1–3)		
1. Data Snooping (Hypothesis Generating)	No	Yes
2. Hypothesis Testing (Static)	Yes	No
3. Replication	No	Yes
(Dynamic Variables Stages 4–5)		
4. Manipulation	Yes	Yes
5. Replication	Yes	Yes

Data Snooping

When one is data snooping, one has little idea about what will be found. The goal of data snooping can be viewed from either of two viewpoints. The first is that the researcher is attempting to build up the R^2 to as close to 1.00 as possible, reasonable, or desirable. The second viewpoint is that the researcher is attempting to reduce the magnitude of the error vector. By identifying variables that are correlated with what was originally considered "error," the criterion can be more accurately predicted and the magnitude of the error reduced. Once again we realize that the error vector is both error of measurement *and* error of specification. By *error of specification* we mean that not all of the correct variables were originally considered in the regression model.

When a large R^2 is observed, one may suspect that the set of variables used to obtain that R^2 is a good set of variables to investigate further. This stage of research can be very exciting because most dramatic discoveries seem to have originated in a stage of this sort; but any discoveries at this stage must be tested on successive samples before one may reasonably say that a discovery was made. Serendipitous findings at this stage can be valuable; one must simply realize that subsequent verification must be made.

Suppose that, for example, a researcher has an interest in why 9th graders vary in their knowledge of and achievement in the area of biological science. She has also chosen a criterion measure of this construct (realizing, of course, that it is not exactly the same as her construct); the criterion measure is a standardized biological science achievement test administered at the end of the 9th grade (after one year of biology instruction). Based on her knowledge of the research area and on discussions with biology teachers and students, she decides to investigate the following variables to see if they will help to predict biology achievement:

gender of student
socioeconomic status of student
time student has spent outdoors in earlier years
number of extra credits student completes
(number of extra credits)^2
size of biology class
gender * size of biology class

The researcher does not have enough information at this point to state hypotheses about the ways that these variables relate to biology achievement; she is data snooping.

She arranges to collect data on the criterion and predictor variables from the 9th-grade students in a particular school. Just before administering the biology achievement test, the school notices that it does not have enough original copies of the test; so the school makes photocopies of the test for the final one-quarter of the students. Our researcher notices that the photocopies are not as legible as the originals. She therefore adds a new variable to her list to consider: legibility of the test.

Because of her data snooping, the researcher finds that the following model, containing only some of her suspected variables, does a good job of predicting biology achievement, attaining an R^2 of .76:

$$(\text{Model } 13.11) \quad B = aU + bS + cA^2 + d(G * X) + e(H * X) + fL + E_{11}$$

where:

B = biology achievement test score;
S = socioeconomic status (SES) of student on 5-point scale;
A^2 = (number of extra credits)^2;
$G * X$ = size of biology class if student is male, 0 otherwise;
$H * X$ = size of biology class if student is female, 0 otherwise; and
L = legibility of the test.

The researcher is ready to state and test her hypotheses.

Hypothesis Testing

A hypothesis cannot be made without the researcher relying on some combination of past research, theoretical orientation, and intuitive insight. Data-snooping results can often aid in developing a supportable hypothesis. As discussed previously, one may in the hypothesis-testing stage compare the difference in R^2 between the Full Model and Restricted Model to some preselected amount of difference, or one may rely on the probability values as proposed in this text. If only one restriction is being made on the Full Model, and if significance is obtained, one will know what is causing that significance. The goal of making an accurate causal statement is enhanced if the hypothesis is tested in a model exhibiting a high R^2. But as can be seen in Figure 13.5, a high R^2 is not viewed as necessary at this stage.

To return to the example, the researcher, based on her data-snooping results, states several hypotheses, of which the following is one; and she collects new data on which to test them. Her Research Hypothesis is:

> For 9th-grade biology students, the score obtained on the biology achievement test will be higher for boys if they are in small classes (fewer than 20) rather than in large classes, and will be higher for girls if they are in large classes (20 and more) rather than in small classes, over and above the effects of SES, number of extra credits squared, and legibility of the test.

The researcher then tests her hypothesis and finds that, at her chosen alpha level, her Statistical Hypothesis can be rejected; so her Research Hypothesis is accepted as tenable. Successively, she finds that all the variables in Model 13.11 are valuable.

Up to this point, no variables have been manipulated, although some are potentially manipulatable. The model containing static (nonmanipulated) variables (Model 13.11) was upheld by the hypothesis testing. The static hypothesis-testing phase now needs to be replicated.

Replication

As can be seen in Figure 13.5, the process of replication appears twice, once in the static phase and once in the manipulation phase. Because it is necessary to proceed from a static model to a dynamic one (containing manipulation) when hunting and testing for causal relationships, the researcher must make a decision concerning the extent and level of successful replication before manipulation. This decision is based largely on the potential cost of the manipulation intended—cost in money, time, and human resources. Rescheduling the large and small biology classes will cost money and may provoke parental, teacher, student, or taxpayer protest. Thus, one may want to replicate the static hypothesis-testing model several times before venturing to manipulate. In the process of replication (before manipulation), one should not be too concerned with probability but with obtaining a high R^2. (It would be ideal if approximately the same weighting coefficients could be obtained, but correlated predictors and small sample sizes usually keep this from happening.)

Manipulation

The first three stages above apply to static data; that is, within-person, focal stimuli, and context variables are measured as they exist, with no effort to change them. Functional relationships are described, so that the criterion variable may be predicted. One should be aware that a variable that is predictive of a criterion in a static model may not be predictive in the same fashion as a dynamic variable. The extreme case would be a variable that simply cannot be manipulated. Another example would be a variable that, when manipulated, causes "side effects" that influence the criterion (either positively or negatively). On the other hand, variables that are found to be not predictive in a static situation may indeed be necessary causers in a dynamic situation. (Hickrod, 1971, discusses this and several other interesting aspects of regression with respect to school-finance concerns.) As a consequence of these realities, some statisticians take a strong stand regarding the analysis of static variables. Werts (1970) states:

> Obviously when partial regression weights are estimated from naturalistic data it will be far from certain that this coefficient will have any resemblance to what would be found under conditions of experimental control. By asking what would happen "if," we are in effect asking the question: what would the data be like if they were not what they are? (p. 130)

We take a more moderate view. Researchers should be aware that what is discovered in a static situation very likely may not hold up in a dynamic situation. But the researcher still may choose to investigate the static situation before the dynamic one because of the potential "cost" of creating the dynamic situation (manipulating variables under experimental control). We also see researchers getting overly enthusiastic in manipulations, often losing sight of the purpose of the manipulation—raising scores on the criterion.

When an R^2 of 1.0 is obtained in a dynamic situation, one can be confident that the causers have been isolated. One then needs to investigate the magnitude of the criterion scores. It may be that the causers of increasing the criterion two units may have been discovered, but a two-unit increase may not be a sufficient increase. The discovered causers may not be sufficient to obtain a three-unit increase. Therefore, other causers must be posited and investigated. The cycle of research has thus started again.

The manipulation step in scientific investigation can be thought of as making a change in subjects, so that predictions can be upset. Johnny is predicted to be a bad reader by the use of the static data, but certain changes can be made, resulting in the prediction becoming incorrect. A certain plot of land is predicted to be of low fertility, but certain changes may make that land desirable for planting crops.

The stage of manipulation contains dynamic variables—those that have been willfully changed by the researcher. As has been emphasized previously, it is desirable to investigate *manipulatable* variables in the snooping and static hypothesis-testing stages so that the move into the manipulation phase will be facilitated.

Our biology researcher wants to manipulate and test each predictor variable in Model 13.11. This is dynamic hypothesis testing, and her hypotheses will, one hopes, be directional since, before manipulation, she has investigated her variables extensively.

Are all of the predictor variables in Model 13.11 manipulatable? Class size and the proportion of males to females in each class can be arranged by the school. Legibility of the test is also directly manipulatable—one needs only to select certain levels of legibility. The number of extra credits may be more difficult to manipulate; some added inducement may be necessary to get students to complete more assignments (if the hypothesized relationship is positive). Or the researcher may feel that it is the content rather than the credit of those assignments that is of benefit and may change those assignments from extra credit to required.

Socioeconomic status is not readily manipulatable—SES is generally based on parental or guardian income, occupation, and education. Changing a child's family circumstances is usually impossible, so the researcher may wish to hypothesize critical elements or aspects of SES that might be varied. Nutrition, self-esteem, need for achievement, and teacher acceptance are surely related to SES and also may affect biology achievement. Indeed, biology achievement might be retarded by an inadequate breakfast and by a lack of desire to achieve. These may be the manipulatable variables that SES summarizes. If so, then a school breakfast program and training that would instill a need for achievement may be manipulated to increase biology achievement for students who would otherwise be predicted to perform poorly without such intervention.

Gender also is not easily manipulated, but gender is placed in the model to interact with a manipulatable variable. It thus performs a useful predictive function without being manipulated. Here, the researcher may not be concerned with gender as a manipulatable variable. She may, however, wish to consider what aspects of gender (underlying dynamic variables) interact with class size to predict biology achievement. Surely simple physiology has an explanatory role—but of unknown value; socialization variables, however, are more likely to be the underlying causative factors that the gender variable accounts for. For example, girls in general may have been taught, "Science is not really for girls." If so, a measure of sex role identification ought to be a better predictor of biology achievement than the binary gender variable. If it is a better predictor, a program designed to break down gender stereotypes ought to reduce the negative influence of being a girl on biology achievement.

If one wishes to go beyond prediction to improvement, then it is important that static variables be examined carefully for their underlying dynamic causative influences on criterion measures (e.g., sex role identification may underlie the effect on achievement of the binary gender variable).

One may not always wish to do away with all nonmanipulatable variables, however. They are valuable predictors in at least two situations: First, nonmanipulatable variables may be used until the underlying dynamic variable is found. Second, nonmanipulatable variables may be unchangeable but may interact with a dynamic variable (e.g., gender interacting with treatment). If one cannot manipulate, then one must rest on the tenuous grounds of theory. Theory always plays a large part because the variables that are included in the analysis were purposefully put there—one may not think initially of that as theory, but it certainly guided one's actions.

Replication of Manipulated Findings

Once manipulations have been found to be successful (resulting in changes in the desired direction), then replications are called for to establish the magnitude of the effect of those manipulations. Researchers should put faith in causers only when manipulations have a consistent effect.

Successful replication by the biology researcher should make her happy. Given that she has had success at each of the previous stages, she can be somewhat assured that the results are valid. Additional replications on new samples of 9th-grade students will give further credence to the findings, for chance may have been operating in all the previously successful results.

What is more likely is that the biology researcher did not perform research at some of the other stages. Because of lack of resources or time, research is often forced to go to the manipulation stage without prior success at the other stages. In these cases, replication is a more crucial stage: But how many successful replications are needed? There is no established answer to that question, although it depends in part upon how crucial it is that the criterion of interest be under control.

Table 13.1 summarized the stages of the proposed research strategy and showed their emphases upon probability and R^2. If success on the goal of either low probability or high R^2 is not obtained at any stage, activity must revert to the initial stage of data snooping.

REPORTING

Once the computer analysis is completed, the researcher needs to convert the computer output into readable form. This information must be communicated to the reader. We first discuss what should be reported. Then we provide several alternatives regarding reporting format. We finish this section with specific instructions for developing plots of the data.

What Should Be Reported?

The information contained in the applied Research Hypothesis should be reported. At the minimum, what should be reported is the Research Hypothesis, the probability of the F test, and the conclusion. If one has a little more space, one could report the Full Model R^2, the adjusted R^2, and the F. Unlimited space (such as in a masters thesis or a doctoral dissertation) would allow for the entire model and the resultant weights to be included.

Reporting Format

Traditionally, correlational analyses have reported correlation coefficients, along with the probability (of the population correlation being zero). ANOVA studies are usually summarized with a source table. Source tables presume that there is interest in all of the sources. In addition, these tables were not designed for directional tests of significance. Consequently, we do not encourage changing the GLM results into such a source table. (If one is using all categorical predictors, then one could use PROC GLM, and such a source table would be produced.)

If journal space is a concern, then all that need be reported is the probability associated with the conclusion. The actual test statistic is not needed, as the test statistic is only an intermediate step to finding the probability. In addition, reporting the actual probability allows readers to interpret the results for themselves. That is, if the researcher's alpha is .05, then a probability value of .03 is significant for the researcher's alpha but is not significant for a reader who has an alpha of .01.

Newman (1973a) suggested a format for reporting results from the GLM. That format is presented in Table 13.2.

Plotting

There are two ways that one can plot the data. One can use SAS to produce the plots, or one can use the weights from the regression model and calculate the plots by hand. For Applied Research Hypothesis 7.1 and Applied Research Hypothesis 9.1, the general plot statement is:

PROC PLOT; PLOT X2 * X3;

This statement produces a scattergram with X2 on the Y axis and X3 on the X axis. One could model the data with the following model statement:

PROC REG: MODEL X2 = X3;

The line of best fit would have an intercept equal to the weight of INTERCEP, and the slope would be the weight of variable X3. To draw this plot by hand, one would have to obtain two predicted points. One easy point is at 0 on the X axis because the predicted score is equal to the intercept, the weight of INTERCEP. The other point should be obtained for a relatively large value of the X-axis variable and can be found from the regression equation:

Predicted X2 = weight of INTERCEP + weight of X3 * X3 score

From the Applied Research Hypothesis 9.1 printout:

Predicted X2 = -12.98 + 2.06 * X3

If we use a value of 10 for X3, then:

Predicted X2 = -12.98 + (2.06 * 10) = 7.62

Once the predicted values have been calculated, they can be connected for the straight line. The curved data in Applied Research Hypothesis 9.1 would require more predicted values, say five, before a smooth second-degree curve could be drawn by hand. The requisite SAS statement to accomplish the same task is in Applied Research Hypothesis 9.1.

When all of the predictor variables are dichotomous, the plotted values are not lines or curves, they are means. Applied Research Hypothesis 7.1 provides an example of how to obtain such a plot with SAS. The means for the four groups can be found from the weights from the Full Model:

Table 13.2
One Suggested Reporting Format

The Complete Regression Model, Which Reflects All of the Empirically Tested Functional Relationships:

$$Y6 = a_0U + a_1X1 + a_2X2 + a_3X3 + a_4X4 + a_5X5 + a_7X7 + a_8X8 + a_9X9 + a_{10}X10 + E$$

where:

Y6	=	the criterion, posttest scores in reading comprehension;
X1	=	1 if in the Multimedia Reading Program, 0 otherwise;
X2	=	1 if in the traditional Basal Text Reading Program, 0 otherwise;
X3	=	1 if male, 0 otherwise;
X4	=	1 if female, 0 otherwise;
X5	=	pretest raw score in reading comprehension as measured by the Ohio Survey Test;
X7	=	1 if male and in the Multimedia Reading Program, 0 otherwise;
X8	=	1 if female and in the Multimedia Reading Program, 0 otherwise;
X9	=	1 if male and in the traditional Basal Text Reading Program, 0 otherwise;
X10	=	1 if female and in the traditional Basal Text Reading Program, 0 otherwise; and
E	=	error vector, difference between predicted score and actual score.

Models, F Ratios, and R^2 for Predicting Posttest Scores of Students

Hypothesis and Models	Model	R^2	df	alpha	F	Computer p	Actual p
Hypothesis 1: The multimedia program is better than the traditional program when covarying pretest scores.							
Model 1 $Y6 = a_0U + a_1X1 + a_5X5 + E$	Full	.50	1,145	.05	.08	.77	.38
Model 2 $Y6 = a_0U + a_5X5 + E$	Restricted	.50					
Hypothesis 2: Females differ from males when covarying the pretest scores.							
Model 3 $Y6 = a_0U + a_3X3 + a_5X5 + E$	Full	.50	1,145	.05	.36	.55	.55
Model 2 $Y6 = a_0U + a_5X5 + E$	Restricted	.50					
Hypothesis 3: There is an interaction between sex and treatment, when covarying pretest scores.							
Model 4 $Y6 = a_0U + a_5X5 + a_7X7 + a_8X8 + a_9X9 + E$	Full	.51	1,144	.05	3.59	.06	.06
Model 5 $Y6 = a_0U + a_1X1 + a_3X3 + a_5X5 + E$	Restricted	.50					

Note. From Newman (1973a).

$$X9 = a_0U + a_{16}X16 + a_{17}X17 + a_{18}X18 + E$$

where:

> $X16 = 1$ if subject in X10 and X13, 0 otherwise;
> $X17 = 1$ if subject in X11 and X12, 0 otherwise; and
> $X18 = 1$ if subject in X11 and X13.

For the subjects in X10 and X13, the Full Model reduces to: $X9 = a_0U + a_{16}X16$, and therefore the mean is $a_0 + a_{16}$, or $101.79 + 8.40 = 110.19$. For the subjects in X11 and X12, the Full Model reduces to: $X9 = a_0U + a_{17}X17$, and therefore the mean is $a_0 + a_{17}$, or $101.79 + 8.55 = 110.34$. For the subjects in X11 and X13, the Full Model reduces to: $X9 = a_0U + a_{18}X18$, and therefore the mean is $a_0 + a_{18}$, or $101.79 + 5.61 = 107.40$. For the subjects in X10 and X12, the Full Model reduces to: $X9 = a_0U$, and therefore the mean is a_0, or 101.79. These means are identified with a plus sign (+) on the Applied Research Hypothesis 7.1 page-5 printout.

FURTHER READING

The GLM approach is an extremely flexible technique that opens up a host of secrets about statistics—once certain mechanical topics are mastered. New practical developments are constantly being made by those using the GLM. More mathematical treatments of the technique than presented here can be found in Maxwell and Delaney (1990), Seber (1984), Ward and Jennings (1973), and Kleinbaum and Kupper (1978). An elementary introduction for hypnotists, written by Starr (1971), contains minimal mathematical concepts.

The Special Interest Group of the American Educational Research Association has been in existence for many years, and their journal, *Multiple Linear Regression Viewpoints*, is a forum for new ideas. Many articles published in that journal have been referenced in this text. *Viewpoints* affords a way for the reader to keep abreast of new regression developments. A nominal annual dues avails one of this journal. Further information on *Viewpoints* can be obtained from the executive secretary, Dr. Steven Spaner, University of Missouri at St. Louis, St. Louis, MO, 63121; or from the editor, Dr. John Pohlmann, Department of Educational Psychology and Special Education, Southern Illinois University, Carbondale, IL, 62901.

APPENDIXES
REFERENCES
INDEX

APPENDIX A:
DATA SET OF SIXTY SUBJECTS

01	4	4	26	8	1	0	82	97	0	1	0	1	80	04	01
02	6	9	28	9	1	0	95	101	0	1	1	0	75	06	02
03	6	8	41	10	0	1	103	104	0	1	0	1	82	07	03
04	8	10	29	10	1	0	109	106	0	1	0	1	84	07	04
05	10	12	40	11	1	0	84	96	0	1	0	1	85	06	05
06	11	13	30	11	1	0	93	100	0	1	1	0	82	06	06
07	12	14	39	12	0	1	98	102	0	1	1	0	84	05	07
08	14	15	32	13	1	0	99	103	0	1	0	1	90	05	08
09	16	15	37	14	0	1	87	98	0	1	0	1	92	06	09
10	17	16	37	15	0	1	106	105	0	1	1	0	92	06	10
11	18	16	36	15	0	1	113	112	0	1	1	0	90	07	11
12	20	17	35	17	0	1	128	125	0	1	0	1	97	11	12
13	22	18	34	18	1	0	89	99	0	1	0	1	100	17	13
14	24	18	33	58	0	1	115	118	0	1	1	0	96	18	14
15	26	20	32	21	0	1	125	124	0	1	1	0	98	18	15
16	27	20	32	22	0	1	91	100	0	1	1	0	100	18	16
17	28	20	31	23	0	1	116	115	0	1	1	0	101	18	17
18	28	21	36	55	1	0	117	117	0	1	0	1	105	18	18
19	29	20	30	24	0	1	123	120	0	1	0	1	106	17	19
20	30	21	29	53	0	1	126	120	0	1	0	1	108	17	20
21	31	21	28	26	0	1	120	116	0	1	1	0	103	17	21
22	32	22	26	52	0	1	129	127	0	1	1	0	105	18	22
23	32	22	37	27	1	0	120	116	0	1	0	1	110	19	23
24	33	22	24	51	0	1	114	113	0	1	0	1	112	25	24
25	34	22	37	52	1	0	100	102	0	1	1	0	107	28	25
26	36	23	38	28	1	0	98	103	0	1	1	0	110	29	26
27	38	24	38	49	1	0	96	102	0	1	0	1	115	30	27
28	40	24	38	48	1	0	85	95	0	1	0	1	120	30	28
29	42	25	39	32	1	0	125	128	0	1	1	0	115	30	29
30	48	25	40	38	1	0	95	102	0	1	1	0	120	30	30
31	5	5	27	8	1	0	84	80	1	0	1	0	75	04	01
32	7	8	27	9	1	0	85	82	1	0	0	1	82	06	02
33	7	8	41	10	0	1	115	123	1	0	0	1	83	07	03

34	8	11	29	11	1	0	87	85	1	0	1	0	80	07	04
35	10	12	41	11	0	1	95	95	1	0	1	0	83	06	05
36	11	12	30	10	1	0	93	92	1	0	1	0	84	06	06
37	12	13	39	11	0	1	99	100	1	0	1	0	85	05	07
38	14	15	32	14	1	0	95	100	1	0	0	1	90	05	08
39	16	16	37	13	0	1	118	128	1	0	0	1	92	06	09
40	17	16	37	14	0	1	88	86	1	0	0	1	93	06	10
41	18	17	36	16	0	1	94	92	1	0	1	0	90	07	11
42	20	18	35	16	0	1	102	102	1	0	1	0	92	11	12
43	21	17	34	19	1	0	89	87	1	0	0	1	98	17	13
44	23	17	33	57	0	1	116	122	1	0	0	1	100	18	14
45	25	19	32	22	0	1	97	97	1	0	0	1	103	18	15
46	27	20	32	21	0	1	103	102	1	0	0	1	105	18	16
47	28	21	31	24	0	1	90	89	1	0	1	0	100	18	17
48	29	22	36	54	1	0	96	96	1	0	1	0	102	18	18
49	30	22	30	24	0	1	108	113	1	0	1	0	103	17	19
50	30	21	39	54	1	0	100	101	1	0	1	0	104	17	20
51	31	22	28	27	0	1	91	89	1	0	0	1	108	17	21
52	32	21	26	51	0	1	112	116	1	0	0	1	110	18	22
53	32	22	37	28	1	0	104	104	1	0	1	0	105	19	23
54	33	21	24	52	0	1	92	90	1	0	0	1	112	25	24
55	33	22	37	51	1	0	120	130	1	0	0	1	110	28	25
56	35	23	38	29	1	0	122	132	1	0	0	1	113	29	26
57	37	24	38	50	1	0	128	140	1	0	1	0	110	30	27
58	39	24	38	49	1	0	124	136	1	0	1	0	113	30	28
59	41	25	39	33	1	0	126	137	1	0	0	1	114	30	29
60	46	25	40	39	1	0	130	142	1	0	0	1	124	30	30

APPENDIX B:
SAS DISCUSSION

Appendix B-1: Necessary SAS statements, along with the minimally required statements to get the answer to Applied Research Hypothesis 4.1 (see SAS Institute, 1990, for a more detailed discussion).

Generic name	Specific statements needed to answer Applied Research Hypothesis 4.1
TITLE	TITLE "APPLIED RESEARCH HYPOTHESIS 4.1";
CMS FILEDEF	CMS FILEDEF NEW DISK BOOK DATA A;
DATA	DATA FIRST;
INFILE	INFILE NEW;
INPUT	INPUT X2 4 - 5 X12 30;
PROCEDURE(S)	PROC REG; MODEL X2 = X12;
OPTION(S)	TEST X12 = 0;

Appendix B-2: Necessary SAS statements, along with the minimally required statements to get the answer to Applied Research Hypothesis 4.2.

Generic name	Specific statements needed to answer Applied Research Hypothesis 4.2
TITLE	TITLE "APPLIED RESEARCH HYPOTHESIS 4.2";
CMS FILEDEF	CMS FILEDEF NEW DISK BOOK DATA A;
DATA	DATA ARH42;
INFILE	INFILE NEW;
INPUT	INPUT X2 4 - 5 X10 28 X11 30 X12 32 X13 34;

	X17 = X10 * X12;
	X18 = X11 * X12;
	X19 = X10 * X13;
	X20 = X11 * X13;
PROCEDURE(S)	PROC REG; MODEL X2 = X17 X18 X19;
OPTION(S)	TEST X17 = 0, X18 = 0, X19 = 0;

The TITLE statement is not really necessary, but it provides that the title will appear at the top of each page of the printout. Any alphabetic or numeric information may appear between the two quotation marks. A semicolon must be placed at the end of each statement, as is the case with most statements.

The CMS FILEDEF statement and its accompanying INFILE statement is a way to access a file from within another file. Thus, the data in Appendix A do not need to be physically inserted into the file that you are currently creating but have been stored in the file "BOOK DATA A." If you do not have the CMS system at your computer installation, you will need to use a different procedure for this operation. Because we will be analyzing the same data on many occasions, less storage space will be required if we do not duplicate the information into the SAS file each time we want to analyze those data.

The DATA statement identifies the name of the data file that has been created. Any set of eight or fewer alphabetic characters is permissible.

The INFILE statement directs the SAS program to make a link to the file defined in the CMS FILEDEF statement. The name of the file (here NEW) can be any six-letter name that is identified in both CMS FILEDEF and INFILE statements.

The INPUT statement specifies the variable names that we want to assign to the variables that are being read from the existing file and the location of those variables. Although there are numerous variables in the file, Applied Research Hypothesis 4.1 requires only one variable in columns 4 and 5 and a second variable in column 30.

The PROCEDURES statement (PROC REG is one example) invokes a specific SAS PROCEDURE (PROCedure). There are numerous PROCedures available in SAS, several of which we refer to in this text. PROCedure REGression will be the PROC that we use most of the time. The word MODEL must appear after the semicolon, followed by the criterion, an equal sign, and the predictor(s). Note that this is a cryptic version of the vector representation of the variables in the Full Model.

The OPTION specifies the TEST restriction(s) that one desires to make on the Full Model. Because the symbol for each weighting coefficient is arbitrary, SAS requires the restrictions be made on the names of the variables themselves. So, in Applied Research Hypothesis 4.1, while we chose to represent the desired restriction as $a_{12} = 0$, the restriction is represented in SAS as X12 = 0. If one had chosen to name the variable in column 30 as MALES instead of X12, then the last two statements would have been:

PROC REG; MODEL X2 = MALES;
TEST MALES = 0;

Appendix B-3: Demonstration of equivalency.

Given:

M1 $=$ new AM;
M2 $=$ new PM;
M3 $=$ comparison AM; and
M4 $=$ comparison PM

show (M1 + M2) + (M1 + M3) + (M4 - M1) is equivalent to:

U + (M1 + M2) + (M1 + M3).

Because (M1 + M2) + (M1 + M3) + (M4 - M1) = U, we can include U in the model and eliminate one of the other vectors, (M4 - M1).

APPENDIX C: ACTIVITIES WITH CIRCLES AND SQUARES

ACTIVITY #1

Research Hypothesis #1: The circumference of a circular object is not the same size as the circumference of a square object.

The Full Model needed to reflect this hypothesis is:

$$\text{MODEL 1: CIRCUM} = aU + bCIRSQ + E_1$$

where:
    CIRCUM = the distance around the object; and
    CIRSQ   = 1 if circle, 0 otherwise.

You are to measure CIRCUM and CIRSQ on each of 18 objects (some may want to measure 9 each, others may want to measure 7 of one kind and 11 of the other). Use a piece of string to measure the distance around the object, and then lay the string next to a ruler to get the actual length. Some may prefer to measure in inches, others in centimeters. Just choose one measuring system and stay with it.
    If the circumference is not different for the two kinds of objects, then one would not need to know which object the measurement came from, and that information would not be needed in the regression model, or the weight of CIRSQ can be reasonably set to zero.

Restriction: b = 0 (want: b ≠ 0)

The Restricted Model would then be:

$$\text{MODEL 2: CIRCUM} = aU + E_2$$

Typical results:

R^2 of the Full Model   = .10 (or very close to .00)
R^2 of the Restricted Model = .00

Using the general F formula for regression:

$$F_{(m1-m2,N-m1)} = \frac{(R_f^2 - R_r^2)/(m1-m2)}{(1-R_f^2)/(N-m1)}$$

where:

m1 = 2;
m2 = 1; and
N = 18.

The results will typically yield an extremely low F value, indicating that the Statistical Hypothesis cannot be rejected and that the Research Hypothesis cannot be accepted. The conclusion is: the circumference of a circular object is not different from the circumference of a square object.

SAS setup:

```
TITLE "ACTIVITY #1";
DATA ALL;
INPUT ID 1-2 CIRCUM 4-9 DIAM 11-16 CIRSQ 18;
CARDS;
01  11.25  3.25 1
02  12.25  3.00 0
.
.
.

PROC REG; MODEL CIRCUM = CIRSQ;
TEST CIRSQ = 0;
```

ACTIVITY #2

Research Hypothesis #2: The circumference of a circular object is a linear function of its diameter.

The Full Model needed to reflect this hypothesis is:

$$\text{MODEL 3: CIRCUM} = aU + b\text{DIAM} + E_3$$

where:

CIRCUM = distance around the circular object; and
DIAM = distance across the circular object.

You already have the required measurements on the circular objects. If the circumference is not a function of the diameter, then the diameter is not needed in the regression model, or the weight of the diameter can be reasonably set to zero.

Restriction: $b = 0$ (want: $b \neq 0$)

The Restricted Model would then be:

$$\text{MODEL 4: CIRCUM } = aU + E_4$$

Typical results:

R^2 of the Full Model   = .99987 (or very close to 1.00)
R^2 of the Restricted Model = .00

Using the general F formula for regression:

$$F_{(m1 - m2, N - m1)} = \frac{(R_f^2 - R_r^2) / (m1 - m2)}{(1 - R_f^2) / (N - m1)}$$

where:
m1 = 2;
m2 = 1; and
N = 9.

The results will typically yield an extremely high F value, indicating that the Statistical Hypothesis can be rejected and that the Research Hypothesis can be accepted. The circumference of a circular object is a linear function of its diameter.

The results will usually depict a slope very close to pi and an intercept very close to zero. The intercept of zero makes conceptual sense in that a circle of zero inches in diameter would be expected to have a circumference of zero inches. Furthermore, you may recognize the value of pi and the resulting simplified equation:

$$C = \pi D$$

There is no need for the error term because the model results in perfect prediction ($R^2 = 1.00$). Because one has employed a model with "perfect predictability," one has discovered a law. Furthermore, only nine subjects were used, with crude measuring sticks, and only one predictor variable.

The SAS setup would be the same as with Activity #1, with these changes. One could delete the square data from Activity #1 or collect new data on 9 circles. An easier approach would be to use SAS to analyze only the circle data. This can be accomplished in a number of ways; we present two.

Option 1: Replace PROC REG statement with:
PROC SORT; BY CIRSQ;
PROC REG; MODEL CIRCUM = DIAM; BY CIRSQ;

Option 2: Replace PROC REG statement with:
PROC REG; MODEL CIRCUM = DIAM;
and include after INPUT statement:
IF CIRSQ = 1;

ACTIVITY #3

Suppose that we are now concerned about whether square objects have the same functional relationship as circular objects.

Research Hypothesis #3: The circumference of a square object is a different linear function of its diameter than is that of a circular object.

We would have to construct a Full Model that would allow a different linear functional relationship for square and circular objects:

$$\text{MODEL 5: CIRCUM} = a_1 UC + b_1 DIAMC + a_2 US + b_2 DIAMS + E_5$$

where:

CIRCUM	=	distance around the object;
UC	=	1 if object is circular;
US	=	1 if object is a square;
DIAMC	=	distance across if object is circle; and
DIAMS	=	distance across if object is square.

The above model is often referred to as the *two-line model* as it allows for two intercepts (a_1 and a_2) and for two slopes (b_1 and b_2). If the Research Hypothesis is not true, then the two kinds of objects follow the same linear function. The restrictions on the Full Model (MODEL 5) needed to get to the one linear function are: $a_1 = a_2$, and $b_1 = b_2$, resulting in:

$$\text{MODEL 6: CIRCUM} = a_3 U + b_3 DIAM + E_6$$

Typical results:

R^2 of the Full Model	=	.99
R^2 of the Restricted Model	=	.95

Using the general F formula for regression:

$$F_{(m1 - m2, N - m1)} = \frac{(R_f^2 - R_r^2) / (m1 - m2)}{(1 - R_f^2) / (N - m1)}$$

where:

m1 = 4;
m2 = 3; and
N = 18.

When MODELS 5 and 6 are compared, one finds that there is a significant difference, as the R^2 for the Full Model is higher than the R^2 of the Restricted Model. Consequently, one can reject the Statistical Hypothesis and accept the Research Hypothesis; one must act as if the circumference of a circular object is a different linear function of the diameter than is that of a square object. Because the distance around the object is a function of the distance across the object, but depending upon whether the object is a square or a circle, we have a case of interaction. The distance across the object interacts with whether or not the object is a square or a circle in the prediction of the distance around the object. We do not usually think of the distance around a square as the circumference, and the fact that there is a different functional relationship for circles and squares may be one of the reasons.

MODEL 5 is preferred over MODEL 6, but there is a more parsimonious model than MODEL 5. The restriction on the preferred model, MODEL 5, would be: $a_1 = a_2 = 0$. Forcing the restrictions onto MODEL 5 results in:

$$\text{MODEL 7: CIRCUM} = b_1\text{DIAMC} + b2\text{DIAMS} + E_7$$

MODEL 7 yields as high an R^2 as does MODEL 5, thus resulting in a new preferred model. It is preferred because it yields the same R^2 and is more parsimonious. MODEL 7, though, still depicts statistical interaction.

SAS setup for activity #3:

```
OPTIONS LINESIZE = 70;
TITLE ACTIVITY 3;
DATA NEW; INPUT CIRCUM DIAM CIRSQ;
US = CIRSQ;
UC = 1 - CIRSQ;
    DIAMC = UC * DIAM;
    DIAMS = US * DIAM;
CARDS;
135      43       0
96       30       0
83.5     26       0
75       25.25    0
65.5     20.5     0
59.5     18.5     0
52.5     16       0
45.5     14.25    0
35.5     11.25    0
145      36       1
126      31.5     1
90       22.25    1
65       16.75    1
51.25    13       1
47       11.75    1
36.25    9        1
10       2.25     1
PROC PRINT;
PROC REG; MODEL CIRCUM = DIAM;
OUTPUT OUT = NEW P = PREDICT R = RESID;
PROC PLOT;
PLOT CIRCUM * DIAM = UC PREDICT * DIAM = '+'/OVERLAY;
TITLE "MODEL 6";
```

```
PROC REG; MODEL CIRCUM = US DIAMC DIAMS;
OUTPUT OUT = NEW P = PRED R = RES;
ONELINE: TEST US = 0, DIAMC = DIAMS;
PROC PLOT;
PLOT CIRCUM * DIAM = UC PRED * DIAM = '+'/OVERLAY;
TITLE "MODEL 5";
PROC REG; MODEL CIRCUM = DIAMC DIAMS;
ZEROINT: TEST INTERCEPT = 0;
TITLE "MODEL 7 WITH INTERCEPT";
PROC REG; MODEL CIRCUM = DIAMC DIAMS / NOINT;
TITLE "MODEL 7 WITH NO INTERCEPT";
```

APPENDIX D:
SAS STATEMENTS FOR
THE VARIOUS APPLIED
RESEARCH HYPOTHESES

OPTIONS LINESIZE = 70;
TITLE "TESTING RESEARCH HYPOTHESES USING MULTIPLE LINEAR REGRESSION";
TITLE2 "APPLIED RESEARCH HYPOTHESIS 4.1";
DATA GLM;
INPUT IDX1 X2 X3 X4 X5 X6 X7 X8 X9 X10 X11 X12 X13 X14;
CARDS;

01	4	4	26	8	1	0	82	97	0	1	0	1	80	04	01
02	6	9	28	9	1	0	95	101	0	1	1	0	75	06	02
03	6	8	41	10	0	1	103	104	0	1	0	1	82	07	03
04	8	10	29	10	1	0	109	106	0	1	0	1	84	07	04
05	10	12	40	11	1	0	84	96	0	1	0	1	85	06	05
06	11	13	30	11	1	0	93	100	0	1	1	0	82	06	06
07	12	14	39	12	0	1	98	102	0	1	1	0	84	05	07
08	14	15	32	13	1	0	99	103	0	1	0	1	90	05	08
09	16	15	37	14	0	1	87	98	0	1	0	1	92	06	09
10	17	16	37	15	0	1	106	105	0	1	1	0	92	06	10
11	18	16	36	15	0	1	113	112	0	1	1	0	90	07	11
12	20	17	35	17	0	1	128	125	0	1	0	1	97	11	12
13	22	18	34	18	1	0	89	99	0	1	0	1	100	17	13
14	24	18	33	58	0	1	115	118	0	1	1	0	96	18	14
15	26	20	32	21	0	1	125	124	0	1	1	0	98	18	15
16	27	20	32	22	0	1	91	100	0	1	1	0	100	18	16
17	28	20	31	23	0	1	116	115	0	1	1	0	101	18	17
18	28	21	36	55	1	0	117	117	0	1	0	1	105	18	18
19	29	20	30	24	0	1	123	120	0	1	0	1	106	17	19
20	30	21	29	53	0	1	126	120	0	1	0	1	108	17	20
21	31	21	28	26	0	1	120	116	0	1	1	0	103	17	21
22	32	22	26	52	0	1	129	127	0	1	1	0	105	18	22
23	32	22	37	27	1	0	120	116	0	1	0	1	110	19	23
24	33	22	24	51	0	1	114	113	0	1	0	1	112	25	24
25	34	22	37	52	1	0	100	102	0	1	1	0	107	28	25
26	36	23	38	28	1	0	98	103	0	1	1	0	110	29	26

```
27  38  24  38  49  1  0   96  102  0  1  0  1  115  30  27
28  40  24  38  48  1  0   85   95  0  1  0  1  120  30  28
29  42  25  39  32  1  0  125  128  0  1  1  0  115  30  29
30  48  25  40  38  1  0   95  102  0  1  1  0  120  30  30
31   5   5  27   8  1  0   84   80  1  0  1  0   75  04  01
32   7   8  27   9  1  0   85   82  1  0  0  1   82  06  02
33   7   8  41  10  0  1  115  123  1  0  0  1   83  07  03
34   8  11  29  11  1  0   87   85  1  0  1  0   80  07  04
35  10  12  41  11  0  1   95   95  1  0  1  0   83  06  05
36  11  12  30  10  1  0   93   92  1  0  1  0   84  06  06
37  12  13  39  11  0  1   99  100  1  0  1  0   85  05  07
38  14  15  32  14  1  0   95  100  1  0  0  1   90  05  08
39  16  16  37  13  0  1  118  128  1  0  0  1   92  06  09
40  17  16  37  14  0  1   88   86  1  0  0  1   93  06  10
41  18  17  36  16  0  1   94   92  1  0  1  0   90  07  11
42  20  18  35  16  0  1  102  102  1  0  1  0   92  11  12
43  21  17  34  19  1  0   89   87  1  0  0  1   98  17  13
44  23  17  33  57  0  1  116  122  1  0  0  1  100  18  14
45  25  19  32  22  0  1   97   97  1  0  0  1  103  18  15
46  27  20  32  21  0  1  103  102  1  0  0  1  105  18  16
47  28  21  31  24  0  1   90   89  1  0  1  0  100  18  17
48  29  22  36  54  1  0   96   96  1  0  1  0  102  18  18
49  30  22  30  24  0  1  108  113  1  0  1  0  103  17  19
50  30  21  39  54  1  0  100  101  1  0  1  0  104  17  20
51  31  22  28  27  0  1   91   89  1  0  0  1  108  17  21
52  32  21  26  51  0  1  112  116  1  0  0  1  110  18  22
53  32  22  37  28  1  0  104  104  1  0  1  0  105  19  23
54  33  21  24  52  0  1   92   90  1  0  0  1  112  25  24
55  33  22  37  51  1  0  120  130  1  0  0  1  110  28  25
56  35  23  38  29  1  0  122  132  1  0  0  1  113  29  26
57  37  24  38  50  1  0  128  140  1  0  1  0  110  30  27
58  39  24  38  49  1  0  124  136  1  0  1  0  113  30  28
59  41  25  39  33  1  0  126  137  1  0  0  1  114  30  29
60  46  25  40  39  1  0  130  142  1  0  0  1  124  30  30
```

```
PROC REG;
MODEL X2 = X12;
DIFF: TEST X12 = 0;
PROC REG;
MODEL X2 = X13;
DIFF: TEST X13 = 0;
*NOTICE BOTH TESTS YIELD SAME F VALUE AND SAME AS FOR 'MODEL'";

OPTIONS LINESIZE = 70;
TITLE "TESTING RESEARCH HYPOTHESES USING MULTIPLE LINEAR
REGRESSION";
TITLE2 "APPLIED RESEARCH HYPOTHESIS 4.2";
CMS FILEDEF NEW DISK BOOK DATA A;
DATA GLM;
INFILE NEW;
INPUT X2 4-5 X6 16 X12 32 X13 34;
```

```
IF X6 = 1 AND X12 = 1 THEN X17 = 1; ELSE X17 = 0;
IF X6 = 1 AND X12 = 0 THEN X18 = 1; ELSE X18 = 0;
IF X6 = 0 AND X12 = 1 THEN X19 = 1; ELSE X19 = 0;
IF X6 = 0 AND X12 = 0 THEN X20 = 1; ELSE X20 = 0;
PROC REG;
MODEL X2 = X17 X18 X19;
TEST X17 = 0, X18 = 0, X19 = 0;

OPTIONS LINESIZE = 70;
TITLE "TESTING RESEARCH HYPOTHESES USING MULTIPLE LINEAR
REGRESSION";
TITLE2 "GENERALIZED RESEARCH HYPOTHESIS 4.2";
TITLE3 "USING DATA FROM";
TITLE4 "APPLIED STATISTICS FOR THE BEHAVIORAL SCIENCES--3RD ED";
TITLE5 "HINKLE, WIERSMA & JURS, 1994";
TITLE6 "COMPUTER EXERCISE PAGE 600";
TITLE7 "CHAPTER 13--COMPUTER EXERCISE 1";
TITLE8 "ANOVA USING PROC REG";
DATA EXER1;
INPUT G1 G2 G3  ANXIETY;
LIST;
CARDS;
1  0  0   8
1  0  0   6
1  0  0   4
1  0  0  12
1  0  0  16
1  0  0  17
1  0  0  12
1  0  0  10
1  0  0  11
1  0  0  13
0  1  0  23
0  1  0  11
0  1  0  17
0  1  0  16
0  1  0   6
0  1  0  14
0  1  0  15
0  1  0  19
0  1  0  10
0  0  1  21
0  0  1  21
0  0  1  22
0  0  1  18
0  0  1  14
0  0  1  21
0  0  1   9
0  0  1  11
PROC REG;
```

```
MODEL ANXIETY = G2 G3;
RESTRICT: TEST G2 = 0, G3 = 0;
TITLE9 "FULL MODEL";
TITLE10 "NOTICE THAT F AND PROBABILITY FOR TEST IS SAME AS FOR
'MODEL'";

OPTIONS LINESIZE = 70;
TITLE "ONE DICHOTOMOUS PREDICTOR...ONE DICHOTOMOUS
CRITERION";
DATA TWO;
INPUT B2 A1 A2;
CARDS;
1   1   0
1   1   0
0   1   0
0   1   0
0   1   0
1   0   1
1   0   1
1   0   1
1   0   1
0   0   1
PROC REG; MODEL B2 = A1;
TEST A1 = 0;
PROC FREQ; TABLES B2 * A1 / CHISQ;

OPTIONS LINESIZE = 70;
TITLE "ARH 4.3  ETA COEFFICIENT FROM HINKLE ET AL., PAGE 527";
DATA ETA;
INPUT ID ANXIETY SCORE;
IF ANXIETY > 0 AND ANXIETY < 4 THEN G1 = 1; ELSE G1 = 0;
IF ANXIETY > 3 AND ANXIETY < 7 THEN G2 = 1; ELSE G2 = 0;
IF ANXIETY > 6 AND ANXIETY < 10 THEN G3 = 1; ELSE G3 = 0;
IF ANXIETY > 9 AND ANXIETY < 13 THEN G4 = 1; ELSE G4 = 0;
IF ANXIETY > 12 AND ANXIETY < 16 THEN G5 = 1; ELSE G5 = 0;
IF ANXIETY > 15 AND ANXIETY < 19 THEN G6 = 1; ELSE G6 = 0;
CARDS;
 1    2    4
 2    3    6
 3    5    8
 4    6   12
 5    7   15
 6    7   15
 7    9   10
 8   11    7
 9   12    7
10   12    5
11   14    6
12   15    8
13   15    9
```

```
14   16   12
15   17   14
16   18   13
PROC PRINT;
PROC REG; MODEL SCORE = G1 G2 G3 G4 G5;
TITLE3 "THIS IS THE MAXIMUM CURVILINEARITY--THE ETA MODEL";
PROC REG; MODEL SCORE = ANXIETY;
TITLE3 "NOTE THAT R-SQUARE FOR THE ETA MODEL IS MUCH HIGHER
THAN FOR";
TITLE4 "THE FOLLOWING LINEAR-FIT MODEL";

OPTIONS LINESIZE = 70;
TITLE "TESTING RESEARCH HYPOTHESES USING MULTIPLE LINEAR
REGRESSION";
TITLE2 "GENERALIZED RESEARCH HYPOTHESIS 4.3";
TITLE3 "USING DATA FROM";
TITLE4 "APPLIED STATISTICS FOR THE BEHAVIORAL SCIENCES--3RD ED";
TITLE5 "HINKLE ET AL., 1994";
TITLE9 "ANOVA USING PROC REG";
TITLE6 "COMPUTER EXERCISE PAGE 602";
TITLE7 "CHAPTER 14--COMPUTER EXERCISE 1";
DATA EXER1;
INPUT L1 L2 L3 L4 SATIS;
GROUP23 = L2 + L3;
CARDS;
1   0   0   0   71
1   0   0   0   74
1   0   0   0   75
1   0   0   0   68
1   0   0   0   74
1   0   0   0   72
1   0   0   0   76
1   0   0   0   73
1   0   0   0   77
0   1   0   0   74
0   1   0   0   81
0   1   0   0   75
0   1   0   0   70
0   1   0   0   74
0   1   0   0   81
0   1   0   0   76
0   0   1   0   83
0   0   1   0   77
0   0   1   0   80
0   0   1   0   76
0   0   1   0   86
0   0   1   0   79
0   0   1   0   86
0   0   1   0   82
0   0   0   1   76
```

```
0  0  0  1  69
0  0  0  1  64
0  0  0  1  76
0  0  0  1  73
0  0  0  1  72
0  0  0  1  68
0  0  0  1  71
PROC PRINT;
PROC MEANS;
PROC REG;
MODEL SATIS =  L2 L3 L4;
RESTRICT: TEST L2 = L3;
TITLE10 "FULL MODEL";
PROC REG;
MODEL SATIS = GROUP23 L4;
TITLE8 "USE REST MOD R-SQUARE AND FULL MOD R-SQUARE TO
CALCULATE F BY";
TITLE9 "HAND TO VERIFY F OF 7.7852";
TITLE10 "RESTRICTED MODEL";

OPTIONS LINESIZE = 70;
TITLE "TESTING RESEARCH HYPOTHESES USING MULTIPLE LINEAR
REGRESSION";
TITLE2 "APPLIED RESEARCH HYPOTHESIS 5.1";
CMS FILEDEF NEW DISK BOOK DATA A;
DATA GLM;
INFILE NEW;
INPUT  X2 4-5 X3 7-8;
PROC REG;
MODEL X2 = X3;
SLOPE: TEST X3 = 0;
TITLE4 "NOTICE F FOR SLOPE TEST (747) IS SAME AS F FOR MODEL";
TITLE5 "AS WELL AS T SQUARED FOR PARAMETER ESTIMATE OF
SLOPE (X3)";
PROC PLOT; PLOT X2 * X3 / HZERO;

OPTIONS LINESIZE = 70;
TITLE "TESTING RESEARCH HYPOTHESES USING MULTIPLE LINEAR
REGRESSION";
TITLE2 "APPLIED RESEARCH HYPOTHESIS 6.1";
CMS FILEDEF NEW DISK BOOK DATA A;
DATA MLR;
INFILE NEW;
INPUT X2 4-5 X4 10-11 X5 13-14 X6 16;
LABEL X2 = 'YOUR VARIABLE #2'
X4 = 'YOUR VARIABLE #4'
X5 = 'YOUR VARIABLE #5'
X6 = 'YOUR VARIABLE #6';
/* LABELS CAN BE UP TO 40 CHARACTERS AND WILL BE PRINTED ON
PRINTOUT. ABOVE IS THE ONLY PLACE THAT YOU NEED TO CHANGE FOR
```

YOUR VARIABLE NAMES. THIS COMMENT WILL NOT BE PRINTED, AND
STARTS WITH /* AND ENDS WITH ASTERISK SLASH.*/
PROC REG; MODEL X2 = X4 X5 X6;
TEST X4 =0, X5 = 0, X6 = 0;

OPTIONS LINESIZE = 70;
TITLE "TESTING RESEARCH HYPOTHESES USING MULTIPLE LINEAR
REGRESSION";
TITLE2 "APPLIED RESEARCH HYPOTHESIS 7.1";
CMS FILEDEF NEW DISK BOOK DATA A;
DATA GLM;
INFILE NEW;
INPUT X9 24-26 X10 28 X11 30 X12 32 X13 34;
X15 = X10 * X12;
X16 = X10 * X13;
X17 = X11 * X12;
X18 = X11 * X13;
IF X10 = 1 THEN X20 = 1;
IF X10 = 0 THEN X20 = -1;
IF X12 = 1 THEN X21 = 1;
IF X12 = 0 THEN X21 = -1;
IF X10 = 1 AND X12 = 1 THEN X22 = 1;
IF X10 = 1 AND X12 = 0 THEN X22 = -1;
IF X10 = 0 AND X12 = 1 THEN X22 = -1;
IF X10 = 0 AND X12 = 0 THEN X22 = 1;
PROC PRINT;
PROC REG; TITLE4 "DIRINT = DIRECTIONAL INTERACTION";
MODEL X9 = X16 X17 X18; OUTPUT OUT = NEW P = PREDICT R = RESID;
DIRINT: TEST X17 = X18 - X16;
PROC PLOT; PLOT X9 * X12 = X10 PREDICT * X12 = '+'/ OVERLAY;
PROC REG; MODEL X9 = X10 X12;
TITLE3 "MODEL BELOW IS RESTRICTED MODEL, GENERATED SO YOU CAN";
TITLE4 "CALCULATE F BY HAND. F WILL BE ABOUT 2.03";
TITLE5 "WITHIN ROUNDING ERROR.";
* ALTERNATIVE SAS PROC FOR ANALYZING STRICTLY CATEGORICAL DATA;
* COMPARE THE LAST CONTRAST F AND PROBABILITY WITH THE
*INTERACTION QUESTION TESTED WITH PROC REG;
PROC GLM;
CLASS X10 X12;
MODEL X9 = X10 X12 X10 * X12;
TITLE3 "MODEL BELOW IS FROM PROC GLM SO YOU CAN COMPARE WITH
THE ABOVE";
TITLE4 "   ";
TITLE5 "NOTE THAT ONLY THE INTERACTION HYPOTHESIS IS OF
INTEREST";
MEANS X10 X12 X10 * X12;
CONTRAST ' METHOD X10 ' X10   1   -1;
CONTRAST ' GROUP X12 '   X12   1   -1;
CONTRAST ' METHOD BY GROUP INT' X10 * X12   1   -1   -1   1;
TITLE3 "HERE ARE THE SPECIFIC CONTRASTS THAT MIMIC EXHIBIT 7.1;

TITLE4 "   ";
TITLE5 "NOTE THAT ONLY THE INTERACTION HYPOTHESIS IS OF INTEREST";
PROC REG; MODEL X9 = X20 X21 X22;
TEST X20 = 0; TEST X21 = 0; TEST X22 = 0;
TITLE3 "HERE IS THE REGRESSION MODEL USING THE CONTRAST COEFFICIENTS";
TITLE4 "WE HAVE TESTED ALL OF THE CONTRASTS, BUT AGAIN";
TITLE5 "NOTE THAT ONLY THE INTERACTION HYPOTHESIS IS OF INTEREST";

OPTIONS LINESIZE = 70;
TITLE "TESTING RESEARCH HYPOTHESES USING MULTIPLE LINEAR REGRESSION";
TITLE2 "APPLIED RESEARCH HYPOTHESIS 7.2";
CMS FILEDEF NEW DISK BOOK DATA A;
DATA GLM;
INFILE NEW;
INPUT X8 20-22 X9 24-26 X10 28 X11 30;
X15 = X10 * X8;
X16 = X11 * X8;
PROC PRINT;
PROC REG; TITLE4 "DIRINT = DIRECTIONAL INTERACTION";
MODEL X9 = X10 X15 X16; OUTPUT OUT = NEW P = PREDICT R = RESID;
DIRINT : TEST X15 = X16;
PROC PLOT; PLOT X9 * X8 = X10 PREDICT * X8 = '+'/OVERLAY;
PROC REG; MODEL X9 = X10 X8;
TITLE3 "MODEL BELOW IS RESTRICTED MODEL, GENERATED SO YOU CAN";
TITLE4 "CALCULATE F BY HAND. F WILL BE ABOUT 327.00";
TITLE5 "WITHIN ROUNDING ERROR";

OPTIONS LINESIZE = 70 PAGESIZE = 60;
TITLE "TESTING RESEARCH HYPOTHESES USING MULTIPLE LINEAR REGRESSION";
TITLE2 "APPLIED RESEARCH HYPOTHESIS 8.1";
CMS FILEDEF NEW DISK BOOK DATA A;
DATA MLR;
INFILE NEW;
INPUT X2 4-5 X12 32 X14 36-39;
PROC REG;
MODEL X2 = X12 X14; OUTPUT OUT = NEW P = PREDICT R = RESID;
RESTRICT: TEST X12 = 0;
TITLE4 'FULL MODEL';
PROC REG; MODEL X2 = X14;
TITLE4 'RESTRICTED MODEL';
TITLE5 "REMEMBER TO CHECK TO SEE IF THE WEIGHT FOR X12 IS POSITIVE";
TITLE6 "AS HYPOTHESIZED, AND IF IT IS, THEN DIVIDE THE COMPUTER";
TITLE7 "PROBABILITY BY 2.";
PROC PLOT; PLOT X2 * X14 = X12 PREDICT * X14 = '+'/OVERLAY;

```
OPTIONS LINESIZE = 70 PAGESIZE = 60;
TITLE "APPLIED RESEARCH HYPOTHESIS 9.1";
CMS FILEDEF NEW DISK BOOK DATA A;
DATA GLM;
INFILE NEW;
INPUT X2 4-5 X3 7-8;
X16 = X3 * X3;
PROC REG; MODEL X2 = X3 X16;
TEST X16 = 0;
PROC REG; MODEL X2 = X3; OUTPUT OUT = NEW P = PREDICT R = RESID;
PROC PLOT; PLOT X2 * X3 PREDICT * X3 = '+' / OVERLAY;

OPTIONS LINESIZE = 70 PAGESIZE = 60;
TITLE2 "APPLIED RESEARCH HYPOTHESIS 9.2";
CMS FILEDEF NEW DISK BOOK DATA A;
DATA GLM;
INFILE NEW;
INPUT X2 4-5 X5 13-14;
X15 = X5 * X5;
PROC REG;
MODEL X2 = X5 X15;
CON1: TEST X5 = 0;
CON2: TEST X15 = 0;
TITLE4 'FULL MODEL';
TITLE6 'CONDITION 1 TEST OF SIGNIFICANCE FOLLOWED BY CONDITION
2 TEST';
PROC REG;
MODEL X2 = X15;
TITLE4 'RESTRICTED MODEL CONDITION 1';
*PROC REG;
*MODEL X2 = X5 X15;
*RESTRICT: TEST X15 = 0;
*TITLE4 'FULL MODEL';
PROC REG;
MODEL X2 = X5;
TITLE4 'RESTRICTED MODEL CONDITION 2';

OPTIONS LINESIZE = 70 PAGESIZE = 60;
TITLE "APPLIED RESEARCH HYPOTHESIS 9.3";
CMS FILEDEF NEW DISK BOOK DATA A;
DATA GLMMLR;
INFILE NEW;
INPUT X2 4-5 X3 7-8;
PROC CORR; VAR X2 X3;
PROC REG;
MODEL X2 = X3;
TITLE4 'FULL MODEL';

OPTIONS LINESIZE = 70 PAGESIZE = 60;
TITLE "APPLIED RESEARCH HYPOTHESIS 9.4";
```

```
CMS FILEDEF NEW DISK BOOK DATA A;
DATA GLM;
INFILE NEW;
INPUT X15 40-41 X16 43-44;
X17 = SIN(X16 * 6.283/12.);
X18 = X15 - (.9724 * X16);
PROC REG;
MODEL X15 = X16 X17;
TEST X17 = 0;
PROC PLOT; PLOT X15 * X16;
PROC REG;
MODEL X18 = X16;
PROC PLOT; PLOT X18 * X16;

OPTIONS LINESIZE = 70;
TITLE "APPLIED RESEARCH HYPOTHESIS 10.1";
CMS FILEDEF NEW DISK BOOK DATA A;
DATA MLR;
INFILE NEW;
INPUT X2 4-5;
X2TEST = X2 - 20;
*NEED TO RUN PROC MEANS FIRST TO GET VALUE OF 23.98333;
X2MIMEAN = (X2 - 23.983333)**2;
X2ADJ = (X2 - 20)**2;
PROC PRINT;
PROC MEANS;
PROC UNIVARIATE; VAR X2TEST;
TITLE1 "ESSF = 60* MEAN OF X2MIMEAN   ESSR = 60 * MEAN OF X2ADJ";
TITLE2 "F = 7.210, WHICH IS THE SQUARE OF THE T FOR TESTING";
TITLE3 "THE HYPOTHESIS THAT X2 - 20 IS DIFFERENT FROM 0, PROVIDED BY";
TITLE4 "PROC UNIVARIATE T OF 2.688.";

OPTIONS LINESIZE = 70 PAGESIZE = 60;
TITLE "APPLIED RESEARCH HYPOTHESIS 10.2";
CMS FILEDEF NEW DISK BOOK DATA A;
DATA GLM;
INFILE NEW;
INPUT X2 4-5 X3 7-8 X12 32;
GAIN = X2 - X3;
PROC REG;
MODEL GAIN = X12;
GAIN: TEST X12 = 0;

OPTIONS LINESIZE = 70 PAGESIZE = 60;
TITLE "APPLIED RESEARCH HYPOTHESIS 10.3";
CMS FILEDEF NEW DISK BOOK DATA A;
DATA GLM;
INFILE NEW;
INPUT X2 4-5 X3 7-8 X12 32;
PROC REG;
```

```
MODEL X2 = X12 X3;
GAIN: TEST X12 = 0;
MODEL X2 = X3;

OPTIONS LINESIZE = 70;
TITLE "APPLIED RESEARCH HYPOTHESIS 10.4";
CMS FILEDEF NEW DISK BOOK DATA A;
PROC FORMAT;
VALUE TIMEF 1 = "PRE" 0 = "POST";
VALUE CELL17F 1 = "PRETREAT" 0 = "OTHER";
VALUE CELL18F 1 = "POSTREAT" 0 = "OTHER";
VALUE CELL19F 1 = "PRECONT" 0 = "OTHER";
VALUE TREATF 1 = "TREAT" 0 = "CONTROL";
DATA GLM;
INFILE NEW;
INPUT  X1 1-2 X2 4-5 X3 7-8 X12 32 X13 34;
X21 = 0; X22 = 0; X23 = 0; X24 = 0; X25 = 0; X26 = 0; X27 = 0;
X28 = 0; X29 = 0; X30 = 0; X31 = 0; X32 = 0; X33 = 0; X34 = 0;
X35 = 0; X36 = 0; X37 = 0; X38 = 0; X39 = 0; X40 = 0; X41 = 0;
X42 = 0; X43 = 0; X44 = 0; X45 = 0; X46 = 0; X47 = 0; X48 = 0;
X49 = 0; X50 = 0; X51 = 0; X52 = 0; X53 = 0; X54 = 0; X55 = 0;
X56 = 0; X57 = 0; X58 = 0; X59 = 0; X60 = 0; X61 = 0; X62 = 0;
X63 = 0; X64 = 0; X65 = 0; X66 = 0; X67 = 0; X68 = 0; X69 = 0;
X70 = 0; X71 = 0; X72 = 0; X73 = 0; X74 = 0; X75 = 0; X76 = 0;
X77 = 0; X78 = 0; X79 = 0; X80 = 0;
IF X1 = 1 THEN X21 = 1;
IF X1 = 2 THEN X22 = 1;
IF X1 = 3 THEN X23 = 1;
IF X1 = 4 THEN X24 = 1;
IF X1 = 5 THEN X25 = 1;
IF X1 = 6 THEN X26 = 1;
IF X1 = 7 THEN X27 = 1;
IF X1 = 8 THEN X28 = 1;
IF X1 = 9 THEN X29 = 1;
IF X1 = 10 THEN X30 = 1;
IF X1 = 11 THEN X31 = 1;
IF X1 = 12 THEN X32 = 1;
IF X1 = 13 THEN X33 = 1;
IF X1 = 14 THEN X34 = 1;
IF X1 = 15 THEN X35 = 1;
IF X1 = 16 THEN X36 = 1;
IF X1 = 17 THEN X37 = 1;
IF X1 = 18 THEN X38 = 1;
IF X1 = 19 THEN X39 = 1;
IF X1 = 20 THEN X40 = 1;
IF X1 = 21 THEN X41 = 1;
IF X1 = 22 THEN X42 = 1;
IF X1 = 23 THEN X43 = 1;
IF X1 = 24 THEN X44 = 1;
IF X1 = 25 THEN X45 = 1;
```

```
IF X1 = 26 THEN X46 = 1;
IF X1 = 27 THEN X47 = 1;
IF X1 = 28 THEN X48 = 1;
IF X1 = 29 THEN X49 = 1;
IF X1 = 30 THEN X50 = 1;
IF X1 = 31 THEN X51 = 1;
IF X1 = 32 THEN X52 = 1;
IF X1 = 33 THEN X53 = 1;
IF X1 = 34 THEN X54 = 1;
IF X1 = 35 THEN X55 = 1;
IF X1 = 36 THEN X56 = 1;
IF X1 = 37 THEN X57 = 1;
IF X1 = 38 THEN X58 = 1;
IF X1 = 39 THEN X59 = 1;
IF X1 = 40 THEN X60 = 1;
IF X1 = 41 THEN X61 = 1;
IF X1 = 42 THEN X62 = 1;
IF X1 = 43 THEN X63 = 1;
IF X1 = 44 THEN X64 = 1;
IF X1 = 45 THEN X65 = 1;
IF X1 = 46 THEN X66 = 1;
IF X1 = 47 THEN X67 = 1;
IF X1 = 48 THEN X68 = 1;
IF X1 = 49 THEN X69 = 1;
IF X1 = 50 THEN X70 = 1;
IF X1 = 51 THEN X71 = 1;
IF X1 = 52 THEN X72 = 1;
IF X1 = 53 THEN X73 = 1;
IF X1 = 54 THEN X74 = 1;
IF X1 = 55 THEN X75 = 1;
IF X1 = 56 THEN X76 = 1;
IF X1 = 57 THEN X77 = 1;
IF X1 = 58 THEN X78 = 1;
IF X1 = 59 THEN X79 = 1;
IF X1 = 60 THEN X80 = 1;
DATA PRE;
SET GLM;
X10 = 1 ; X2AND3 = X2;
DATA POST;
SET GLM;
X10 = 0; X2AND3 = X3;
DATA COMBINED;
SET PRE POST;
X11 = 1 - X10;
X17 = 0; X18 = 0; X19 = 0; X20 = 0;
X17 = X12 * X10;
X18 = X12 * X11;
X19 = X13 * X10;
PROC PRINT; VAR X1 X2 X3 X12 X21 X22 X10 X2AND3 X17 X18 X19;
FORMAT X17 CELL17F. X18 CELL18F. X19 CELL19F. X10 TIMEF. X12
```

```
TREATF.;
PROC REG;
MODEL X2AND3 = X17 X18 X19 X21-X77 X79;
INT: TEST X18 - X17 + X19 = 0;
MODEL X2AND3 = X10 X12 X21-X77 X79;

OPTIONS LINESIZE = 70 PAGESIZE = 60;
TITLE "APPLIED RESEARCH HYPOTHESIS 11.1";
CMS FILEDEF NEW DISK BOOK DATA A;
DATA GLM;
INFILE NEW;
INPUT ID 1-2 X2 4-5 X12 32 X14 36-39;
IF ID < 31;
*ABOVE STATEMENT COULD BE REPLACED WITH: OBS = 30;
PROC REG;
MODEL X2 = X12 X14;

OPTIONS LINESIZE = 70 PAGESIZE = 60;
TITLE "SECOND STEP OF CROSS-VALIDATION FROM CHAPTER 11";
CMS FILEDEF NEW DISK BOOK DATA A;
DATA GLM;
INFILE NEW;
INPUT ID 1-2 X2 4-5 X12 32 X14 36-39;
IF ID > 30;
X15 = 4.62768*X12 + .92690*X14 - 69.82533;
PROC CORR; VAR X2 X15;

OPTIONS LINESIZE = 70;
TITLE "TESTING RESEARCH HYPOTHESES USING MULTIPLE LINEAR
REGRESSION";
TITLE2 "APPENDIX 11-3 STEPWISE FORWARD, BACKWARD, AND ALL
POSSIBLE";
CMS FILEDEF NEW DISK BOOK DATA A;
DATA GLM;
INFILE NEW;
INPUT X2 4-5 X4 10-11 X5 13-14 X8 20-22 X9 24-26 X10 28 X12 32;
PROC REG; TITLE4 "VARIOUS FISHING TECHNIQUES";
MODEL X2 = X4 X5 X8 X9 X10 X12 / SELECTION = F;
TITLE6 "OUTPUT BELOW IS FORWARD STEPWISE";
PROC REG;
MODEL X2 = X4 X5 X8 X9 X10 X12 / SELECTION = B;
TITLE6 "OUTPUT BELOW IS BACKWARD STEPWISE";
PROC REG;
MODEL X2 = X4 X5 X8 X9 X10 X12 / SELECTION = RSQUARE;
TITLE6 "OUTPUT BELOW IS ALL POSSIBLE COMBINATIONS";

OPTIONS LINESIZE = 70;
TITLE "APPLIED RESEARCH HYPOTHESIS 12.1";
CMS FILEDEF NEW DISK BOOK DATA A;
DATA MLR;
```

INFILE NEW;
INPUT X2 4-5 X6 16;
PROC REG; MODEL X2 = X6;
TEST X6 = 3;

OPTIONS LINESIZE = 70;
TITLE "APPLIED RESEARCH HYPOTHESIS 12.2";
CMS FILEDEF NEW DISK BOOK DATA A;
DATA GLM;
INFILE NEW;
INPUT X2 4-5 X3 7-8;
X2F = X2 - 2 * X3;
*   ;
*WILL HAVE TO RUN MODEL FIRST TO GET THE NUMBERS BELOW;
*   ;
X2FULL = (X2 - 2.060944 * X3+ 12.976262) **2;
X2ADJ2 = (X2 - 2 * X3 +11.8833333) **2;
PROC REG; MODEL X2 = X3;
TEST X3 = 2;
PROC MEANS;
TITLE1 "ESS FORMULA IS F = (EESR - ESSF) / (M1 - M2)";
TITLE2 "   _____   ";
TITLE3 "   ESSF / (N - M1)   ";
TITLE5 "FOR THIS DATA ((9.4363 * 60) - (9.3313 * 60))/(2 - 1) ";
TITLE6 "   _____ ";
TITLE7 "   (9.3313 * 60) / 58   ";
TITLE9 "EQUAL TO .6526, WITHIN ROUNDING ERROR OF THE VALUE OF
.6532";

OPTIONS LINESIZE = 70;
TITLE "APPLIED RESEARCH HYPOTHESIS 12.3";
CMS FILEDEF NEW DISK BOOK DATA A;
DATA GLM;
INFILE NEW;
INPUT ID 1-2 X2 4-5 X3 7-8;
SEST = 11.4766 * (1 - .9820) **.5;
OVERUND = 1.64 * SEST;
PROC REG; MODEL X2 = X3; OUTPUT OUT = NEWER R = RESIDUAL;
PROC PLOT; PLOT X2 * X3; PROC PLOT; PLOT RESIDUAL * X3;
PROC PRINT; VAR X2 X3 RESIDUAL OVERUND;

APPENDIX E:
ANOVA SOURCE TABLES

SOURCE TABLE FOR APPLIED RESEARCH HYPOTHESIS 4.1

Source	df	Sum of Squares	Mean Square	F
X13	1	2.00669	2.00669	.015
Error	58	7,768.97664	133.94787	
Total	59	7,770.98333		

$R^2 = .0003 = (2.00669 / 7,770.98333)$

SOURCE TABLE FOR APPLIED RESEARCH HYPOTHESIS 7.1

Source	df	Sum of Squares	Mean Square	F
X10	1	124.144	124.144	.52
X12	1	111.890	111.890	.47
X10 * X12	1	480.745	480.745	2.03
Error	56	13,249.727	236.602	
Total	59	13,942.733		

Using Type I sum of squares from the PROC GLM output:

$R^2 = .00727 = (101.400 / 13,942.733)$ (for X10)
$R^2 = .00795 = (110.859 / 13,942.733)$ (for X12)
$R^2 = .03448 = (480.745 / 13,942.733)$ (for X10 * X12)
$R^2 = .04970$ (for the sum of the three separate effects)

SAS does not automatically provide the R^2 of the Restricted Model. One can obtain the R^2 of the Restricted Model by actually having SAS calculate the Restricted Model by inserting another PROC REG statement. Alternatively, one could use the information from the Full Model if there was only one restriction; that is, if one wanted to know how much variance was accounted for by a particular variable. Given the formula for the general F test:

$$F = \frac{(R_f^2 - R_r^2)/(1)}{(1 - R_f^2)/(N - m1)}$$

If both sides of the equation are multiplied by:

$(1 - R_f^2)/(N - m1)$

we have:

$F * (1 - R_f^2)/(N - m1) = (R_f^2 - R_r^2)$

If R_f^2 is subtracted from both sides, we have:

$-R_f^2 + F * (1 - R_f^2)/(N - m1) = -R_r^2$

If both sides are multiplied by -1, we have:

$R_f^2 - F * (1 - R_f^2)/(N - m1) = R_r^2$

APPENDIX F:
EQUIVALENCY OF *F*-TEST FORMULAE

The *F* test is now developed in terms of R^2 from Equation 5.4. The R^2 is the proportion of variance accounted for.

Equation 5.4:

$$F_{(df_p, df_w)} = \frac{\hat{v}_p}{\hat{v}_w} = \frac{(ESS_r - ESS_f)/df_p}{(ESS_f)/df_w}$$

$$R^2 = \frac{SS_p / N}{SS_t / N}$$

or the proportion of sum of squares accounted for:

$$R^2 = \frac{SS_p}{SS_t}$$

$1 - R^2$ would then be the proportion of sum of squares unaccounted for:

$$1 - R^2 = \frac{ESS}{SS_t}$$

thus:

$$\frac{ESS_f}{SS_t} = 1 - R_f^2$$

and

$$\frac{ESS_r}{SS_t} = 1 - R_r^2$$

Dividing equation 5.4 by SS_t yields:

$$\frac{\left[\dfrac{ESS_r}{SS_t} - \dfrac{ESS_f}{SS_t} / df_p\right]}{\dfrac{ESS_f}{SS_t} / df_w}$$

Substituting the above derived equalities:

$$F = \frac{[(1 - R_r^2) - (1 - R_f^2)] / df_p}{(1 - R_f^2) / df_w}$$

By rearranging terms, the above formulation can be simplified to:

$$F_{(df_p, df_w)} = \frac{[R_f^2 - R_r^2] / df_p}{(1 - R_f^2) / df_w}$$

APPENDIX G: POWER TABLES

Table G.1
Power as a Function of **L** and **U** at $\alpha = .01$

U	\multicolumn{12}{c}{L}
	2.00	4.00	6.00	8.00	10.00	12.00	14.00	16.00	18.00	20.00	25.00	30.00
1	12	28	45	60	72	81	88	92	95	97	99	*
2	08	20	35	49	61	72	80	87	91	94	98	99
3	07	16	29	42	54	65	74	82	87	91	97	99
4	06	14	25	37	49	60	69	77	84	89	96	98
5	05	12	22	33	44	55	65	74	80	86	94	98
6	05	11	19	30	41	51	61	70	77	83	93	97
7	04	10	18	27	37	48	58	67	74	81	91	96
8	04	09	16	25	35	45	55	64	72	78	90	96
9	04	08	15	23	33	42	52	61	69	76	88	95
10	03	08	14	22	31	40	49	58	66	74	87	94
11	03	07	13	20	29	38	47	56	64	71	85	93
12	03	07	12	19	27	36	45	54	62	69	83	92
13	03	06	12	18	26	34	43	52	60	67	82	91
14	03	06	11	17	25	33	41	50	58	65	80	90
15	03	06	10	16	23	31	40	48	56	64	79	89
16	03	06	10	16	22	30	38	46	54	62	77	88
20	02	05	08	13	19	26	33	41	48	56	72	84
24	02	04	07	12	17	22	29	36	43	51	67	80
28	02	04	07	10	15	20	26	32	39	46	62	76
32	02	04	06	09	13	18	22	29	32	42	58	72
40	02	03	05	08	11	15	20	25	30	36	51	65
50	02	03	05	07	09	13	16	21	25	31	44	58
60	02	03	04	06	08	11	14	18	22	26	39	52
80	02	02	03	05	06	09	11	14	17	21	31	43
100	01	02	03	04	06	07	09	11	14	17	26	36

* Power greater than .995.

Note. From *Statistical power analysis for the behavioral sciences*, Rev. Ed. (pp. 416–418, 440–442) by J. Cohen, 1977, New York: Academic Press.

Table G.2

Power as a Function of **L** and **U** at $\alpha = .05$

						L						
U	2.00	4.00	6.00	8.00	10.00	12.00	14.00	16.00	18.00	20.00	25.00	30.00
1	29	52	69	81	89	93	96	98	99	99	*	*
2	23	42	58	72	82	88	93	96	97	99	*	*
3	19	36	52	65	76	84	90	93	96	98	99	*
4	17	32	47	60	72	80	87	91	94	96	99	*
5	16	29	43	56	68	77	84	89	93	95	98	*
6	15	27	40	53	64	74	81	87	91	94	98	99
7	14	25	38	50	61	71	79	85	89	93	97	99
8	13	24	36	48	59	68	77	83	88	92	97	99
9	13	23	34	45	56	66	74	81	86	90	96	99
10	12	21	32	43	54	64	72	79	85	89	96	98
11	12	21	31	42	52	64	70	78	83	88	95	98
12	11	20	30	40	50	60	69	76	82	87	94	98
13	11	19	29	39	49	58	67	74	80	85	93	97
14	11	18	28	37	47	57	65	73	79	84	93	97
15	11	18	27	36	46	55	64	71	78	83	92	97
16	10	17	26	35	45	54	62	70	76	82	91	96
20	10	16	23	31	40	49	57	65	72	78	88	94
24	09	15	21	29	37	45	53	60	67	74	85	92
28	09	14	20	27	34	42	49	57	64	70	82	91
32	08	13	18	25	32	39	46	53	60	67	80	88
40	08	12	17	22	28	40	41	48	55	61	74	84
50	08	11	15	20	25	31	37	43	49	55	69	80
60	07	10	14	18	23	28	33	39	45	50	64	75
80	07	09	12	16	20	24	28	33	38	43	56	67
100	07	09	11	14	18	21	25	29	34	38	50	61

* Power greater than .995.

Note. From *Statistical power analysis for the behavioral sciences*, by J. Cohen, 1977.

Table G.3
Power as a Function of **L** and **U** at $\alpha = .10$

						L						
U	2.00	4.00	6.00	8.00	10.00	12.00	14.00	16.00	18.00	20.00	25.00	30.00
1	41	64	79	88	94	97	98	99	*	*	*	*
2	33	54	70	81	89	93	96	98	99	99	*	*
3	30	48	64	76	85	90	94	97	98	99	*	*
4	27	44	60	72	81	88	92	95	97	98	*	*
5	25	41	56	69	78	85	90	94	96	98	99	*
6	24	39	53	66	75	83	89	92	95	97	99	*
7	23	37	51	63	73	81	87	91	94	96	99	*
8	22	35	49	61	71	79	85	90	93	95	98	*
9	21	34	47	58	69	77	84	88	92	95	98	99
10	20	33	45	57	67	75	82	87	91	94	98	99
11	20	31	43	55	65	74	80	86	90	93	97	99
12	19	31	42	53	63	72	79	85	89	92	97	99
13	19	30	41	52	62	70	78	84	88	92	97	99
14	19	29	40	50	60	69	76	82	87	91	96	99
15	18	28	39	49	59	68	75	81	86	90	96	98
16	18	28	38	48	58	66	74	80	85	89	95	98
20	17	26	35	44	53	62	70	76	82	86	93	97
24	16	24	32	41	50	58	66	72	78	83	92	96
28	16	23	31	39	47	55	62	69	75	80	90	95
32	15	22	29	37	45	52	60	66	72	78	88	94
40	15	20	27	34	41	48	55	61	67	73	84	91
50	14	19	25	31	37	44	50	56	62	68	79	88
60	14	18	23	29	34	40	46	52	58	64	75	84
80	13	17	21	26	31	36	41	46	52	57	69	78
100	13	16	20	24	28	32	37	42	47	52	63	73

* Power greater than .995.

Note. From *Statistical power analysis for the behavioral sciences*, by J. Cohen, 1977.

Table G.4

L as a Function of Power and **U** at $\alpha = .01$

U	.25	.50	.60	2/3	.70	.75	.80	.85	.90	.95	.99
1	3.62	6.64	8.00	9.03	9.61	10.56	11.68	13.05	14.88	17.81	24.03
2	4.67	8.19	9.75	10.92	11.57	12.64	13.88	15.40	17.43	20.65	27.42
3	5.44	9.31	11.01	12.27	12.97	14.12	15.46	17.09	19.25	22.67	29.83
4	6.07	10.23	12.04	13.38	14.12	15.34	16.75	18.47	20.74	24.33	31.80
5	6.63	11.03	12.94	14.34	15.12	16.40	17.87	19.66	22.03	25.76	33.50
6	7.13	11.75	13.74	15.21	16.01	17.34	18.87	20.73	23.18	27.04	35.02
7	7.58	12.41	14.47	15.99	16.83	18.20	19.79	21.71	24.24	28.21	36.41
8	8.01	13.02	15.15	16.73	17.59	19.00	20.64	22.61	25.21	29.29	37.69
9	8.40	13.59	15.79	17.41	18.30	19.75	21.43	23.46	26.12	30.30	38.89
10	8.78	14.13	16.39	18.05	18.96	20.46	22.18	24.25	26.98	31.26	40.02
11	9.14	14.64	16.96	18.67	19.60	21.13	22.89	25.01	27.80	32.16	41.09
12	9.48	15.13	17.50	19.25	20.20	21.77	23.56	25.73	28.58	33.02	42.11
13	9.80	15.59	18.03	19.81	20.78	22.38	24.21	26.42	29.32	33.85	43.09
14	10.12	16.04	18.53	20.35	21.34	22.97	24.83	27.08	30.03	34.64	44.02
15	10.42	16.48	19.01	20.86	21.88	23.53	25.43	27.72	30.72	35.40	44.93
16	10.72	16.90	19.48	21.37	22.40	24.08	26.01	28.34	31.39	36.14	45.80
20	11.81	18.45	21.21	23.22	24.32	26.11	28.16	30.63	33.85	38.86	49.03
24	12.80	19.86	22.78	24.90	26.06	27.94	30.10	32.69	36.07	41.32	51.93
28	13.70	21.15	24.21	26.44	27.65	29.62	31.88	34.59	38.11	43.58	54.60
32	14.55	22.35	25.55	27.87	29.13	31.19	33.53	36.35	40.01	45.67	57.08
40	16.10	24.54	27.99	30.48	31.84	34.04	36.55	39.56	43.46	49.49	61.57
50	17.83	27.00	30.72	33.40	34.86	37.22	39.92	43.14	47.31	53.74	66.59
60	19.39	29.21	33.18	36.04	37.59	40.10	42.96	46.38	50.79	57.58	71.12
80	22.18	33.15	37.55	40.72	42.43	45.21	48.36	52.11	56.96	64.39	79.13
100	24.63	36.62	41.40	44.84	46.70	49.70	53.10	57.16	62.38	70.37	86.18

Note. From *Statistical power analysis for the behavioral sciences*, by J. Cohen, 1977.

Table G.5

L as a Function of Power and **U** at $\alpha = .05$

					Power						
U	.25	.50	.60	2/3	.70	.75	.80	.85	.90	.95	.99
1	1.65	3.84	4.90	5.71	6.17	6.94	7.85	8.98	10.51	13.00	18.37
2	2.26	4.96	6.21	7.17	7.70	8.59	9.64	10.92	12.65	15.44	21.40
3	2.71	5.76	7.15	8.21	8.79	9.76	10.90	12.30	14.17	17.17	23.52
4	3.08	6.42	7.92	9.05	9.68	10.72	11.94	13.42	15.40	18.57	25.24
5	3.41	6.99	8.59	9.79	10.45	11.55	12.83	14.39	16.47	19.78	26.73
6	3.70	7.50	9.19	10.44	11.14	12.29	13.62	15.26	17.42	20.86	28.05
7	3.97	7.97	9.73	11.04	11.77	12.96	14.35	16.04	18.28	21.84	29.25
8	4.22	8.40	10.24	11.60	12.35	13.59	15.02	16.77	19.08	22.74	30.36
9	4.45	8.81	10.71	12.12	12.89	14.17	15.65	17.45	19.83	23.59	31.39
10	4.67	9.19	11.15	12.60	13.40	14.72	16.24	18.09	20.53	24.38	32.36
11	4.88	9.56	11.58	13.07	13.89	15.24	16.80	18.70	21.20	25.14	33.29
12	5.08	9.90	11.98	13.51	14.35	15.74	17.34	19.28	21.83	25.86	34.16
13	5.28	10.24	12.36	13.93	14.80	16.21	17.85	19.83	22.44	26.54	35.00
14	5.46	10.55	12.73	14.34	15.22	16.67	18.34	20.36	23.02	27.20	35.81
15	5.64	10.86	13.09	14.73	15.63	17.11	18.81	20.87	23.58	27.84	36.58
16	5.81	11.16	13.43	15.11	16.03	17.53	19.27	21.37	24.12	28.45	37.33
20	6.46	12.26	14.71	16.51	17.50	19.11	20.96	23.20	26.13	30.72	40.10
24	7.04	13.26	15.87	17.78	18.82	20.53	22.49	24.85	27.94	32.76	42.59
28	7.57	14.17	16.93	18.94	20.04	21.83	23.89	26.36	29.60	34.64	44.86
32	8.07	15.02	17.91	20.02	21.17	23.04	25.19	27.77	31.14	36.37	46.98
40	8.98	16.58	19.71	21.99	23.23	25.25	27.56	30.33	33.94	39.54	50.83
50	10.00	18.31	21.72	24.19	25.53	27.71	30.20	33.19	37.07	43.07	55.12
60	10.92	19.88	23.53	26.17	27.61	29.94	32.59	35.77	39.89	46.25	58.98
80	12.56	22.67	26.75	29.70	31.29	33.88	36.83	40.34	44.89	51.89	65.83
100	14.00	25.12	29.59	32.80	34.54	37.36	40.56	44.37	49.29	56.85	71.84

Note. From *Statistical power analysis for the behavioral sciences*, by J. Cohen, 1977.

Table G.6
L as a Function of Power and U at $\alpha = .10$

U	Power										
	.25	.50	.60	·2/3	.70	.75	.80	.85	.90	.95	.99
1	.91	2.70	3.60	4.30	4.70	5.38	6.18	7.19	8.56	10.82	15.77
2	1.27	3.56	4.65	5.50	5.97	6.77	7.71	8.88	10.46	13.02	18.56
3	1.55	4.18	5.41	6.36	6.88	7.76	8.80	10.08	11.80	14.57	20.51
4	1.78	4.69	6.04	7.06	7.63	8.57	9.68	11.05	12.88	15.83	22.09
5	1.98	5.14	6.58	7.66	8.27	9.27	10.45	11.89	13.82	16.91	23.44
6	2.16	5.53	7.06	8.20	8.84	9.90	11.13	12.64	14.65	17.87	24.65
7	2.33	5.90	7.50	8.70	9.36	10.47	11.75	13.32	15.41	18.75	25.74
8	2.49	6.24	7.91	9.16	9.85	10.99	12.32	13.95	16.11	19.55	26.76
9	2.63	6.55	8.29	9.58	10.30	11.48	12.86	14.54	16.77	20.31	27.70
10	2.77	6.85	8.65	9.99	10.73	11.95	13.37	15.10	17.39	21.02	28.58
11	2.90	7.13	8.99	10.37	11.13	12.39	13.85	15.62	17.97	21.69	29.42
12	3.03	7.40	9.31	10.73	11.52	12.81	14.30	16.12	18.53	22.33	30.22
13	3.15	7.66	9.62	11.08	11.89	13.21	14.74	16.60	19.06	22.94	30.99
14	3.26	7.91	9.92	11.42	12.24	13.59	15.16	17.06	19.57	23.53	31.72
15	3.37	8.15	10.21	11.74	12.58	13.96	15.56	17.50	20.06	24.09	32.42
16	3.48	8.38	10.49	12.05	12.91	14.32	15.95	17.93	20.54	24.64	33.10
20	3.88	9.24	11.53	13.21	14.14	15.65	17.40	19.51	22.30	26.66	35.62
24	4.25	10.02	12.46	14.25	15.24	16.85	18.70	20.94	23.88	28.48	37.88
28	4.58	10.73	13.32	15.21	16.25	17.95	19.90	22.25	25.33	30.14	39.95
32	4.89	11.39	14.11	16.10	17.19	18.97	21.01	23.46	26.68	31.69	41.87
40	5.46	12.60	15.57	17.73	18.90	20.83	23.03	25.68	29.13	34.50	45.37
50	6.10	13.95	17.19	19.54	20.82	22.91	25.29	28.15	31.87	37.64	49.27
60	6.68	15.18	18.66	21.18	22.55	24.78	27.33	30.38	34.34	40.47	52.78
80	7.71	17.34	21.26	24.09	25.62	28.11	30.95	34.34	38.72	45.49	58.99
100	8.61	19.26	23.55	26.65	28.32	31.04	34.13	37.81	42.58	49.90	64.45

Note. From *Statistical power analysis for the behavioral sciences*, by J. Cohen, 1977.

APPENDIX H:
POLICY-CAPTURING ACTIVITY

Please rate 10 of the following applicants for the new position in the university's Department of Educational Administration. Please use the following rating scale.

Number of candidates

Low	=	1	1
		2	2
Med	=	3	4
		4	2
High	=	5	1

Therefore, only one candidate will receive a "1" and one a "5," two will receive a "2" and two a "4," and 4 will receive a "3."

1 ID
__ RATING
50 AGE
F SEX
3 YEARS OF RELEVANT EXPERIENCE
1 TYPE OF EXPERIENCE 1=COMMUNITY COLLEGE 0=NO COMMUNITY COLLEGE
3 LETTERS OF REFERENCE (ON A 1–5 SCALE) 1=BAD 3=MEDIUM 5=GREAT

2 ID
__ RATING
50 AGE
F SEX
3 YEARS OF RELEVANT EXPERIENCE
1 TYPE OF EXPERIENCE 1=COMMUNITY COLLEGE 0=NO COMMUNITY COLLEGE
5 LETTERS OF REFERENCE (ON A 1–5 SCALE) 1=BAD 3=MEDIUM 5=GREAT

3 ID
__ RATING
50 AGE

F SEX
3 YEARS OF RELEVANT EXPERIENCE
0 TYPE OF EXPERIENCE 1=COMMUNITY COLLEGE 0=NO COMMUNITY COLLEGE
3 LETTERS OF REFERENCE (ON A 1–5 SCALE) 1=BAD 3=MEDIUM 5=GREAT

4 ID
__ RATING
50 AGE
F SEX
3 YEARS OF RELEVANT EXPERIENCE
0 TYPE OF EXPERIENCE 1=COMMUNITY COLLEGE 0=NO COMMUNITY COLLEGE
5 LETTERS OF REFERENCE (ON A 1–5 SCALE) 1=BAD 3=MEDIUM 5=GREAT

5 ID
__ RATING
30 AGE
F SEX
3 YEARS OF RELEVANT EXPERIENCE
1 TYPE OF EXPERIENCE 1=COMMUNITY COLLEGE 0=NO COMMUNITY COLLEGE
3 LETTERS OF REFERENCE (ON A 1–5 SCALE) 1=BAD 3=MEDIUM 5=GREAT

6 ID
__ RATING
30 AGE
F SEX
3 YEARS OF RELEVANT EXPERIENCE
1 TYPE OF EXPERIENCE 1=COMMUNITY COLLEGE 0=NO COMMUNITY COLLEGE
5 LETTERS OF REFERENCE (ON A 1–5 SCALE) 1=BAD 3=MEDIUM 5=GREAT

7 ID
__ RATING
30 AGE
F SEX
3 YEARS OF RELEVANT EXPERIENCE
0 TYPE OF EXPERIENCE 1=COMMUNITY COLLEGE 0=NO COMMUNITY COLLEGE
3 LETTERS OF REFERENCE (ON A 1–5 SCALE) 1=BAD 3=MEDIUM 5=GREAT

8 ID
__ RATING
30 AGE
F SEX
3 YEARS OF RELEVANT EXPERIENCE
0 TYPE OF EXPERIENCE 1=COMMUNITY COLLEGE 0=NO COMMUNITY COLLEGE
5 LETTERS OF REFERENCE (ON A 1–5 SCALE) 1=BAD 3=MEDIUM 5=GREAT

9 ID
__ RATING
50 AGE
M SEX
3 YEARS OF RELEVANT EXPERIENCE

1 TYPE OF EXPERIENCE 1=COMMUNITY COLLEGE 0=NO COMMUNITY COLLEGE
3 LETTERS OF REFERENCE (ON A 1–5 SCALE) 1=BAD 3=MEDIUM 5=GREAT

10 ID
__ RATING
50 AGE
M SEX
3 YEARS OF RELEVANT EXPERIENCE
1 TYPE OF EXPERIENCE 1=COMMUNITY COLLEGE 0=NO COMMUNITY COLLEGE
5 LETTERS OF REFERENCE (ON A 1–5 SCALE) 1=BAD 3=MEDIUM 5=GREAT

11 ID
__ RATING
50 AGE
M SEX
3 YEARS OF RELEVANT EXPERIENCE
0 TYPE OF EXPERIENCE 1=COMMUNITY COLLEGE 0=NO COMMUNITY COLLEGE
3 LETTERS OF REFERENCE (ON A 1–5 SCALE) 1=BAD 3=MEDIUM 5=GREAT

12 ID
__ RATING
50 AGE
M SEX
3 YEARS OF RELEVANT EXPERIENCE
0 TYPE OF EXPERIENCE 1=COMMUNITY COLLEGE 0=NO COMMUNITY COLLEGE
5 LETTERS OF REFERENCE (ON A 1–5 SCALE) 1=BAD 3=MEDIUM 5=GREAT

13 ID
__ RATING
30 AGE
M SEX
3 YEARS OF RELEVANT EXPERIENCE
1 TYPE OF EXPERIENCE 1=COMMUNITY COLLEGE 0=NO COMMUNITY COLLEGE
3 LETTERS OF REFERENCE (ON A 1–5 SCALE) 1=BAD 3=MEDIUM 5=GREAT

14 ID
__ RATING
30 AGE
M SEX
3 YEARS OF RELEVANT EXPERIENCE
1 TYPE OF EXPERIENCE 1=COMMUNITY COLLEGE 0=NO COMMUNITY COLLEGE
5 LETTERS OF REFERENCE (ON A 1–5 SCALE) 1=BAD 3=MEDIUM 5=GREAT

15 ID
__ RATING
30 AGE
M SEX
3 YEARS OF RELEVANT EXPERIENCE
0 TYPE OF EXPERIENCE 1=COMMUNITY COLLEGE 0=NO COMMUNITY COLLEGE
3 LETTERS OF REFERENCE (ON A 1–5 SCALE) 1=BAD 3=MEDIUM 5=GREAT

16 ID
__ RATING
30 AGE
M SEX
3 YEARS OF RELEVANT EXPERIENCE
0 TYPE OF EXPERIENCE 1=COMMUNITY COLLEGE 0=NO COMMUNITY COLLEGE
5 LETTERS OF REFERENCE (ON A 1–5 SCALE) 1=BAD 3=MEDIUM 5=GREAT

17 ID
__ RATING
30 AGE
M SEX
10 YEARS OF RELEVANT EXPERIENCE
1 TYPE OF EXPERIENCE 1=COMMUNITY COLLEGE 0=NO COMMUNITY COLLEGE
3 LETTERS OF REFERENCE (ON A 1–5 SCALE) 1=BAD 3=MEDIUM 5=GREAT

18 ID
__ RATING
30 AGE
M SEX
10 YEARS OF RELEVANT EXPERIENCE
1 TYPE OF EXPERIENCE 1=COMMUNITY COLLEGE 0=NO COMMUNITY COLLEGE
5 LETTERS OF REFERENCE (ON A 1–5 SCALE) 1=BAD 3=MEDIUM 5=GREAT

19 ID
__ RATING
30 AGE
M SEX
10 YEARS OF RELEVANT EXPERIENCE
0 TYPE OF EXPERIENCE 1=COMMUNITY COLLEGE 0=NO COMMUNITY COLLEGE
3 LETTERS OF REFERENCE (ON A 1–5 SCALE) 1=BAD 3=MEDIUM 5=GREAT

20 ID
__ RATING
30 AGE
M SEX
10 YEARS OF RELEVANT EXPERIENCE
0 TYPE OF EXPERIENCE 1=COMMUNITY COLLEGE 0=NO COMMUNITY COLLEGE
5 LETTERS OF REFERENCE (ON A 1–5 SCALE) 1=BAD 3=MEDIUM 5=GREAT

21 ID
__ RATING
50 AGE
M SEX
10 YEARS OF RELEVANT EXPERIENCE
1 TYPE OF EXPERIENCE 1=COMMUNITY COLLEGE 0=NO COMMUNITY COLLEGE
3 LETTERS OF REFERENCE (ON A 1–5 SCALE) 1=BAD 3=MEDIUM 5=GREAT

22 ID
__ RATING

50 AGE
M SEX
10 YEARS OF RELEVANT EXPERIENCE
1 TYPE OF EXPERIENCE 1=COMMUNITY COLLEGE 0=NO COMMUNITY COLLEGE
5 LETTERS OF REFERENCE (ON A 1–5 SCALE) 1=BAD 3=MEDIUM 5=GREAT

23 ID
__ RATING
50 AGE
M SEX
10 YEARS OF RELEVANT EXPERIENCE
0 TYPE OF EXPERIENCE 1=COMMUNITY COLLEGE 0=NO COMMUNITY COLLEGE
3 LETTERS OF REFERENCE (ON A 1–5 SCALE) 1=BAD 3=MEDIUM 5=GREAT

24 ID
__ RATING
50 AGE
M SEX
10 YEARS OF RELEVANT EXPERIENCE
0 TYPE OF EXPERIENCE 1=COMMUNITY COLLEGE 0=NO COMMUNITY COLLEGE
5 LETTERS OF REFERENCE (ON A 1–5 SCALE) 1=BAD 3=MEDIUM 5=GREAT

25 ID
__ RATING
30 AGE
F SEX
10 YEARS OF RELEVANT EXPERIENCE
1 TYPE OF EXPERIENCE 1=COMMUNITY COLLEGE 0=NO COMMUNITY COLLEGE
3 LETTERS OF REFERENCE (ON A 1–5 SCALE) 1=BAD 3=MEDIUM 5=GREAT

26 ID
__ RATING
30 AGE
F SEX
10 YEARS OF RELEVANT EXPERIENCE
1 TYPE OF EXPERIENCE 1=COMMUNITY COLLEGE 0=NO COMMUNITY COLLEGE
5 LETTERS OF REFERENCE (ON A 1–5 SCALE) 1=BAD 3=MEDIUM 5=GREAT

27 ID
__ RATING
30 AGE
F SEX
10 YEARS OF RELEVANT EXPERIENCE
0 TYPE OF EXPERIENCE 1=COMMUNITY COLLEGE 0=NO COMMUNITY COLLEGE
3 LETTERS OF REFERENCE (ON A 1–5 SCALE) 1=BAD 3=MEDIUM 5=GREAT

28 ID
__ RATING
30 AGE
F SEX

10 YEARS OF RELEVANT EXPERIENCE
0 TYPE OF EXPERIENCE 1=COMMUNITY COLLEGE 0=NO COMMUNITY COLLEGE
5 LETTERS OF REFERENCE (ON A 1–5 SCALE) 1=BAD 3=MEDIUM 5=GREAT

29 ID
__ RATING
50 AGE
F SEX
10 YEARS OF RELEVANT EXPERIENCE
1 TYPE OF EXPERIENCE 1=COMMUNITY COLLEGE 0=NO COMMUNITY COLLEGE
3 LETTERS OF REFERENCE (ON A 1–5 SCALE) 1=BAD 3=MEDIUM 5=GREAT

30 ID
___ RATING
50 AGE
F SEX
10 YEARS OF RELEVANT EXPERIENCE
1 TYPE OF EXPERIENCE 1=COMMUNITY COLLEGE 0=NO COMMUNITY COLLEGE
5 LETTERS OF REFERENCE (ON A 1–5 SCALE) 1=BAD 3=MEDIUM 5=GREAT

31 ID
__ RATING
50 AGE
F SEX
10 YEARS OF RELEVANT EXPERIENCE
0 TYPE OF EXPERIENCE 1=COMMUNITY COLLEGE 0=NO COMMUNITY COLLEGE
3 LETTERS OF REFERENCE (ON A 1–5 SCALE) 1=BAD 3=MEDIUM 5=GREAT

32 ID
__ RATING
50 AGE
F SEX
10 YEARS OF RELEVANT EXPERIENCE
0 TYPE OF EXPERIENCE 1=COMMUNITY COLLEGE 0=NO COMMUNITY COLLEGE
5 LETTERS OF REFERENCE (ON A 1–5 SCALE) 1=BAD 3=MEDIUM 5=GREAT

APPENDIX I: MICROCOMPUTER SETUPS FOR SELECTED APPLIED RESEARCH HYPOTHESES

We would like to thank Jim Salzman, a doctoral student at the University of Akron, for his work in 1994 on these SPSS setups. Please note that the job setups will likely differ from computer to computer. In addition, the setups are very similar to those needed on a mainframe. For instance, the first two lines of each of these problems are not needed at the New Mexico State University mainframe computer.

```
//NEWMAN JOB 94522,'SALZMAN',CLASS=H
// EXEC SPSS
TITLE 'TESTING RESEARCH HYPOTHESES USING MLR'
SUBTITLE 'APPLIED RESEARCH HYPOTHESIS 4.1'
DATA LIST /ID 1-2 X2 4-5 X3 7-8 X4 12-13 X5 15-16 X6 18
      X7 19 X8 21-23 X9 24-26 X10 28 X11 29 X12 30
      X13 31 X14 33-35
BEGIN DATA
01   04   04     26   08   10    082097   0101   080
02   06   09     28   09   10    095101   0110   075
03   06   08     41   10   01    103104   0101   082
04   08   10     29   10   10    109106   0101   084
05   10   12     40   11   01    084096   0101   085
06   11   13     30   11   10    093100   0110   082
07   12   14     39   12   01    098102   0110   084
08   14   15     32   13   10    099103   0101   090
09   16   15     37   14   01    087098   0101   092
10   17   16     37   15   01    106105   0110   092
11   18   16     36   15   01    113112   0110   90
12   20   17     35   17   01    128125   0101   97
13   22   18     34   18   10    89 99    0101   100
14   24   18     33   58   01    115118   0110   96
15   26   20     32   21   01    125124   0110   98
16   27   20     32   22   01    91100    0110   100
17   28   20     31   23   01    116115   0110   101
18   28   21     36   55   10    117117   0101   105
19   29   20     30   24   01    123120   0101   106
```

20	30	21	29	53	01	126120	0101	108
21	31	21	28	26	01	120116	0110	103
22	32	22	26	52	01	129127	0110	105
23	32	22	37	27	10	120116	0101	110
24	33	22	24	51	01	114113	0101	112
25	34	22	37	52	10	100102	0110	107
26	36	23	38	28	10	98103	0110	110
27	38	24	38	49	10	96102	0101	115
28	40	24	38	48	10	85 95	0101	120
29	42	25	39	32	10	125128	0110	115
30	48	25	40	38	10	95102	0110	120
31	5	5	27	8	10	84 80	1010	75
32	7	8	27	9	10	85 82	1001	82
33	7	8	41	10	01	115123	1001	83
34	8	11	29	11	10	87 85	1010	80
35	10	12	41	11	01	95 95	1010	83
36	11	12	30	10	10	93 92	1010	84
37	12	13	39	11	01	99100	1010	85
38	14	15	32	14	10	95100	1001	90
39	16	16	37	13	01	118128	1001	92
40	17	16	37	14	01	88 86	1001	93
41	18	17	36	16	01	94 92	1010	90
42	20	18	35	16	01	102102	1010	92
43	21	17	34	19	10	89 87	1001	98
44	23	17	33	57	01	116122	1001	100
45	25	19	32	22	01	97 97	1001	103
46	27	20	32	21	01	103102	1001	105
47	28	21	31	24	01	90 89	1010	100
48	29	22	36	54	10	96 96	1010	102
49	30	22	30	24	01	108113	1010	103
50	30	21	39	54	10	100101	1010	104
51	31	22	28	27	01	91 89	1001	108
52	32	21	26	51	01	112116	1001	110
53	32	22	37	28	10	104104	1010	105
54	33	21	24	52	01	92 90	1001	112
55	33	22	37	51	10	120130	1001	110
56	35	23	38	29	10	122132	1001	113
57	37	24	38	50	10	128140	1010	110
58	39	24	38	49	10	124136	1010	113
59	41	25	39	33	10	126137	1001	114
60	46	25	40	39	10	130142	1001	124

```
END DATA.
REGRESSION  VARIABLES=X2 X12 X13/
   DEPENDENT=X2/
   TEST (X12)/
   DEPENDENT=X2/
   TEST (X13)/


//R2JAS1 JOB 94522,'SALZMAN',CLASS=H
// EXEC SPSS
```

TITLE 'ONE DICHOTOMOUS PREDICTOR--ONE DICHOTOMOUS CRITERION'
DATA LIST / B2 1 A1 3 A2 5.
BEGIN DATA

```
1   1   0
1   1   0
0   1   0
0   1   0
0   1   0
1   0   1
1   0   1
1   0   1
1   0   1
0   0   1
```

END DATA.
REGRESSION VARIABLES=B2 A1/
   DEPENDENT=B2/
   TEST (A1)/
CROSSTABS TABLES = B2 BY A1 A2
OPTIONS 4 14 15
STATISTICS 1

//NEWMAN JOB 94522,'SALZMAN',CLASS=H
// EXEC SPSS
TITLE 'TESTING RESEARCH HYPOTHESES USING MLR'
SUBTITLE 'APPLIED RESEARCH HYPOTHESIS 4.2'
DATA LIST /ID 1-2 X2 4-5 X3 7-8 X4 12-13 X5 15-16 X6 18
       X7 19 X8 21-23 X9 24-26 X10 28 X11 29 X12 30
       X13 31 X14 33-35
BEGIN DATA

```
01   04   04    26   08   10   082097   0101   080
02   06   09    28   09   10   095101   0110   075
03   06   08    41   10   01   103104   0101   082
04   08   10    29   10   10   109106   0101   084
05   10   12    40   11   01   084096   0101   085
06   11   13    30   11   10   093100   0110   082
07   12   14    39   12   01   098102   0110   084
08   14   15    32   13   10   099103   0101   090
09   16   15    37   14   01   087098   0101   092
10   17   16    37   15   01   106105   0110   092
11   18   16    36   15   01   113112   0110   90
12   20   17    35   17   01   128125   0101   97
13   22   18    34   18   10    89 99   0101   100
14   24   18    33   58   01   115118   0110   96
15   26   20    32   21   01   125124   0110   98
16   27   20    32   22   01    91100   0110   100
17   28   20    31   23   01   116115   0110   101
18   28   21    36   55   10   117117   0101   105
19   29   20    30   24   01   123120   0101   106
20   30   21    29   53   01   126120   0101   108
```

21	31	21	28	26	01	120116	0110	103
22	32	22	26	52	01	129127	0110	105
23	32	22	37	27	10	120116	0101	110
24	33	22	24	51	01	114113	0101	112
25	34	22	37	52	10	100102	0110	107
26	36	23	38	28	10	98103	0110	110
27	38	24	38	49	10	96102	0101	115
28	40	24	38	48	10	85 95	0101	120
29	42	25	39	32	10	125128	0110	115
30	48	25	40	38	10	95102	0110	120
31	5	5	27	8	10	84 80	1010	75
32	7	8	27	9	10	85 82	1001	82
33	7	8	41	10	01	115123	1001	83
34	8	11	29	11	10	87 85	1010	80
35	10	12	41	11	01	95 95	1010	83
36	11	12	30	10	10	93 92	1010	84
37	12	13	39	11	01	99100	1010	85
38	14	15	32	14	10	95100	1001	90
39	16	16	37	13	01	118128	1001	92
40	17	16	37	14	01	88 86	1001	93
41	18	17	36	16	01	94 92	1010	90
42	20	18	35	16	01	102102	1010	92
43	21	17	34	19	10	89 87	1001	98
44	23	17	33	57	01	116122	1001	100
45	25	19	32	22	01	97 97	1001	103
46	27	20	32	21	01	103102	1001	105
47	28	21	31	24	01	90 89	1010	100
48	29	22	36	54	10	96 96	1010	102
49	30	22	30	24	01	108113	1010	103
50	30	21	39	54	10	100101	1010	104
51	31	22	28	27	01	91 89	1001	108
52	32	21	26	51	01	112116	1001	110
53	32	22	37	28	10	104104	1010	105
54	33	21	24	52	01	92 90	1001	112
55	33	22	37	51	10	120130	1001	110
56	35	23	38	29	10	122132	1001	113
57	37	24	38	50	10	128140	1010	110
58	39	24	38	49	10	124136	1010	113
59	41	25	39	33	10	126137	1001	114
60	46	25	40	39	10	130142	1001	124

END DATA.
COMPUTE X17 = 0
COMPUTE X18 = 0
COMPUTE X19 = 0
COMPUTE X20 = 0
IF (X6=1 AND X12=1) X17=1
IF (X6=1 AND X12=0) X18=1
IF (X6=0 AND X12=1) X19=1
IF (X6=0 AND X12=0) X20=1

REGRESSION VARIABLES=X2 X17 X18 X19 /
    DEPENDENT=X2/
    TEST (X17 X18 X19)/

//R2JAS1 JOB 94522,'SALZMAN',CLASS=H
// EXEC SPSS
TITLE 'TESTING RESEARCH HYPOTHESES USING MLR'
SUBTITLE 'GENERALIZED RESEARCH HYPOTHESIS 4.2'
DATA LIST / G1 1 G2 3 G3 5 ANXIETY 7-8.
BEGIN DATA
1   0   0   08
1   0   0   06
1   0   0   04
1   0   0   12
1   0   0   16
1   0   0   17
1   0   0   12
1   0   0   10
1   0   0   11
1   0   0   13
0   1   0   23
0   1   0   11
0   1   0   17
0   1   0   16
0   1   0   06
0   1   0   14
0   1   0   15
0   1   0   19
0   1   0   10
0   0   1   21
0   0   1   21
0   0   1   22
0   0   1   18
0   0   1   14
0   0   1   21
0   0   1   09
0   0   1   11
END DATA.
REGRESSION VARIABLES=G2 G3 ANXIETY/
    DEPENDENT=ANXIETY/
    TEST (G2) (G3) (G2, G3)/

//R2JAS1 JOB 94522,'SALZMAN',CLASS=H
// EXEC SPSS
TITLE 'ETA COEFFICIENT FROM HINKLE, ET AL., 1994, PAGE 525'
DATA LIST / ID 1-2 ANXIETY 4-5 SCORE 7-8
BEGIN DATA
01   02   04
02   03   06
03   05   08

```
04   06   12
05   07   15
06   07   15
07   09   10
08   11   07
09   12   07
10   12   05
11   14   06
12   15   08
13   15   09
14   16   12
15   17   14
16   18   13
END DATA.
COMPUTE G1=0
COMPUTE G2=0
COMPUTE G3=0
COMPUTE G4=0
COMPUTE G5=0
COMPUTE G6=0
IF (ANXIETY GT 0 AND ANXIETY LT 4) G1=1
IF (ANXIETY GT 3 AND ANXIETY LT 7) G2=1
IF (ANXIETY GT 6 AND ANXIETY LT 10) G3=1
IF (ANXIETY GT 9 AND ANXIETY LT 13) G4=1
IF (ANXIETY GT 12 AND ANXIETY LT 16) G5=1
IF (ANXIETY GT 15 AND ANXIETY LT 19) G6=1
REGRESSION VARIABLES=SCORE G1 G2 G3 G4 G5 ANXIETY/
    DEPENDENT=SCORE/
    TEST (G1,G2,G3,G4,G5)/
    DEPENDENT=SCORE/
    TEST (ANXIETY)/


//R2JAS1 JOB 94522,'SALZMAN',CLASS=H
// EXEC SPSS
TITLE 'TESTING RESEARCH HYPOTHESES USING MLR'
SUBTITLE 'GENERALIZED RESEARCH HYPOTHESIS 4.3'
COMMENT USING DATA FROM HINKLE ET AL., 1994, CHAPTER 14,
    PAGE 602.
DATA LIST /L1 1 L2 3 L3 5 L4 7 SATIS 9-10.
BEGIN DATA
1   0   0   0   71
1   0   0   0   74
1   0   0   0   75
1   0   0   0   68
1   0   0   0   74
1   0   0   0   72
1   0   0   0   76
1   0   0   0   73
1   0   0   0   77
0   1   0   0   74
```

```
0   1   0   0   81
0   1   0   0   75
0   1   0   0   70
0   1   0   0   74
0   1   0   0   81
0   1   0   0   76
0   0   1   0   83
0   0   1   0   77
0   0   1   0   80
0   0   1   0   76
0   0   1   0   86
0   0   1   0   82
0   0   0   1   76
0   0   0   1   69
0   0   0   1   64
0   0   0   1   76
0   0   0   1   73
0   0   0   1   72
0   0   0   1   68
0   0   0   1   71
END DATA.
COMPUTE GROUP23 = L2 + L3
REGRESSION VARIABLES = L2 L3 L4 SATIS/
    DEPENDENT=SATIS/ENTER/
REGRESSION VARIABLES = L4 GROUP23 SATIS/
    DEPENDENT=SATIS/ENTER/
```

APPENDIX J:
INDEX FOR SAS

Below is an alphabetical list of the various nuances that can be used to make the SAS printout more interpretable or to facilitate a particular analysis. Numbers refer to Applied Research Hypotheses.

/* */ 6.1
all variables 4.1
both group membership vectors 4.1
chi-square 4.6
combining files 10.4
comment 12.2
data transformations 7.1
different name for data file 9.3
ESS formula for F 10.1 and 10.2
formula in title 12.2
gain score created 10.2
generated predicted score 11.1
generating restricted model 7.1
generation of Full Model and Restricted Model Generalized Research Hypothesis 4.3
Hinkle example 4.2
if-else statements 4.3
label for test statement 10.3
labels 6.1
multiple restrictions 6.1
nonzero restriction 12.1
page size 8.1
person vectors 10.4
plot Hzero 5.1
plot overlay 7.1, 8.1, and 9.4
printing of new variables 10.1
PROC CORR 9.3
PROC FREQ 4.6
PROC MEANS 4.3 and 10.1

PROC PRINT 4.3 and 7.1
PROC UNIVARIATE 10.1
raw data 4.1
repeated measures 10.4
restrict analysis to first N subjects 11.1
SELECT IF 11.1
selected data 5.1
sine function 9.4
test of mean not equal to 0 10.1
TEST statement 4.1
title cards 4.2
title with space and override titles 4.3
titles 4.1
two successive restrictions on one model 9.2

REFERENCES

Andrews, F., Morgan, J., & Sonquist, J. (1967). *Multiple classification analysis.* Ann Arbor: The University of Michigan, Institute for Social Research.

Bender, J. A., Kelly, F. J., Pierson, J. K., & Kaplan, H. (1968). Analysis of the comparative advantages of unlike exercises in relation to prior individual strength levels. *The Research Quarterly, 39,* 443–448.

Boneau, C. A. (1960). The effects of violations of assumptions underlying the *t*-test. *Psychological Bulletin, 57,* 49–64.

Bottenberg, R. A., & Ward, J. H. (1963). *Applied multiple linear regression.* Lackland Air Force Base, TX: Aerospace Medical Division, AD 413128.

Bray, J. H., & Maxwell, S. E. (1985). *Multivariate analysis of variance.* Beverly Hills, CA: Sage.

Brown, R. (1992). The Beta or not the Beta: What is the research question? *Multiple Linear Regression Viewpoints, 19* (1), 1–6.

Byrne, J. (1974). The use of regression equations to demonstrate causality. *Multiple Linear Regression Viewpoints, 5* (1), 11–22.

Campbell, D. T., & Stanley, J. C. (1966). *Experimental and quasiexperimental designs for research.* Chicago: Rand McNally.

Castaneda, A., Palermo, D. S., & McCandless, B. (1956). Complex learning and performance as a function of anxiety in children and task difficulty. *Child Development, 27,* 328–332.

Christal, R. E. (1968). Selecting a harem—and other applications of the policy capturing model. *The Journal of Experimental Education, 36* (1), 24–27.

Cohen, J. (1970). Approximate power and sample size determination for common one-sample and two-sample hypothesis tests. *Educational and Psychological Measurement, 30,* 811–832.

Cohen, J. (1977). *Statistical power analysis for the behavioral sciences* (2nd ed.). New York: Academic Press.

Cohen, J. (1990). Things I have learned (so far). *American Psychologist, 45* (12), 1304–1312.

Cohen, J., & Cohen, P. (1975). *Applied multiple regression/correlation analysis for the behavioral sciences.* New York: Halstead.

Colliver, J. A., Verhulst, S. J., & Kolm, P. (1987). A simple multiple linear regression test for differential effects of a given independent variable on several dependent measures. *Multiple Linear Regression Viewpoints, 15* (2), 134–141.

Connett, W. E., Houston, S. R., & Shaw, D. G. (1972). The use of factor regression in data analysis. *Multiple Linear Regression Viewpoints, 2* (4), 46–49.

Constas, M. A., & Francis, J. D. (1992). A graphical method for selecting the best sub-set regression model. *Multiple Linear Regression Viewpoints, 19* (1), 16–25.

DuCette, J., & Wolk, S. (1972). Ability and achievement as moderating variables of student satisfaction and teacher preparation. *The Journal of Experimental Education, 41* (1), 12–17.

Fraas, J. W., & Drushal, M. E. (1987). The use of MLR models to analyze partial interaction: An educational application. *Multiple Linear Regression Viewpoints, 15* (2), 85–96.

Galileo, G. (1953). *The two chief world systems* (S. Drake, Trans.). Berkeley: University of California Press. (Original work published 1632.)

Glass, G. V., & Smith, M. L. (1979). Meta-analysis of research on class size and achievement. *Educational Evaluation and Policy Analysis, 1,* 2–16.

Goldberg, L. R. (1972). Parameters of personality inventory construction and utilization: A comparison of prediction strategies and tactics. *Multivariate Behavioral Research Monographs, 72,* 2.

Hand, D. J., & Taylor, C. C. (1987). *Multivariate analysis of variance and repeated measures: A practical approach for behavioral scientists.* New York: Chapman & Hall.

Harris, R. J. (1985). *A primer of multivariate statistics* (2nd ed.). Orlando, FL: Academic Press.

Harris, R. J. (1992, April). *Structure coefficients versus scoring coefficients as bases for interpreting emergent variables in multiple regression and related techniques.* Paper presented at the meeting of the American Educational Research Association, San Francisco.

Harris, R. J. (1993). "Beta" weights should be used to interpret regression variates and to assess in-context variable importance. *Mid-western Educational Researcher, 6* (1), 11–14.

Haupt, C. C. (1993). Facilitation effects: Determinants of drug use among Hispanic and White non-Hispanic fourth, fifth, and sixth grade public school students. *Dissertation Abstracts International, 53 (7), 530A.* (University Microfilms No. DA9232434).

Hickrod, G. A. (1971). Local demand for education: A critique of school finance and economic research circa 1959–1969. *Review of Educational Research, 41,* 35–50.

Hinkle, D. E., Wiersma, W., & Jurs, S. G. (1994). *Applied statistics for the behavioral sciences* (3rd ed.). Boston: Houghton Mifflin.

Houston, S. (1987). The use of judgement analysis and a modified canonical JAN in evaluation methodology. *Multiple Linear Regression Viewpoints, 15* (2), 49–84.

Interuniversity Consortium for Political and Social Research. (1993). *Guide to resources and services 1991-1992.* Ann Arbor: Author.

Jennings, E. E. (1967). Fixed effects analysis of variance by regression analysis. *Multivariate Behavioral Research, 2,* 95–108.

Kelly, F. J., Beggs, D. L., McNeil, K. A., Eichelberger, T., & Lyon, J. (1969). *Research design in the behavioral sciences: Multiple regression approach.* Carbondale: Southern Illinois University Press.

Kelly, F. J., McNeil, K. A., & Newman, I. (1973). Suggested inferential statistical models for research in behavior modification. *The Journal of Experimental Education, 41* (4), 54–63.

Kerlinger, F. N., & Pedhazur, E. J. (1973). *Multiple regression in behavioral research.* New York: Holt, Rinehart & Winston.

Kinion E. (1990). Corporate manager's leadership style and existence of employee health promotion programs. *Multiple Linear Regression Viewpoints, 17* (2), 15–30.

Klein, M., & Newman, I. (1974). Estimated parameters of three shrinkage estimate formuli. *Multiple Linear Regression Viewpoints, 4* (4), 6–11.

Kleinbaum, D. G., & Kupper, L. L. (1978). *Applied regression analysis and other multivariable methods.* North Scitvate, MA: Duxbury Press.

Leitner, D. (1979). Using multiple regression to interpret chi-square contingency table analysis. *Multiple Linear Regression Viewpoints, 10* (1), 39–45.

Levine, D. U., & Stephenson, R. S. (1988). Differing policy implications of alternate multiple regressions using the same set of student background variables to predict academic achievement. *Multiple Linear Regression Viewpoints, 16* (2), 94–104.

Lewis-Beck, M. S. (1980). *Applied regression: An introduction.* Beverly Hills, CA: Sage.

Lunneborg, C. E. (1992). A case for interpreting regression weights. *Multiple Linear Regression Viewpoints, 19* (1), 26–36.

Maxwell, S. E., & Delaney, H. D. (1990). *Designing experiments and analyzing data: A model comparison perspective.* Belmont, CA: Wadsworth.

McNeil, K. A. (1970a). Meeting the goals of research with multiple regression analysis. *Multivariate Behavioral Research, 5,* 375–386.

McNeil, K. A. (1970b). The negative aspects of the eta coefficient as an index of curvilinearity. *Multiple Linear Regression Viewpoints, 1* (1), 7–17.

McNeil, K. A. (1974). The multiple linear regression approach to "Chi Square" hypotheses. *The Journal of Experimental Education, 43* (2), 53–55.

McNeil, K. A. (1990). The case against interpreting regression weights. *Multiple Linear Regression Viewpoints, 18* (1), 1–7.

McNeil, K. A. (1991a). The case for non-zero restrictions in statistical analysis. *Multiple Linear Regression Viewpoints, 18* (1), 47–54.

McNeil, K. A. (1991b, January). *Reasons for not interpreting regression coefficients.* Paper presented at the meeting of the Southwest Educational Research Association. San Antonio, TX.

McNeil, K. A. (1992). Response to Lunneborg: The conditions for interpretation of regression weights. *Multiple Linear Regression Viewpoints, 19* (1), 37–43.

McNeil, K. A. (1993). Cautions and conditions for interpreting weighting coefficients. *Mid-western Educational Researcher, 6* (1), 11–14.

McNeil, K. A., & Beggs, D. L. (1969, February). *The mathematical equivalence of the test of significance of the point biserial correlation coefficient and the t-test for the difference between means.* Paper presented at the meeting of the American Educational Research Association, Los Angeles.

McNeil, K. A., Evans, J., & McNeil, J. T. (1979, April). *Non-linear transformation of the criterion.* Paper presented at the meeting of the American Educational Research Association, San Francisco.

McNeil, K. A., & Findlay, E. (1980). Evaluating Title I early childhood programs: Problems, the applicability of Model C, and several evaluation plans. *Multiple Linear Regression Viewpoints, 10* (4), 41–50.

McNeil, K. A., & Kelly, F. J. (1970). Express functional relationships among data rather than assume "intervalness." *The Journal of Experimental Education, 39* (2), 43–48.

McNeil, K. A., Kelly, F. J., & McNeil, J. T. (1975). *Testing research hypotheses using multiple linear regression.* Carbondale: Southern Illinois University Press.

McNeil, K. A., & McShane, M. (1974). Complexity in behavioral research, as viewed within the multiple linear regression approach. *Multiple Linear Regression Viewpoints, 4* (4), 12–15.

McNeil, K. A., & Spaner, S. D. (1971). A defense for including highly correlated predictor variables in multiple regression models. *Multivariate Behavioral Research, 6,* 117–125.

Mendenhall, W. (1968). *Introduction to linear models and the design and analysis of experiments.* Belmont, CA: Wadsworth.

Morgan, J. (1971, June). Can a computer beat the horses? *Look, 35,* 33.

Mosteller, F., & Tukey, J. W. (1968). Data analysis, including statistics. In G. Lindzey & E. Aronson (Eds.), *The handbook of social psychology: Vol. 2* (pp. 88–203). Reading, MA: AddisonWesley.

Mueller, R. O. (1990). Teaching ANCOVA: The importance of random assignment. *Multiple Linear Regression Viewpoints, 17* (2), 1–14.

Newman, I. (1972). Some further considerations of using factor regression analysis. *Multiple Linear Regression Viewpoints, 3* (2), 39–41.

Newman, I. (1973a). A revised "suggested format for the presentation of multiple regression analyses." *Multiple Linear Regression Viewpoints, 4* (1), 45–47.

Newman, I. (1973b). Variations between shrinkage estimation formulas and the appropriateness of their interpretation. *Multiple Linear Regression Viewpoints, 4* (2), 45–48.

Newman, I. (1988, October). *There is no such thing as multivariate analysis: All analyses are univariate.* Paper presented at the meeting of the Midwestern Educational Research Association, Chicago.

Newman, I., & Benz, C. (1980). An estimate of power for intact groups and individual subjects. *Multiple Linear Regression Viewpoints, 10* (4), 51–60.

Newman, I., Benz, C., & Williams, J. D. (1990). Alternative in analyzing the Solomon four group design. *Multiple Linear Regression Viewpoints, 17* (2), 91–103.

Newman, I., Lewis, E., & McNeil, K. A. (1973). Multiple linear regression models which more closely reflect Bayesian concerns. *Multiple Linear Regression Viewpoints, 3* (3), 71–77.

Newman, I., Lewis, E., McNeil, K. A., & Fugita, S. S. (1978). Multiple linear regression models which more closely reflect Bayesian concerns. *The Ohio Journal of Science, 1,* 47–50.

Newman, I., McNeil, K. A., Seymour, G. A., & Garver, T. K. (1978, March). *A Monte Carlo evaluation of estimated parameters of five shrinkage estimate formula.* Paper presented at the meeting of the American Educational Research Association, San Francisco.

Novick, M. R., & Jackson, P. H. (1970). Bayesian guidance technology. *Review of Educational Research, 40,* 459–494.

Pedazur, E. J. (1982). *Multiple regression in behavioral research: Explanation and prediction* (2nd ed.). New York: Holt, Rinehart & Winston.

Pohlmann, J. P. (1973). Incorporating cost information into the selection of variables in multiple regression analysis. *Multiple Linear Regression Viewpoints, 4* (2), 18–26.

Pohlmann, J. P. (1993). Some nonstandard applications of the analysis of covariance model. *Multiple Linear Regression Viewpoints, 20* (1), 20–24.

Presley, R. J., & Huberty, C. (1988). Predicting statistics achievement: A prototypical regression analysis. *Multiple Linear Regression Viewpoints, 16* (1), 36–77.

Rosenthal, R. (1984). *Meta-analytic procedures for social research.* Beverly Hills, CA: Sage.

SAS Institute. (1990). *SAS Procedures Guide, Version 6* (3rd ed.). Cary, NC: SAS Institute.

Saunders, D. R. (1956). Moderator variables in prediction. *Educational and Psychological Measurement, 16,* 209–222.

Seber, G. A. F. (1984). *Multivariate observations.* New York: John Wiley & Sons.

Smith, G., McNeil, K. A., & Mitchell, N. (1986). Regression and Model C for evaluation. *Multiple Linear Regression Viewpoints, 16* (1), 75–89.

Spaner, S. (1970). *Application of multivariate techniques to developmental data: The derivation, verification, and validation of a predictive model of twenty-four month cognitive data.* Unpublished doctoral dissertation, Southern Illinois University, Carbondale.

Starr, F. H. (1971). The remarriage of multiple regression and statistical inference: A promising approach for hypnosis researchers. *The American Journal of Clinical Hypnosis, 13,* 175–197.

Stevens, J. (1986). *Applied multivariate statistics for the social sciences.* Hillsdale, NJ: Lawrence Erlbaum Associates.

Tallmadge, G. K., & Wood, C. T. (1981). *User's guide: ESEA Title I evaluation and reporting system.* Mountain View, CA: RMC Research.

Tatsuoka, M. M. (1988) *Multivariate analysis: Techniques for educational and psychological research* (2nd ed.). New York: Macmillan.

Thayer, J. D. (1990). Implementing variable selection techniques in regression. *Multiple Linear Regression Viewpoints, 17* (2), 67–90.

Tracz, S. M., & Leitner, D. W. (1989). The meta-analysis of the effect of class size on achievement: A secondary analysis. *Multiple Linear Regression Viewpoints, 17* (1), 16–29.

Trochim, W. M. K., & Stanley, T. D. (1991). "Regression discontinuity design" by any other name might be less problematic. *Evaluation Review, 15* (5), 605–624.

Ward, J., & Jennings, E. (1973). *Introduction to linear models.* Englewood Cliffs, NJ: Prentice Hall.

Webster, W. J., & Eichelberger, R. T. (1972). A use of the regression model in educational evaluation. *The Journal of Experimental Education, 40* (3), 91–96.

Webster, W., & Mendro, R. (1992, April). *Measuring the effects of schooling: Expanded school effects indices.* Paper presented at the meeting of the American Educational Research Association, San Francisco.

Werts, C. E. (1970). The partitioning of variance in school effects studies: A reconsideration. *American Educational Research Journal, 7,* 127–132.

Williams, J. D. (1974a). *Regression analysis in educational research.* New York: MSS Information Corporation.

Williams, J. D. (1974b). Regression solutions to the AxBxS design. *Multiple Linear Regression Viewpoints, 5* (2), 3–9.

Williams, J. D. (1987). The use of nonsense coding with ANOVA situations. *Multiple Linear Regression Viewpoints, 15* (2), 29–39.

Williams, J. D., & Lindem, A. C. (1974). Regression computer programs for setwise regression and three related analysis of variance techniques. *Multiple Linear Regression Viewpoints, 4* (4), 30–46.

Williams, J. D., & Newman, I. (1982, March). *Using multiple linear models to simultaneously analyze a Solomon four group design.* Paper presented at the meeting of the American Educational Research Association, New York.

Williams, J. D., Williams, J. A., & Roman, S. J. (1988). A ten year study of salary differential by sex through a regression methodology. *Multiple Linear Regression Viewpoints, 16* (1), 91–107.

Winer, B. J. (1971) *Statistical principles in experimental design* (2nd ed.). New York: McGraw-Hill.

Woehlke, P., Elmore, P. B., & Spearing, D. L. (1990). Testing assumptions in multiple regression: Comparison of procedures available in SAS and SPSSX. *Multiple Linear Regression Viewpoints, 17* (2), 48–66.

Wood, D. A., & Langevin, M. J. (1972). Moderating the prediction of grades in freshman engineering. *Journal of Educational Measurement, 9,* 311–320.

INDEX

addition of two vectors, 36

adjusted R^2, 154, 300, 307

analysis of covariance, 137, 149–65

analysis of variance, 3, 19–27, 48–78, 117

a priori comparison, 61

assumptions, 26–27, 158–66, 174

attenuation, 257

automatic unit vector analogue, 53, 58, 61

backward stepwise, 112–13, 261

Bayesian statistics, 275

beta weights, 84

blocking variable, 151, 231, 258

b weights, 84

categorical vectors, 35

causality, 7, 208, 216–18, 300–302

ceiling effect, 200

cell means model, 124

chi-square, 172

coefficient of determination, 102

coefficient of nondetermination, 102

competing explainers, 7, 149, 217, 223, 274, 302

complexity, 46, 155, 192, 297

confidence intervals, 282

confidence level, 17

configural variable, 147

contingency coefficient, 76

contrasts, 61–64, 66–68, 102, 117–18

control: experimental, 239–74; goal of research, 8–9, 297–302, 305–7; of individual differences, 250; statistical, 149–51

correlated predictor variables, 189, 301

correlated *t* test, 232

covariate, 5, 140, 153–54, 158–60, 162, 164–67, 173, 194, 227, 258, 274

criterion–referenced testing, 218, 274

cross validation, 257

curvilinear interaction, 291, 293

data snooping, 4, 196–97, 253, 259–64, 287, 302–3, 307

degrees of freedom, 3, 5, 20–23, 29–30, 64, 69, 76, 84–89, 93, 111, 121, 125–27, 132, 147–48, 154, 157, 163, 181, 219–23, 232, 253, 256, 296, 299

dependent *t* test, 232

description, 6, 17, 79, 196

dichotomous variable, 48, 54, 76–77, 116

directional hypothesis, 52–54, 64, 72, 119, 134, 137–38, 162, 164, 216–17, 219

directional probability, 55, 60, 122, 138, 160, 186–87, 190, 215, 239, 277–78

disordinal interaction, 126

dynamic variable, 305–6

error of specification, 303

error sum of squares, 86, 92, 219, 276, 279, 297

error variance, 15, 87–89, 102, 154, 258, 281

eta coefficient, 170, 193

evaluation, 268–86

external validity, 215

focal stimulus variable, 13

forward stepwise, 112–13, 261

gain scores, 226

generalization, 7, 19, 231, 290–93

goals of research, 47, 54, 287
grand mean, 22, 28–30
group membership vectors, 35, 43–47, 50, 59, 296
guessing effect, 200

homogeneity, 137, 159, 162–63, 165–67, 174, 280
homogeneity of regression lines, 159
homoscedasticity, 160, 174–78
hypothesis seeking, 193, 196

improvement, 6–9, 306
individual differences, 168, 222–24, 228–32, 249–51
internal validity, 215
interval scale, 254

Johnson-Neyman technique, 134

linear combination of vectors, 41, 43
linear relationship, 27, 79–97, 181, 186–89, 287–88

main effect, 116–27, 258
manipulation, 7–9, 216, 218, 239, 302, 304–7
measurement, 13, 18, 32–33, 35, 197, 205, 254, 257,
measurement error, 15, 18, 102, 292, 303
missing data, 265–66
moderator variable, 147
most general regression model, 294
multicolinearity, 255
multiple comparisons, 71, 122, 260
multiple correlation, 98–109
mutually exclusive vectors, 43–44, 50, 153, 165

Newton's law, 174, 258, 287–94
nonlinear interaction, 208, 291–93
nonlinear relationship, 114, 178, 197
nonmanipulatable variables, 301, 306
nonmanipulated, 54, 304
nonparallel slopes, 162
nonsense coding, 68
nonzero restriction, 276–80
null hypothesis, 27, 85
null vector, 38, 41–42

one-degree-of-freedom questions, 73
ordinal interaction, 125–26, 176
outliers, 266, 273

over and above, 110–15, 127, 146, 151–67, 173, 180–82, 186–87, 190, 226–30, 295–98, 304
overfitting, 253, 261

parameter estimates, 55, 133
parsimony, 287–93
path analysis, 301
person variables, 10–13, 16, 166
person vectors, 151, 222–23, 228–32, 250–51
piece of information, 3, 5, 31, 33–34, 45–47, 84, 87, 113–15, 124–25, 132, 147, 219, 284
planned comparison, 72
point-biserial correlation, 54
point change, 232
policy capturing, 283–86
polynomials, 178–96, 202–5, 208–15, 255–56, 286, 288–94
post hoc comparisons, 4, 48, 61–68, 71–72
power, 26, 76, 168–73, 222–26, 253
power analysis, 26, 168–172
prediction, 6–8, 11–17, 46–47, 49, 82–84, 107–9, 194, 214–17, 256–57, 267, 271–75, 300, 305
properties of vectors, 38–41

random sampling, 19, 22, 24, 217, 252
raw score weight, 84
regression constant, 55, 81–83, 97, 153–54, 186–90, 288
reliability, 205–6, 257
repeated measures, 166, 212–15, 221–24, 228–30, 237–51
replication, 155, 181, 196, 210, 215, 217, 253, 257–59, 261, 263, 266, 290, 302, 304, 305, 307
rescaling, 197–208
research strategy, 178, 287–94, 302–7
residual, 281
response surface, 208

sample size, 26, 77–78, 168–72, 253, 298–99
sampling error, 18–19, 22, 30–31, 81, 130, 132, 137
sampling variation, 7–8, 19, 22–24, 27, 82, 88, 93, 154, 158, 243–44
sequence, 162, 247–49, 258–61
serendipitous findings, 262–64, 303
setwise regression, 261
simple effect, 125
single population mean, 218, 274, 294, 297

slope of the line, 83, 158
smart vectors, 35
snooping, 4, 192–97, 253, 258–64, 302–7
source table, 31, 307
Spearman rank correlation, 202
square root transformation, 202
standard error of estimate, 280–82
standard partial regression weight, 84
standard weight, 84
stepwise regression, 4, 112–13, 260–63, 284–85
straight-line hypothesis, 96
subject-by-treatment interaction, 250
subtraction of two vectors, 37
suppressor variable, 255

third-degree model, 193, 211
time series, 212, 251
treatment-by-aptitude interaction, 175
trend analysis, 66
two-line model, 321

two-parallel-line model, 132
Type I error, 24, 26, 168, 260–61
Type II error, 24, 26, 169

unbiased estimate, 18, 21
understanding, 1–2, 56, 72, 88, 138, 155, 163, 174, 203, 257, 302
unique variance, 87–89, 107–10
U-shaped curve, 181, 190 ·

validity, 205, 254, 290, 293
validity generalization, 287, 290, 293
validity of assumptions, 27
vector notation, 32–38
Venn diagram, 15–16, 56–57, 95, 102–5, 108, 110, 113–14, 119–20
Viewpoints, 310

Y-intercept, 80–88, 94–97, 129–34, 145–46, 159, 167, 183, 187, 191, 233–36, 241–44, 248, 290, 308

Keith McNeil received his PhD in educational psychology from the University of Texas at Austin in 1967. He began teaching the GLM at Southern Illinois University at Carbondale that year. He ran his own consulting business for 2 years, worked in a state department of education for 1 year, worked 8 years on a federal contract providing evaluation assistance to state and local educational agencies, and worked for 5 years as an evaluator in the Dallas public schools. He rejoined higher education in 1989 at New Mexico State University, where he teaches statistics and research design. He has been chair of the Special Interest Group in Multiple Linear Regression on four different occasions. He recently coauthored *Research for the Helping Professions*.

Isadore Newman received his PhD in educational psychology with a specialty in statistics and measurement from Southern Illinois University at Carbondale in 1971. He has been a professor at Akron University since 1971. During his professional career, he has served on over 300 dissertation committees and has presented hundreds of papers at state, national, and international meetings. He has written 9 books and monographs and has served on many editorial boards, in addition to being the editor of *Multiple Linear Regression Viewpoints* and the *Midwestern Educational Researcher*.

Francis J. Kelly received his PhD from the University of Texas at Austin in 1963. He has been teaching the GLM and developmental psychology at Southern Illinois University at Carbondale since then. He has mentored many doctoral students and was the primary author of *Research Design in the Behavioral Sciences: Multiple Regression Approach* (1969). He was also the primary author of an introductory educational psychology textbook, *Educational Psychology: A Behavioral Approach* (1969) and a genetic psychology monograph, *The Dialectics of Social Adaptation and Individual Constructivism* (1984).